193
Advances in Polymer Science

Editorial Board:
A. Abe · A.-C. Albertsson · R. Duncan · K. Dušek · W. H. de Jeu
J.-F. Joanny · H.-H. Kausch · S. Kobayashi · K.-S. Lee · L. Leibler
T. E. Long · I. Manners · M. Möller · O. Nuyken · E. M. Terentjev
B. Voit · G. Wegner · U. Wiesner

Advances in Polymer Science

Recently Published and Forthcoming Volumes

Surface-Initiated Polymerization II
Volume Editor: Jordan, R.
Vol. 198, 2006

Surface-Initiated Polymerization I
Volume Editor: Jordan, R.
Vol. 197, 2006

Conformation-Dependent Design of Sequences in Copolymers II
Volume Editor: Khokhlov, A. R.
Vol. 196, 2006

Conformation-Dependent Design of Sequences in Copolymers I
Volume Editor: Khokhlov, A. R.
Vol. 195, 2006

Enzyme-Catalyzed Synthesis of Polymers
Volume Editors: Kobayashi, S., Ritter, H., Kaplan, D.
Vol. 194, 2006

Polymer Therapeutics II
Polymers as Drugs, Conjugates and Gene Delivery Systems
Volume Editors: Satchi-Fainaro, R., Duncan, R.
Vol. 193, 2006

Polymer Therapeutics I
Polymers as Drugs, Conjugates and Gene Delivery Systems
Volume Editors: Satchi-Fainaro, R., Duncan, R.
Vol. 192, 2006

Interphases and Mesophases in Polymer Crystallization III
Volume Editor: Allegra, G.
Vol. 191, 2005

Block Copolymers II
Volume Editor: Abetz, V.
Vol. 190, 2005

Block Copolymers I
Volume Editor: Abetz, V.
Vol. 189, 2005

Intrinsic Molecular Mobility and Toughness of Polymers II
Volume Editor: Kausch, H.-H.
Vol. 188, 2005

Intrinsic Molecular Mobility and Toughness of Polymers I
Volume Editor: Kausch, H.-H.
Vol. 187, 2005

Polysaccharides I
Structure, Characterization and Use
Volume Editor: Heinze, T.
Vol. 186, 2005

Advanced Computer Simulation Approaches for Soft Matter Sciences II
Volume Editors: Holm, C., Kremer, K.
Vol. 185, 2005

Crosslinking in Materials Science
Vol. 184, 2005

Phase Behavior of Polymer Blends
Volume Editor: Freed, K.
Vol. 183, 2005

Polymer Analysis/Polymer Theory
Vol. 182, 2005

Interphases and Mesophases in Polymer Crystallization II
Volume Editor: Allegra, G.
Vol. 181, 2005

Interphases and Mesophases in Polymer Crystallization I
Volume Editor: Allegra, G.
Vol. 180, 2005

Polymer Therapeutics II

Polymers as Drugs, Conjugates and Gene Delivery Systems

Volume Editors: Ronit Satchi-Fainaro · Ruth Duncan

With contributions by

V. Y. Alakhov · S. Allen · C. M. Barnes · E. V. Batrakova
Y. T. A. Chim · M. C. Davies · R. Duncan · J. S. Ellis
J. Fang · K. Greish · A. V. Kabanov · K. Kataoka
H. Maeda · N. Nishiyama · C. J. Roberts
R. Satchi-Fainaro · S. Sherman · S. J. B. Tendler

The series *Advances in Polymer Science* presents critical reviews of the present and future trends in polymer and biopolymer science including chemistry, physical chemistry, physics and material science. It is adressed to all scientists at universities and in industry who wish to keep abreast of advances in the topics covered.

As a rule, contributions are specially commissioned. The editors and publishers will, however, always be pleased to receive suggestions and supplementary information. Papers are accepted for *Advances in Polymer Science* in English.

In references *Advances in Polymer Science* is abbreviated *Adv Polym Sci* and is cited as a journal.

Springer WWW home page: http://www.springer.com
Visit the APS content at http://www.springerlink.com/

Library of Congress Control Number: 2005933609

ISSN 0065-3195
ISBN-10 3-540-29211-X Springer Berlin Heidelberg New York
ISBN-13 978-3-540-29211-1 Springer Berlin Heidelberg New York
DOI 10.1007/11547785

This work is subject to copyright. All rights are reserved, whether the whole or part of the material is concerned, specifically the rights of translation, reprinting, reuse of illustrations, recitation, broadcasting, reproduction on microfilm or in any other way, and storage in data banks. Duplication of this publication or parts thereof is permitted only under the provisions of the German Copyright Law of September 9, 1965, in its current version, and permission for use must always be obtained from Springer. Violations are liable for prosecution under the German Copyright Law.

Springer is a part of Springer Science+Business Media

springer.com

© Springer-Verlag Berlin Heidelberg 2006
Printed in Germany

The use of registered names, trademarks, etc. in this publication does not imply, even in the absence of a specific statement, that such names are exempt from the relevant protective laws and regulations and therefore free for general use.

Cover design: *Design & Production* GmbH, Heidelberg
Typesetting and Production: LE-TEX Jelonek, Schmidt & Vöckler GbR, Leipzig

Printed on acid-free paper   02/3141 YL – 5 4 3 2 1 0

Volume Editors

Dr. Ronit Satchi-Fainaro
Harvard Medical School
and Children's Hospital
Boston Department of Surgery
Vascular Biology Program
1 Blackfan Circle
Boston, MA 02115, USA
ronit.satchi-fainaro@childrens.harvard.edu

Prof. Ruth Duncan
Welsh School of Pharmacy
Cardiff University
Redwood Building
King Edward VII Avenue
Cardiff CF 10 3XF, UK
DuncanR@cf.ac.uk

Editorial Board

Prof. Akihiro Abe
Department of Industrial Chemistry
Tokyo Institute of Polytechnics
1583 Iiyama, Atsugi-shi 243-02, Japan
aabe@chem.t-kougei.ac.jp

Prof. A.-C. Albertsson
Department of Polymer Technology
The Royal Institute of Technology
10044 Stockholm, Sweden
aila@polymer.kth.se

Prof. Ruth Duncan
Welsh School of Pharmacy
Cardiff University
Redwood Building
King Edward VII Avenue
Cardiff CF 10 3XF, UK
DuncanR@cf.ac.uk

Prof. Karel Dušek
Institute of Macromolecular Chemistry,
Czech
Academy of Sciences of the Czech Republic
Heyrovský Sq. 2
16206 Prague 6, Czech Republic
dusek@imc.cas.cz

Prof. W. H. de Jeu
FOM-Institute AMOLF
Kruislaan 407
1098 SJ Amsterdam, The Netherlands
dejeu@amolf.nl
and Dutch Polymer Institute
Eindhoven University of Technology
PO Box 513
5600 MB Eindhoven, The Netherlands

Prof. Jean-François Joanny
Physicochimie Curie
Institut Curie section recherche
26 rue d'Ulm
75248 Paris cedex 05, France
jean-francois.joanny@curie.fr

Prof. Hans-Henning Kausch
Ecole Polytechnique Fédérale de Lausanne
Science de Base
Station 6
1015 Lausanne, Switzerland
kausch.cully@bluewin.ch

Prof. Shiro Kobayashi
R & D Center for Bio-based Materials
Kyoto Institute of Technology
Matsugasaki, Sakyo-ku
Kyoto 606-8585, Japan
kobayash@kit.ac.jp

Prof. Kwang-Sup Lee
Department of Polymer Science &
Engineering
Hannam University
133 Ojung-Dong Daejeon,
306-791, Korea
kslee@hannam.ac.kr

Prof. L. Leibler
Matière Molle et Chimie
Ecole Supérieure de Physique
et Chimie Industrielles (ESPCI)
10 rue Vauquelin
75231 Paris Cedex 05, France
ludwik.leibler@espci.fr

Prof. Timothy E. Long
Department of Chemistry
and Research Institute
Virginia Tech
2110 Hahn Hall (0344)
Blacksburg, VA 24061, USA
telong@vt.edu

Prof. Ian Manners
School of Chemistry
University of Bristol
Cantock's Close
BS8 1TS Bristol, UK
Ian.Manners@bristol.ac.uk

Prof. Martin Möller
Deutsches Wollforschungsinstitut
an der RWTH Aachen e.V.
Pauwelsstraße 8
52056 Aachen, Germany
moeller@dwi.rwth-aachen.de

Prof. Oskar Nuyken
Lehrstuhl für Makromolekulare Stoffe
TU München
Lichtenbergstr. 4
85747 Garching, Germany
oskar.nuyken@ch.tum.de

Prof. E. M. Terentjev
Cavendish Laboratory
Madingley Road
Cambridge CB 3 OHE, UK
emt1000@cam.ac.uk

Prof. Brigitte Voit
Institut für Polymerforschung Dresden
Hohe Straße 6
01069 Dresden, Germany
voit@ipfdd.de

Prof. Gerhard Wegner
Max-Planck-Institut
für Polymerforschung
Ackermannweg 10
Postfach 3148
55128 Mainz, Germany
wegner@mpip-mainz.mpg.de

Prof. Ulrich Wiesner
Materials Science & Engineering
Cornell University
329 Bard Hall
Ithaca, NY 14853, USA
ubw1@cornell.edu

Advances in Polymer Science
Also Available Electronically

For all customers who have a standing order to Advances in Polymer Science, we offer the electronic version via SpringerLink free of charge. Please contact your librarian who can receive a password or free access to the full articles by registering at:

springerlink.com

If you do not have a subscription, you can still view the tables of contents of the volumes and the abstract of each article by going to the SpringerLink Homepage, clicking on "Browse by Online Libraries", then "Chemical Sciences", and finally choose Advances in Polymer Science.

You will find information about the

- Editorial Board
- Aims and Scope
- Instructions for Authors
- Sample Contribution

at springeronline.com using the search function.

Contents

Polymer Therapeutics for Cancer:
Current Status and Future Challenges
R. Satchi-Fainaro · R. Duncan · C. M. Barnes   1

Nanostructured Devices Based on Block Copolymer Assemblies
for Drug Delivery: Designing Structures for Enhanced Drug Function
N. Nishiyama · K. Kataoka 67

The EPR Effect and Polymeric Drugs:
A Paradigm Shift for Cancer Chemotherapy in the 21st Century
H. Maeda · K. Greish · J. Fang 103

Molecular-Scale Studies on Biopolymers Using Atomic Force Microscopy
J. S. Ellis · S. Allen · Y. T. A. Chim · C. J. Roberts · S. J. B. Tendler ·
M. C. Davies 123

Polymer Genomics
A. V. Kabanov · E. V. Batrakova · S. Sherman · V. Y. Alakhov 173

Author Index Volumes 101–193 199

Subject Index 223

Contents of Volume 192

Polymer Therapeutics I

Volume Editors: Ronit Satchi-Fainaro, Ruth Duncan
ISBN: 3-540-29210-1

Polymer Therapeutics: Polymers as Drugs, Drug and Protein Conjugates and Gene Delivery Systems: Past, Present and Future Opportunities
R. Duncan · H. Ringsdorf · R. Satchi-Fainaro

Polymers as Drugs
P. K. Dhal · S. R. Holmes-Farley · C. C. Huval · T. H. Jozefiak

Domino Dendrimers
R. J. Amir · D. Shabat

PEGylation of Proteins as Tailored Chemistry for Optimized Bioconjugates
G. Pasut · F. M. Veronese

Gene Delivery Using Polymer Therapeutics
E. Wagner · J. Kloeckner

Polymer Therapeutics for Cancer: Current Status and Future Challenges

Ronit Satchi-Fainaro[1] (✉) · Ruth Duncan[2] · Carmen M. Barnes[1]

[1] Harvard Medical School, Children's Hospital Boston, Department of Surgery, Vascular Biology Program, Karp Family Research Laboratories 12.007D, 1 Blackfan Circle, Boston, MA 02115, USA
ronit.satchi-fainaro@childrens.harvard.edu

[2] Centre for Polymer Therapeutics Welsh School of Pharmacy, Cardiff University, Redwood Building, King Edward VII Avenue, Cardiff CF10 3XF, UK

1	Introduction	4
2	Passive or active targeting?	7
3	Targeting tumour cells or tumour vasculature?	9
4	Parenteral drug targeting	11
5	**Polymer-protein conjugates**	11
5.1	PEG-adenosine deaminase	13
5.2	PEG-L-asparaginase	14
5.3	PEG-granulocyte-colony stimulating factor (PEG-G-CSF)	15
5.4	PEG-interferon-α (IFN-α)	16
5.5	Styrene-co-maleic anhydride-neocarzinostatin (SMANCS)	17
5.6	Preclinical polymer-protein conjugates	18
6	**Polymer-drug conjugates**	18
6.1	HPMA copolymer-Gly-Phe-Leu-Gly-doxorubicin (PK1, FCE28068)	23
6.2	HPMA copolymer-Gly-Phe-Leu-Gly-doxorubicin-galactosamine (PK2, FCE28069)	25
6.3	HPMA copolymer-antibody-doxorubicin conjugates	26
6.4	HPMA copolymer-paclitaxel (PNU166945)	27
6.5	HPMA copolymer-camptothecin (MAG-CPT; PNU 166148)	28
6.6	HPMA copolymer-platinate (AP5280)	31
6.7	HPMA copolymer-DACH platinate (AP5346)	32
6.8	Poly-L-glutamic(PG)-paclitaxel (CT-2103, XYOTAX)	33
6.9	Poly-L-glutamic(PG)-camptothecin (CT-2106)	35
6.10	PEG-camptothecin (PROTHECAN)	36
6.11	PEG-paclitaxel	37
6.12	Dextran-doxorubicin (AD-70, DOX-OXD)	37
6.13	Polymeric micelles	37
6.14	Brain tumour implants – local delivery of chemotherapy	38
7	**Other compounds in preclinical stage**	39
7.1	DE-310	39
7.2	Carboxymethyl dextran-CPT analogue (T-2513) conjugate (T-0128)	39

7.3	Polyacetal-diethylstilboestrol	40
7.4	HPMA copolymer-1,5-diazaanthraquinone	42
8	**Targeting tumour vasculature**	**42**
8.1	Use of targeting moieties to deliver drugs to the tumour vasculature	43
8.1.1	Targeting tumour vessels using markers of angiogenesis	43
8.1.2	Drug targeting to angiogenic vessels using peptide motifs	45
8.2	HPMA copolymer-TNP-470 (caplostatin)	46
8.3	Delivery schedules and vehicles to target angiogenesis	48
8.4	Related technologies: PEGylated-liposomes to target tumour vasculature	49
8.4.1	PEGylated-liposome-Raf mutant	49
8.4.2	Liposomal-PEG-Ala-Pro-Arg-Pro-Gly (DSPE-PEG-APRPG)	49
9	**Combination of polymer therapeutics**	**50**
9.1	Polymer directed enzyme prodrug therapy (PDEPT)	50
9.2	Polymer-enzyme liposome therapy (PELT)	52
9.3	Combination of polymer therapeutics inducing oxidative stress	54
9.3.1	PEG-XO	54
9.3.2	PEG-DAO	54
9.3.3	PEG-ZnPP	55
9.3.4	PEG-DAC and PEG-ZnPP combination	55
10	**Conclusions**	**56**
References		**56**

Abstract Drug delivery systems for cancer therapeutics have revolutionized medicine. Delivery systems have improved the efficacy and reduced the toxicity of current therapies and resulted in the development of new ones. Today, millions of cancer patients have directly benefited from drug delivery systems, and polymers have been at the frontline of these technological advances. Targeted delivery systems of chemotherapeutics to the tumour compartment can be achieved systemically, either passively or actively. Polymer conjugation radically changes the pharmacokinetics of the bound drug, and conjugates with prolonged circulation times target tumours passively via the enhanced permeability and retention (EPR) effect. Polymer conjugates can also be modified with moieties to directly target the tumour cells or the tumour vasculature. In this chapter, we review the successful clinical application of polymer–protein conjugates, and promising clinical results arising from trials with polymer–anticancer-drug conjugates. Over the last decade more than twelve polymer-drug conjugates have entered Phase I/II clinical trial as intravenously injectable anticancer agents. Only one of the polymer conjugates that has reached clinical trial directly targets tumour cells, while another one targets the tumour vasculature. Conjugation to polymers may save the fate of the many promising drug/peptide chemotherapies that fail each year due to high toxicity or poor pharmacokinetics. Yet, these technologies have not been exploited to their full potential. Only a few combinations of a limited number of chemotherapeutic drugs and polymer delivery systems are being tested in clinical and preclinical trials today. Furthermore, genomics and proteomics research is producing novel peptides, proteins and oligonucleotides that lack effective delivery systems. Thus, the full potential for drug delivery systems based on NCEs (new chemical entities), such as "polymer therapeutics", lies ahead.

Keywords Angiogenesis · Drug targeting · EPR effect · HPMA copolymer · PEG · Polymer therapeutics

Abbreviations

Amino-DAQ	1,5-diazaanthraquinone derivative
ASCO	American Society of Clinical Oncology
ASGP	Asialoglycoprotein
ASGPR	Asialoglycoprotein receptor
ATWLPPR	Alanine-threonine-tryptophan-leucine-proline-proline-arginine
AUC	Area under the curve
BBB	Blood brain barrier
BCNU	1,3-bis(2-chloroethyl)-1-nitrosourea
bFGF	Basic fibroblast growth factor
CM	Carboxymethyl
CPT	Camptothecin
Da	Daltons
DAO	D-amino acid oxidase
DES	Diethylstilboestrol
DLT	Dose limiting toxicity
DMXAA	Dimethyl-xanthenone-4-acetic acid
DOTA	1,4,7,10-tetraazacyclododecane-N,N',N'',N'''-tetraacetic acid
Dox	Doxorubicin
DSPE	Distearoylphosphatidylethanolamine
EC	Endothelial cell
EGF	Epidermal growth factor
en	Ethylenediamine
EPR effect	Enhanced permeability and retention effect
FDA	Food and Drug Administration
FPLC	Fast protein liquid chromatography
HIV/AIDS	Human immunodefficiency virus/Acquired immunodefficiency syndrome
HO	Heme oxygenase
HPLC	High-pressure liquid chromatography
HPMA	N-(2-hydroxypropyl)methacrylamide
HuIg	Human immunoglobulin
i.p.	Intraperitonealy
i.v.	Intravenously
IFL	Irinotecan, fluorouracil, and [calcium folinate] leucovorin
IFN-α	Interferon-α
IFN-β	Interferon-β
IgG	Immunoglobulin
IL-6	Interleukin-6
LAK cells	Lymphokine-activated killer cell
LD$_{10}$	Dose of drug lethal to 10% of animals
MA	Methacryloyil
mAb	Monoclonal antibody
MAG	HPMA: methacryloyl-glycine (MA-Gly)-ONp 95:5 or 90:10
MMP-2	Matrix metalloproteinase-2
MMP-9	Matrix metalloproteinase-2
mPEG	monoPEG

MTD	Maximum tolerated dose
NCE	New chemical entities
NGR	Asparagine-glycine-arginine
NK	Natural killer cells
NSCLC	Non-small-cell lung carcinoma
O_2^-	Superoxide anion
ONp	*p*-Nitrophenyl
PAAm	Polyacrylamide
PCT	Paclitaxel
PDAAm	Polydimethylacrylamide
PDEPT	Polymer directed enzyme prodrug therapy
PEG	Polyethyleneglycol
PEG-G-CSF	PEGylated recombinant methionyl human granulocyte colony stimulating factor
PEI	Poly(ethyleneimine)
PELT	Polymer enzyme liposome therapy
PGA	Poly-*L*-glutamic acid
PLC	Phospholipase C
POG	Pediatric Oncology Group
PS2	Poor performance status 2
PVA	Polyvinyl alcohol
PVP	Polyvinylpyrrolidone
RES	Reticuloendothelial system
RGD	Arginine-glycine-aspartate
ROS	Reactive oxygen species
s.c.	Subcutaneously
ScFv	Single-chain Fv antibody fragment
SCID	Severe combined immunodeficiency disease
SCLC	Small-cell lung cancer
SMANCS	Styrene-co-maleic anhydride-neocarzinostatin
SS-NH-PEG	Succinimidyl ester of PEG
STELLAR	Selective targeting for efficacy in lung cancer, Lower adverse reaction
TBA	Thiobutylamidine
TEM	Tumor endothelial marker
TNFα	Tumor necrosis factor-α
VEGF/VPF	Vascular endothelial growth factor/Vascular permeability factor
VEGFR	Vascular endothelial growth factor receptor
VTA	Vascular targeting agents
XO	Xanthine oxidase
ZnPP	Zinc protoporphyrin

1
Introduction

Chemotherapeutic treatment of neoplastic diseases is often restricted by adverse systemic toxicity, which limits the dose of drug that can be administered, or by the appearance of drug resistance. Lack of selectivity is only one (albeit a major) obstacle hindering the optimisation of drug effectiveness. Others include inaccessibility of target, premature drug metabolism and allergic reactions [1]. There is a great demand for innovative drug delivery systems that can better target antitumour drugs and that can overcome resistance in its many forms. The question is: how can we meet these challenges?

A great deal of research has concentrated on ways to develop new cancer therapeutics that specifically target tumour cells compared with normal cells, exploiting the differences between neoplastic and normal tissues. These targeted therapies should be more effective and decrease toxicity to normal tissues.

Several systems have been developed in order to restrict the delivery of the chemotherapeutic agent to the tumour site. With the identification of cell-specific receptor/antigens on tumour cells [2] and tumour endothelial cells [3], it has been possible to actively target chemotherapeutic or antiangiogenic agents using ligand- or antibody-bearing delivery systems. Alternatively, the drug can be loaded into high-capacity drug carriers such as liposomes or entrapped in degradable polymers for sustained drug release and localized chemotherapy systems [4]. In the controlled polymer drug delivery systems, the active molecule is released continuously at therapeutic levels by polymer degradation and diffusion through the polymer pores. Clinical approved examples include Zoladex [5, 6], Lupron Depot, and Decapeptyl [7], which are injectable polymer rods or microspheres of luteinizing hormone-releasing hormone (LHRH) analogues for the treatment of advanced prostate cancer [4, 8]. Localized chemotherapy systems have been particularly appealing for the brain, where the presence of the blood–brain barrier limits delivery of therapeutics by blood. Gliadel, an implantable polymer wafer that locally delivers carmustine, has been used successfully for the treatment of malignant gliomas after surgery [9]. Interestingly, we found that HPMA copolymer-TNP-470 (caplostatin) [10] was able to treat orthotopic intracranial U87 human glioblastoma in mice [11], even though it does not cross the blood brain barrier, a fact that eliminated the neurotoxicity associated with the unconjugated TNP-470. This can be attributed to the leakiness of blood vessels in some brain tumours, allowing polymer conjugates to target these tumours by the EPR effect.

Drugs can also be conjugated to polymer carriers, named "polymer therapeutics" [12], that can be either directly conjugated to targeting proteins/peptides or derivatised with adapters conjugated to a targeting moiety. "Polymer therapeutics" [13] is a term used to describe polymeric drugs [14], polymer-drug conjugates [15], polymer-protein conjugates [16], polymeric micelles to which a drug is covalently bound [17], and multi-component polyplexes that are being developed as nonviral vectors [18] (Fig. 1). All subclasses consist of at least three parts: (a) a specific water-soluble polymer, either as the bioactive itself or as an inert functional part of a multifaceted construct for improved drug, protein or gene delivery; (b) a biodegradable polymer-drug linker, and; (c) the bioactive antitumour drug.

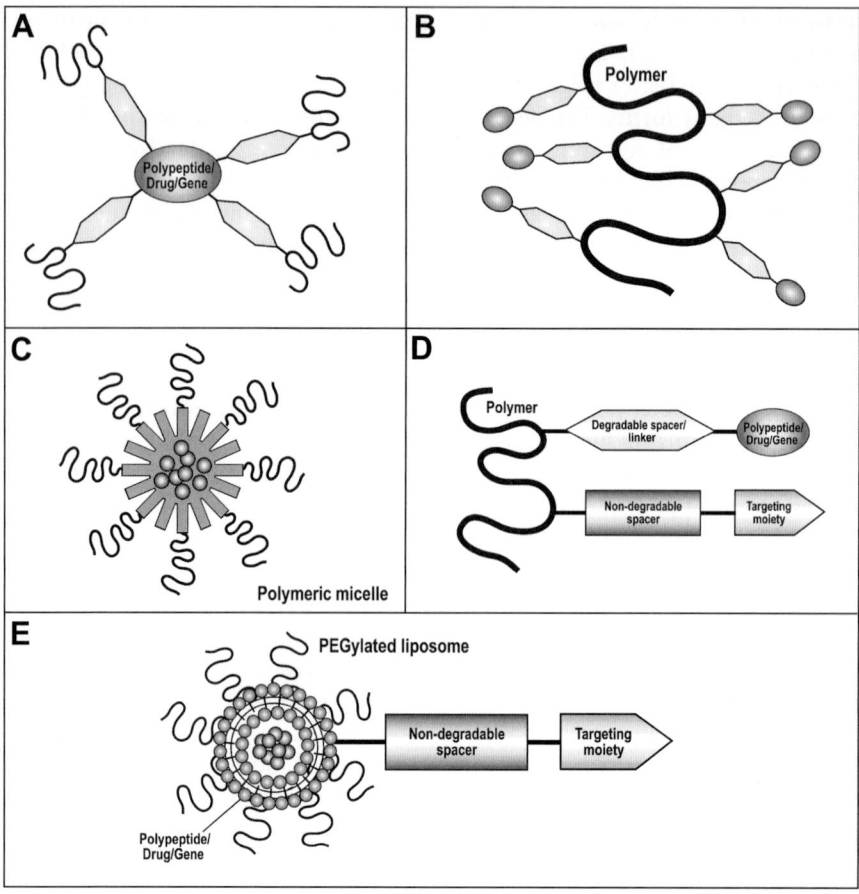

Fig. 1 Schematic diagram of possible combinations of actively targeted conjugates: **A** Soluble polymer-protein conjugate (20 nm) or polyplex: hydrophilic polymers bearing a cationic block-DNA complex (40–60 nm); **B** Soluble polymeric drug (5–15 nm) carrier (polymer therapeutics, modified from [87]); **C** Polymeric micelle (60–100 nm) – amphiphilic block entrapping a drug; **D** Soluble polymeric drug carrier bearing a targeting moiety (5–15 nm); **E** PEGylated stealth liposome carrying the active entity conjugated to a targeting moiety (200–500 nm)

Because in polymer therapeutics the drugs are chemically conjugated, they differ from controlled drug delivery systems in that they are more like new chemical entities (NCE). Not only is their pharmacokinetic profile distinct from that of the parent drug, but the route of cellular uptake may also differ, as the polymer-drug can only enter cells by the endocytic route, leading to lysosomotropic drug delivery. Several conjugates can release drug intracellularly while others release it extracellularly, depending on the polymer-drug linker and the activating moiety. While polymer therapeutics share many features with other macromolecular drugs and prodrugs (proteins, antibodies, and oligonucleotides, and immunoconjugates), their chemistry makes them amenable to flexible tailoring, for example of their molecular weight, number and types of drugs per polymer, targeting moieties and even bioresponsive elements [12]. Polymer-protein conjugates have made it to the clinic since the early 1990s, with the approval of polyethylene glycol (PEG)-adenosine deaminase, PEG-L-asparaginase and styrene maleic anhydride (SMANCS) [19]. During the past two decades, the field of polymer therapeutics has continued to grow due to the advances in both polymer chemistry and biological sciences, and promising results from clinical trials involving polymer-anticancer-drug conjugates [12]. With the emergence of hybrid biotechnologies, which combine the synthesis of innovative polymers with biological macromolecules (proteins, oligonucleotides, antibodies), a number of compounds have been developed that are suitable for clinical development and use (Tables 1, 2, and 3).

It is surprising, however, that with the abundance of novel drugs and targets offered in the post-genomic era and novel sophisticated chemistry available, only four drugs (doxorubicin, camptothecin, paclitaxel and platinate) and four polymers (HPMA copolymer, Poly-L-glutamic acid, PEG, and Dextran) are repeatedly used to develop these promising new polymer therapeutics. Therefore, we will examine here future directions and challenges in this field. The purpose of this chapter is to compare different therapeutic targeted delivery systems and strategies for chemotherapeutic and antiangiogenic agents, focusing on those polymer therapeutics that have been approved by the FDA or that are undergoing clinical and preclinical trials. The rationale for the design of preclinical lead compounds is summarised, and the challenges for effective and clinical development of these complex macromolecular prodrugs are discussed.

2
Passive or active targeting?

Targeting can be achieved either actively, by specifically including a recognition moiety into the carrier ("active targeting"), or passively, as a result of some physical or chemical characteristics of the carrier ("passive targeting") [20] (Fig. 2). The active approach relies upon the selective localisation of a ligand at a cell-specific receptor. Passive targeting refers to the exploitation of the natural (passive) distribution pattern of a drug-carrier in vivo. The latter is based upon mechanical entrapment of the carrier by shape or size or uptake by the cells of the reticuloendothelial system (RES). Maeda called the passive targeting phenomenon the "enhanced permeability and retention (EPR) effect" [21], and attributed it to two factors: the disorganised pathology of angiogenic tumour vasculature with its discontinuous endothelium,

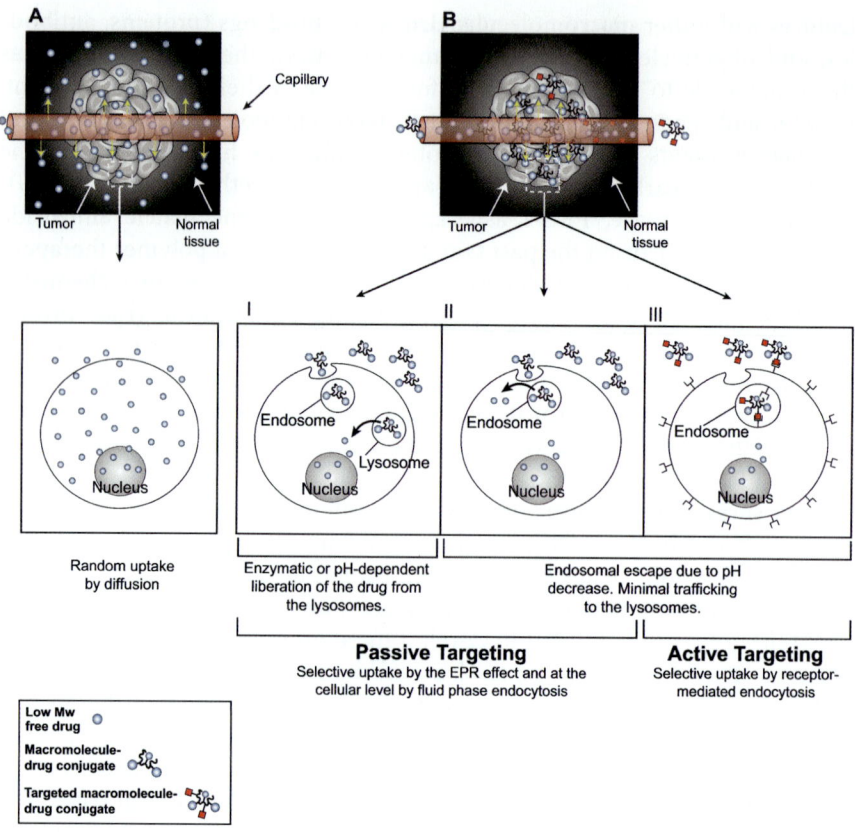

Fig. 2 Schematic representation of internalization of polymer-anticancer conjugates into cells. At the *top*, panel **A** represents low molecular weight (MW) drugs able to diffuse through normal vasculature to any tissue and internalise into cells. Panel **B** represents high molecular weight macromolecules unable to fenestrate through the normal vasculature, but able to extravasate through the leaky and permeable vasculature in the tumour tissue. At the *bottom*, low molecular weight drugs diffuse throughout the cells, while the macromolecule-drug conjugates internalize into the cells through passive targeting by the enhanced permeability and retention (EPR) effect. The conjugates can internalize by fluid phase pinocytosis through (I) the lysosomotropic pathway, releasing the drug in the lysosome by diffusion out to the cytoplasm and to its target; or (II) the endosomotropic pathway, where the bioresponsive polymer changes conformation and causes an increase in endosomal membrane permeability which allows the macromolecular drugs to escape into the cytoplasm; or (III) through active targeting by receptor-mediated endocytosis to cells presenting the target antigen

leading to hyperpermeability to circulating macromolecules, and the lack of effective tumour lymphatic drainage, which leads to subsequent macromolecular accumulation (Fig. 2). This concept is described in more detail elsewhere in this volume (Maeda, Greish and Fang). Long-circulating macromolecules – including albumin and polymer conjugates, polymeric micelles and liposomes – accumulate passively in solid tumour tissue by the EPR effect, and intravenously-administered drug delivery systems can increase the tumour concentration of antitumour drugs up to 70-fold. However, vasculature permeability and intertumoural hydrostatic pressure vary during tumour progression. These features, together with irregular blood flow, can lead to heterogeneous distribution of macromolecular medicines in tumour tissue [22]. Furthermore, some drugs cause a decrease in tumour vessel hyperpermeability [23] by themselves. Therefore, there is an increasing need for novel targeting moieties to target the tumour endothelial cells since the polymer conjugation causes a siginificant increase in the half-life of the free drug.

3
Targeting tumour cells or tumour vasculature?

The process of angiogenesis (Fig. 3), new capillary blood vessel growth from pre-existing vasculature [24], is now recognised as an important control point in cancer. Most tumours do not start out angiogenic, but remain as small, dormant tumours for years or a lifetime. They cannot grow until they can recruit new blood vessels. As a result, the microvascular endothelial cell recruited by a tumour has become an important second target in cancer therapy. A basis for this concept has been that tumour vascular endothelial cells are genetically stable unlike tumour cells. However, it has been recently reported that tumour endothelial cells could also acquire cytogenetical abnormality in the tumour microenvironment [25]. It is becoming clear that tumour vasculature is much more complex than initially thought.

Angiogenesis involves many growth factors and their receptors, cytokines, proteases and adhesion molecules [26, 27]; thus, multiple targets for therapeutic intervention and targeting opportunities for antiangiogenic therapy for cancer exist. Treating both the cancer cell and the endothelial cell in a tumour may be more effective than treating the cancer cell alone. Table 4 summarises the advantages of targeting the vessels of the tumour instead of, or in addition to treating the tumour itself.

Angiogenesis inhibitors are emerging as a new class of drugs. In the US there are currently more than 40 angiogenesis inhibitors in various clinical trials for late stage cancer. Members of this family of drugs differ by their targets and vary from low MW molecules to polypeptides and antibodies. The fact that the development of a functional vasculature within a tumour is a requisite for its growth and progression [28] has led to the design of therapies directed toward the tumour vasculature. These therapies aim either to prevent the formation of new vessels (antiangiogenic cytostatic agents such as Endostatin, Angiostatin and TNP-470, vascular endothelial growth

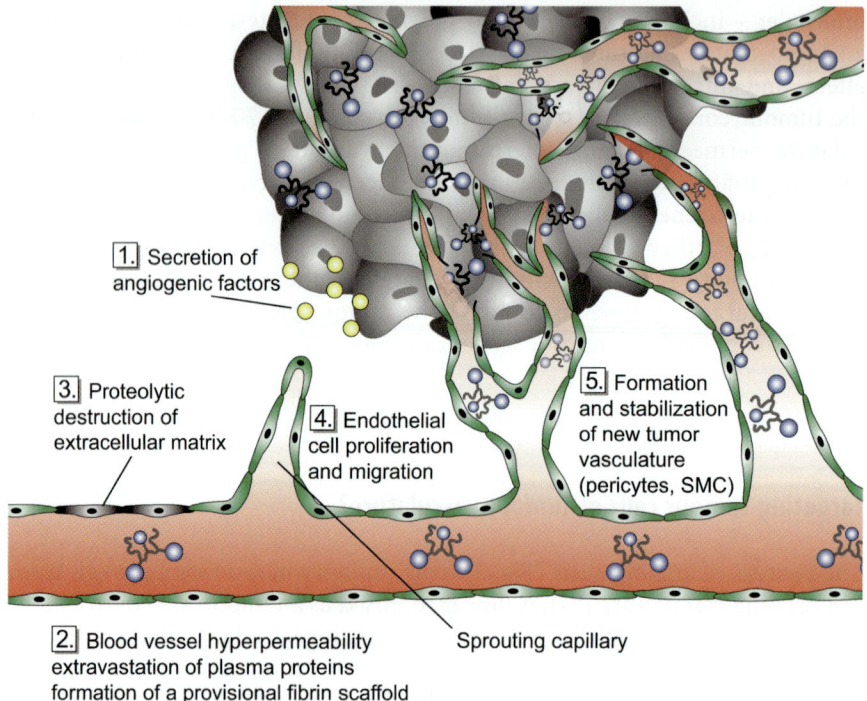

Fig. 3 Angiogenic steps. Schematic representation of the angiogenesis process. Polymer therapeutics can be used to target each one of the steps

factor (VEGF) antagonists or VEGF receptor (VEGFR) inhibitors); or to damage existing vessels (cytotoxic vascular targeting agents (VTA)). VTAs allow rapid destruction of existing blood vessels in tumours containing activated EC. They consist of antitubulin agents such as combretastatin [29] analogues and colchicine analogues. Other drugs such as flavone acetic acid analogue and dimethyl-xanthenone-4-acetic acid (DMXAA) induce TNF-α and serotonin and inhibit blood flow.

Despite the differences between various angiogenesis inhibitors, the common ground is that all can benefit from specific targeting. A proper delivery system would enable optimisation of their pharmacokinetic profiles. Currently, antiangiogenic proteins are delivered into the circulation and achieve their high therapeutic index by selective inhibition of proliferating and migrating EC in an angiogenic focus, without having a similar effect on quiescent endothelium in the remaining vasculature. If these direct angiogenesis inhibitors, such as thrombospondin, angiostatin and endostatin [30], could be targeted to the angiogenic focus in a tumour, potency could be potentially enhanced. However, for those angiogenesis inhibitors where dose cannot be increased because of side-effects (such as TNP-470, a Fumagillin analogue [31]), or for those whose efficacy is not improved by dose escalation, targeting to the microvascular endothelium in a tumour bed may greatly increase the usefulness of the inhibitor. Conjugating these novel antianigogenic agents and vascular targeting agents to polymers will significantly improve their efficacy by (i) increasing their tumour specificity and solubility; (ii) improving their pharmacokinetics; and (iii) reducing their toxicity. Here we will review some of the polymer therapeutics targeted to the vasculature and other agents which can benefit from being conjugated to polymers in the future.

4
Parenteral drug targeting

Modern approaches to parenteral drug targeting include liposomes, immunoconjugates, polymeric microparticles and biodegradable polymeric implants designed for localised or sustained-controlled release [2, 8]. Each has advantages and disadvantages. Immunoconjugates provide selective targeting, but antibodies can be immunogenic, their pharmacokinetics are governed by molecular weight, and they have a limited drug-carrying capacity. Liposomes have a high drug-carrying capacity, but stability can be an issue (either releasing the drug too quickly or entrapping too strongly), and they are prone to reticuloendothelial system capture. Polymer conjugates can be synthesised to specific molecular weight and composition, but their drug-carrying capacity is relatively low, and they can present challenges for characterisation. The past decade has seen the realisation that the ideal platform for drug delivery will marry the benefits of these three components into hybrid nanotechnolgies for each application. A PEGylated liposome DOXIL (containing doxorubicin) [32] is a successful anticancer treatment, and many polymer conjugates use antibodies to mediate cell specificity [33]. Another interesting hybrid is a polymerised cationic liposome that has been linked to an endothelial targeting ligand and used to deliver a mutant Raf gene [34]. Polymer-protein and polymer drug conjugates share many common features, but the biological rationale for their design is very different.

5
Polymer-protein conjugates

Recombinant DNA and monoclonal antibody technology have created a biotech revolution that is providing a growing number of peptide-, protein- and antibody-based drugs [35, 36]. Most of these proteins are limited in their clinical applications because of unexpectedly low therapeutic effects. The reason for this limitation is that these proteins are immediately decomposed by various proteases in vivo and are rapidly excreted from the blood circulation, leading to a short plasma half-life. Furthermore, they are limited by poor stability and, for proteins, immunogenicity. Consequently, frequent administration at an excessively high dose is required to reveal their therapeutic effects in vivo. As a result, homeostasis is destroyed, and unexpected side effects occur. Therefore, there has been a continuing search for improved alternatives. Bioconjugation with water-soluble polymers improves the plasma clearance and body distribution, resulting in increased therapeutic effects and decreased side effects. In the 1970s, pioneering research by Davis, Abuchowski and colleagues foresaw the potential of the conjugation of polyethylene glycol (PEG) to proteins [37]. This concept is described in more detail elsewhere in this volume (Pasut and Veronese). This technique is now well established and is called PEGylation [38]. PEGylation increases protein solubility and stability, and hence prolongs plasma half-life, and can reduce protein immunogenicity [39, 40]. Moreover, by preventing rapid renal clearance of small proteins and receptor-mediated protein uptake by cells of the reticuloendothelial system, PEGylation can be used to prolong plasma half-life. The resultant need for less frequent dosing is of great benefit to the patient and encourages compliance. PEG is a particularly attractive polymer for conjugation. It is widely used as a pharmaceutical excipient, and the flexible, highly water-soluble polymer chain extends to give a hydrodynamic radius that is some 5–10 times greater than that of a globular protein of equivalent molecular weight. There are three requirements for optimised synthesis of a polymer-protein conjugate: a semitelechelic polymer–that is, one with a single reactive group at one terminal end to avoid protein crosslinking during conjugation; the ability to introduce a linker that will not generate toxic or immunogenic by-products and that will provide appropriate stability characteristics (dependent on the protein being bound); and an approach that will provide reproducible site-specific protein modification. Of paramount importance is the maintenance of the biological activity of the protein after PEGylation and covalent binding to the polymer. Recently, improved conjugation techniques have been developed, including site-specific modification following protein mutagenesis [41, 42], the use of the enzyme transglutaminase to PEGylate selectively at glutamine in the protein [43], and the design of degradable PEG-protein linkages to maximize the return of protein bioactivity [43]. With increasingly sophisticated conjugate design, many of the early challenges for the clinical development of polymer-protein conjugates are being met [44]. Although there are few available polymers for conjugation such as HPMA copolymers, poly(ethyleneimine) (PEI), linear polyamidoamines, polyvinylpyrrolidone (PVP), polyacrylamide (PAAm), polydimethylacrylamide (PDAAm), polyvinyl alcohol (PVA), chitosan, dextrin and dextran, PEG is still the most popular option since the clinical value of PEGylation is now well established. PVP, PAAm, and PDAAm could be functionalised by introducing various comonomers upon radical polymerisation. PVA and dextran have many primary OH groups that can be used for bioconjugation on the side chain. Introduction of other polymers for protein conjugation include the following conjugates: HPMA copolymer-β-lactamase [45], HPMA copolymer-cathepsin B [46], polyvinylpyrrolidone (PVP)-TNFα [47, 48], PVP-IL-6 [49], and so on.

5.1
PEG-adenosine deaminase

PEG-adenosine deaminase (ADAGEN; Enzon) was the first PEGylated protein to enter the market, in 1990 [50]. It is used to treat adenosine deaminase-deficient X-linked severe combined immunodeficiency disease (SCID), commonly known as the "bubble boy disease". It is an alternative to bone marrow transplantation and enzyme replacement by gene therapy. Since the introduction of ADAGEN, a large number of PEGylated-protein and -peptide pharmaceuticals have followed (Table 1).

Table 1 Polymer-protein conjugates in preclinical, early clinical trials or on the market

Polymer-protein	Name	Indication	Status	Refs.
PEG-adenosine deaminase	Adagen	SCID syndrome	1990	[50]
Styrene maleic anhydride-neocarzinostatin (SMANCS)	Zinostatin, Stimalmer	Hepatocellular carcinoma	1993 (Japan)	[240]
PEG-L-asparaginase	Oncaspar	Acute lymphoblastic leukemia	1994	[241]
PEG-interferon-α 2b	PEGINTRON	Hepatitis C	2000	[79]
PEG-interferon-α 2b	PEGINTRON	Cancer (Renal cell carcinoma, hemangiomas, angioblastomas, giant cell tumours), multiple sclerosis, HIV/AIDS	Various clinical trials	[79, 80, 161]
PEG-interferon-α 2a	PEGASYS	Hepatitis C	2002	[242, 243]
PEG-human growth hormone (HGR)	Pegvisomant	Acromegaly	2002 (EU)	[244]
PEG-granulocyte-colony stimulating factor (G-CSF)	PEG-filgrastim, Neulasta	Prevention of chemotherapy-associated neutropenia	2002	[55]

Table 1 (continued)

Polymer-protein	Name	Indication	Status	Refs.
PEG-anti-TNFα Fab	CD870	Rheumatoid arthritis	Phase II	[42]
PEG-insulin		Diabetes	Pre-clinical	[245, 246]
EGF-N-hydroxysuccinimide-PEG3400-biotin to OX26/streptavidin. (Transferrin)		Imaging human brain tumors	Pre-clinical	[84]
HPMA copolymer-GLy-Gly-cathepsin B		Activating moiety in a PDEPT combination	Pre-clinical	[221]
HPMA copolymer-Gly-Gly-β-lactamase		Activating moiety in a PDEPT combination	Pre-clinical	[45]
PVP-IL-6		Enhancement of thrombopoietic activity	Pre-clinical	[49]
PVP-TNFα		Sarcoma-180	Pre-clinical	[47, 48]

5.2
PEG-L-asparaginase

The chemotherapy agent L-asparaginase has been an important part of acute lymphoblastic leukemia therapy for over 30 years. Leukemia cells have a requirement for the amino acid L-asparagine. Two of the main disadvantages of the drug are the need for frequent intramuscular injection and a very high rate of allergic reactions. Because of this, L-asparaginase seemed like an ideal target for PEGylation and PEG-L-asparaginase was developed in the 1970s and 1980s. PEG-L-asparaginase (ONCASPAR; Enzon) was the first antitumour PEGylated protein to be approved for clinical use in 1994 as a treatment for acute lymphoblastic leukaemia. The conjugate contains multiple PEG chains of MW $\sim$ 5000 Da linked to the enzyme, and it has successfully overcome many of the problems associated with the use of L-asparaginase derived from *E. coli*. The conjugate has undergone extensive testing and appears to retain its antileukemic effectiveness while allowing less frequent administration than the native compound. Compared with the native enzyme, PEG-L-asparaginase has the advantages of reduced hypersensitivity, a longer plasma half-life (8–30 h to ∼ 14 days) and slower total clearance [51]. Consequently, PEG-*L*-asparaginase can be administered by a 1 h infusion every two weeks, instead of the 2–3 times per week for four weeks required for the native enzyme to ensure depletion of circulating levels of *L*-asparagine. *L*-Asparagine levels fall below the limit of detection within an hour and this effect is sustained throughout the two-week interval [51]. Recently, PEG-*L*-asparaginase was administered weekly based upon the results of the Pediatric Oncology Group (POG) 9310, which demonstrated the superiority of weekly administration compared with every other week PEG-*L*-asparaginase in the induction for relapsed ALL [52]. While the actual cost to patients for PEG-*L*-asparaginase is greater than that of multiple injections of other *L*-asparaginases, the reduced need for physician visits and treatment of complications of therapy may make overall treatment costs considerably less than that of the conventional *L*-asparaginases. Most importantly, PEGylation of *L*-asparaginase decreases hypersensitivity reactions to only 8% of patients and the conjugate can be used in combination with chemotherapy to treat patients that are hypersensitive to the native enzyme. Recently, a Phase I-II trial of PEG-*L*-asparaginase was conducted in patients with multiple myeloma [53]. In the 17 patients who were evaluable for response, a complete response was observed in one patient after four doses, and stable disease was observed in eight patients.

5.3
PEG-granulocyte-colony stimulating factor (PEG-G-CSF)

Recombinant methionyl G-CSF (filgrastim) is a protein of MW 19 000 Da obtained by bacterial fermentation from *E. coli*. It was developed by Amgen Inc. as an adjunct to chemotherapy. Administration of G-CSF moderates the neutropenia frequently induced by cytotoxic chemotherapy. Recently, PEGylated recombinant methionyl human granulocyte colony stimulating factor (PEG-G-CSF; pegfilgrastim; PEG-rmetHuG-CSF; Neulasta; Amgen) was approved for clinical use [54], and is used to prevent severe cancer chemotherapy-induced neutropenia [55]. The covalent attachment of PEG to the *N*-terminal amine group of the parent molecule was attained using site-directed reductive alkylation. In this case, a single chain of PEG of MW ∼ 20 000 Da is bound to the protein. PEGylation increases the size of filgrastim so that it becomes too large for renal clearance. Consequently, neutrophil-mediated clearance predominates during elimination of the drug. Conjugation to PEG extends the median serum half-life of pegfilgrastim to 42 hours, compared with 3.5–3.8 hours for filgrastim, though in fact the half-life is variable, depending on the absolute neutrophil count, which in turn reflects the ability of pegfilgrastim to sustain production of those same cells. Because of this, PEG-G-CSF has the benefit of less frequent administration, being given by a single subcutaneous injection (100 µg/kg s.c.) on day 2 of each chemotherapy cycle.

The native G-CSF must be given daily (5 μg/kg/day s.c.) for two weeks to achieve the same protection. Although some allergic reactions have been observed following administration of GCSF, none were observed in clinical trials using the PEG-G-CSF, and the major side-effect observed with both the free form and the conjugate is bone pain [56, 57].

5.4
PEG-interferon-α (IFN-α)

IFNs are multifunctional regulatory cytokines involved in the control of cell function and replication, and IFN-α and IFN-β directly inhibit the proliferation of tumour cells of different histological origins [58, 59]. IFN-α is also an antiangiogenic molecule. The first evidence that IFN-α had antiendothelial activity was reported in 1980 when it was found to inhibit the motility of vascular endothelial cells in vitro in a dose-dependent and reversible manner [60], and subsequently found to inhibit angiogenesis in vivo [58]. Recent studies indicate that IFN-α and IFN-β can also down-regulate the expression of proangiogenic molecules, such as basic fibroblast growth factor (bFGF) [61], IL-8 [62], and the metalloproteases, matrix metalloprotease-2 (MMP-2) and MMP-9 [63], and to activate host effector cells [64].

IFN-α has been widely used not only as an antiviral agent to treat chronic hepatitis, but also as a cytotoxic agent to treat certain leukaemias and some bladder cancers [58]. Experimental studies in mice showed that the antiangiogenic efficacy of IFN-α is optimal at low doses and declines at higher doses [65]. At low doses, IFN-α inhibits tumour-cell production of bFGF [61, 66], as well as endothelial-cell motility, and therefore can be considered to have both direct and indirect antiangiogenic activity [60]. For therapy of human bladder carcinoma grown in athymic nude mice, the optimal biological dose (the dose that exerts a maximal antiangiogenic effect) is relatively low; for example 10 000 units administered on a daily schedule [65]. Low dose IFN-α is the most successful therapy against life-threatening hemangiomas [67, 68], where new blood vessel growth is associated with increased expression of bFGF [69–72]. Certain tumours, such as giant-cell tumours of the bone and angioblastomas, only or mainly produce bFGF, making them ideal candidates for interferon therapy. When patients with these tumours were treated with interferon (IFN)-α at low daily doses, drug resistance was not observed with therapy of 1–3.5 years duration [69, 70].

Pharmacokinetic studies have demonstrated that the half-life of IFNs in the circulation of patients is on the order of minutes [73], thus making therapeutic levels difficult to sustain, which may in turn compromise the effectiveness of IFN in the therapy of bFGF producing solid tumours [73, 74]. PEGylation of IFN delays its clearance and reduces its immunogenicity [74]. Therapeutic levels can be maintained through once-weekly dosing [75]. Recent clinical trials have shown that PEG-IFN-α is significantly more effective than free-form IFN-α in the treatment of chronic hepatitis C [76, 77], and differences in clinical outcome have been related to the ability of PEGylation to sustain absorption, restrict the volume of distribution, and reduce clearance of IFN-α [76]. Two PEG-interferon-α conjugates (IFN-α-2a and IFN-α-2b), Pegasys (Roche) [78] and PEG-Intron (Schering) [79] have been approved as treatments for hepatitis C [38]. IFN-α-2a and IFN-α-2b display similar biological activities and only differ in respect to a single amino acid, but the molecular weight of PEG used for conjugation and the linker employed is very different in each product. Pegasys is a conjugate containing Roche's IFN-α-2a linked to a 40 000 Da branched PEG. PEG-Intron is a conjugate of Schering-Plough's IFN-α-2b linked to a 12 000 Da PEG. The latter is synthesised using degradable carbamate linker to histidine residues in the protein, designed to minimize loss of protein activity. Consequently, Pegasys has a higher specific activity in vitro and a longer plasma half-life than PEG-Intron. Both conjugates have shown clinically superior antiviral activity compared to IFN-α. PEG-IFN-α is also under clinical evaluation in other indications, including cancer, multiple sclerosis and HIV/AIDS. Efficacy of IFN-α in the treatment of melanoma and renal cell carcinoma is well established, but there are problems, including toxic side effects (mild to moderate nausea, anorexia, fatigue and depression) and a short plasma half-life ($t_{1/2}$ = 2.3 h) that necessitate administration three times per week. In a Phase I/II study, PEGylated IFN-α-2b was given by subcutaneous injection (0.75 to 7.5 µg/kg s.c.) once per week for twelve weeks to patients with advanced solid tumours (primarily, renal cell carcinoma). PEGylated IFN-α-2b had a maximum tolerated dose (MTD) of 6.0 µg/kg/week, and produced an objective response rate of 14% in 44 previously untreated renal-cell carcinoma patients and was well-tolerated [80].

5.5
Styrene-co-maleic anhydride-neocarzinostatin (SMANCS)

Maeda and colleagues characterized a conjugate of two polymer chains of styrene maleic anhydride (SMA) covalently bound to the antitumour protein neocarzinostatin (NCS) [81]. It was originally synthesised to increase NCS lipid solubility and thus facilitate therapeutic artery administration (via femoral artery access) in the phase-contrast agent Lipidol and to increase the plasma half-life of NCS. SMANCS demonstrated significant antitumour activity in a number of animal models and moreover the remarkable tumour/blood ratio for SMANCS in Lipidol of > 2500 measured in a rabbit liver tumour model demonstrated much higher targeting than reported for any other system [82]. The first clinical evaluation of SMANCS was reported in 1983. 44 patients, mostly with nonresectable hepatoma, were treated with SMANCS, 86% demonstrated decreased α-fetoprotein levels and 95% demonstrated a decrease in tumour size [83]. SMANCS (3–4 mg) can be administered every 3–4 weeks, while X-ray imaging confirmed selective retention of SMANCS in the tumour. Subsequent and more rigorous evaluation has validated these early trends, and a multi-center phase II study involving 400 patients with primary hepatoma has been completed in Japan [83]. In this study the sizes of the tumors were measurable in 322 out of 400 patients, and of these 322, 308 patients displayed regression. One year after the initial administration of the drug, all tumours were reduced to < 50% of the initial size [83]. SMANCS received market approval in Japan in 1990 for hepatocellular carcinoma. SMANCS is described in more detail elsewhere in this volume (Maeda, Greish and Fang).

5.6
Preclinical polymer-protein conjugates

PEGylated construct of ^{125}I-EGF and a MAb to the transferrin receptor

Human brain gliomas overexpress the receptor for epidermal growth factor (EGF), and radiolabeled EGF is a potential peptide radiopharmaceutical for imaging human brain tumours, should this peptide be made transportable through the blood-brain barrier (BBB) in vivo. To achieve this goal, peptide drug delivery to the brain has been explored by conjugating a peptide radiopharmaceutical ^{125}I-EGF to a BBB drug delivery vector [84]. The OX26 monoclonal antibody (MAb) undergoes receptor-mediated transcytosis through the BBB via the brain capillary endothelial transferrin receptor, and is therefore a good delivery vector to the brain. In this study, the radiopharmaceutical was conjugated to the delivery vector via the streptavidin/biotin complex using either a short linker or a PEGylated linker. EGF was monobiotinylated with either N-hydroxysuccinimide-bis(aminohexanoyl)-biotin or hydroxysuccinimide-PEG3400-biotin, which uses poly(ethylene glycol) (PEG) of 3400 Da molecular mass (PEG3400), and both constructs were evaluated. Attachment of [^{125}I]EGF-PEG3400-biotin to the OX26/streptavidin conjugate did not impair binding of the construct to the EGF receptor in C6 glioma cells [84]. These studies demonstrate that the use of the extended PEG linker releases steric hindrance of MAb transport vectors upon binding of EGF to its cognate receptor on glioma cells. Attachment of EGF peptide radiopharmaceuticals to BBB drug delivery systems such as the OX26 MAb using extended PEG linkers allows for retention of the bifunctionality of the conjugate with binding to both EGF and transferrin receptors.

6
Polymer-drug conjugates

Cancer drug targeting is a rapidly developing research discipline that has recently yielded a number of different drug-delivery approaches. One such strategy has been to couple small-molecular-weight cancer drugs to polymers. This coupling results in an altered biodistribution of the drugs following intravenous administration that favors concentration of the drug in tumours. Normally, low-molecular-weight anticancer drugs will nonselectively penetrate most tissues, because they pass rapidly through cell membranes. This results in a relatively rapid distribution of the drug with no tumour selectivity. In the case of polymer-drug conjugates, however, the polymer-drug linkages are designed to be stable in the bloodstream. This feature, along with the fact that the higher molecular weight polymer-drug can only gain entry into cells via endocytosis, results in circulation of the polymer-drug for a longer period than with the drug alone. Because most normal tissues have nonleaky microvasculatures, the polymer-drug accumulates more in tumour tissue, which has a notoriously leaky vascular supply. Conjugates with prolonged circulation times target tumours by the EPR effect. Once in the tumour interstitium the polymer-drug can only enter cells by the endocytic route, leading to lysosomotropic drug delivery. Several conjugates have peptidyl polymer-drug linkers amenable to cleavage by lysosomal thiol-dependent proteases. In this case, prodrug activation occurs intracellularly. In contrast, other conjugates that contain an ester link between drug and polymer can release drug by chemical hydrolysis or esterase degradation extracellularly. There are currently more than 12 different polymer-drug conjugates at various stages of clinical trials and further compounds are reported in preclinical development (Table 2). The first synthetic polymer-drug conjugate to enter clinical evaluation was HPMA copolymer-doxorubicin (PK1, FCE28068) in 1994 (Fig. 4). Since then five other anticancer compounds and two gamma camera imaging agents (^{131}I or ^{123}I-labelling) conjugated to this polymer have been evaluated clinically.

Tumour-specific polymer-drug conjugates can also be created by adding specific targeting moieties to the polymer to aid in treatment of specific tumours (for example by adding galactosamine to target hepatocellular carcinoma). HPMA copolymer-Gly-Phe-Leu-Gly-doxorubicin containing galactosamine (PK2, FCE28069) is the only polymer-drug conjugate bearing a target ligand to be tested clinically [85] (Fig. 5). PK2 is tumour-specific in the sense it targets a receptor as hepatocellular carcinoma. Although most of the conjugate was localised to normal liver hepatocytes, it was estimated that the doxorubicin concentration in hepatoma would still be 12–50-fold higher than could be achieved by the administration of free drug. With a better understanding of biological targets combined with advances in synthetic chemistry, novel and improved polymer therapeutics for cancer treatment are in constant development and should soon be on the market [13].

Also in the 1970s, the combination of De Duve's realization that the endocytic pathway might be useful for "lysosomotropic drug delivery" [86] (Fig. 2) and Ringsdorf's vision of the idealized polymer chemistry for drug conjugation [87] produced the concept of targetable polymer-drug conjugates (Fig. 1). Whereas protein PEGylation was born from the desire to improve the

Table 2 Polymer-drug conjugates in early clinical trials as anticancer agents

Polymer-drug (linker)	Name	Company	Status	Refs.
Dextran-doxorubicin	AD-70, DOX-OXD		Phase I	[137]
HPMA copolymer-doxorubicin (Amide)	PK1, FCE 28068	UK Cancer Research Campaign/Pfizer Inc.	Phase II	[15, 96]
HPMA copolymer-doxorubicin-galactosamine (Amide)	PK2, FCE 28069	UK Cancer Research Campaign/Pfizer Inc.	Phase I/II	[15, 85]
HPMA copolymer-paclitaxel (Ester)	PNU166945	Pfizer Inc.	Phase I	[109]
HPMA copolymer-camptothecin (Ester)	MAG-CPT, PCNU 166148	Pfizer Inc.	Phase I	[247]
HPMA copolymer-platinate (Malonate)	AP5280	Access Pharmaceuticals Inc.	Phase I/II	[91, 117, 118]
PGA-paclitaxel (Ester)	CT-2103, XYOTAX	Cell Therapeutics Inc./Chugai Pharmaceutical Co Ltd.	Phase II/III	[119, 120, 125]
PGA-camptothecin (Ester)	CT-2106	Cell Therapeutics Inc.	Phase I	[128]
PEG-camptotecin (Ester)	Prothecan	Enzon Inc.	Phase II	[133]
PEG-paclitaxel			Phase I	[115]
Polysaccharide (CM-dextran-polyalcohol)-camptothecin	DE-310	Daiichi Pharmaceutical Co., Ltd.	Phase I	[151]
PEG-aspartic acid-doxorubicin micelle (Amide/free drug)	NK911	National Cancer Institute Japan	Phase I	[140, 248]

Table 2 (continued)

Polymer-drug (linker)	Name	Company	Status	Refs.
HPMA-DACH platinate	AP5346	Access Pharmaceuticals	Phase I	Proceedings of the 16th EORTC-NCI-AACR Symposium, October 2004, Geneva.
Paclitaxel-incorporating micellar nanoparticle formulation	NK105	National Cancer Institute Japan	Phase I	Matsumura (personal communication)

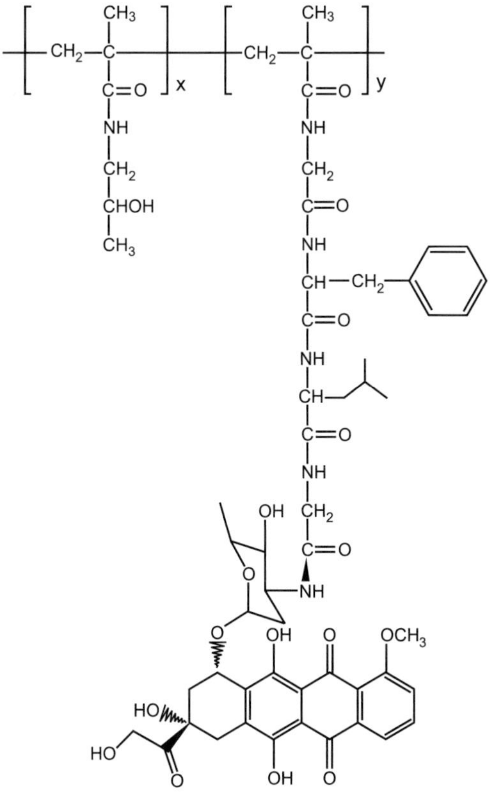

Fig. 4 HPMA copolymer-doxorubicin (PK1, FCE28068)

Table 3 Antiangiogenic versus conventional therapy

	Endothelial cells (EC)	Tumour cells (TC)
Effect	1 EC supplies O_2 and nutrients for 50–100 TC. Applicable to all solid tumours	All TC
Accessibility	EC are in direct contact with the circulation	"Hidden" in the tumour tissue
Targeting therapy	Tumour associated – EC possess unique phenotypic characteristics	Few identified specific tumour cell antigens for selective tumour types
Side effects/Toxicity	Few or no side effects	Yes
Duration of treatment	Long	Short
Expected regression	Slow	Fast
Goal	Stable disease Regression to avascular State	Eradication of Tumour Cells

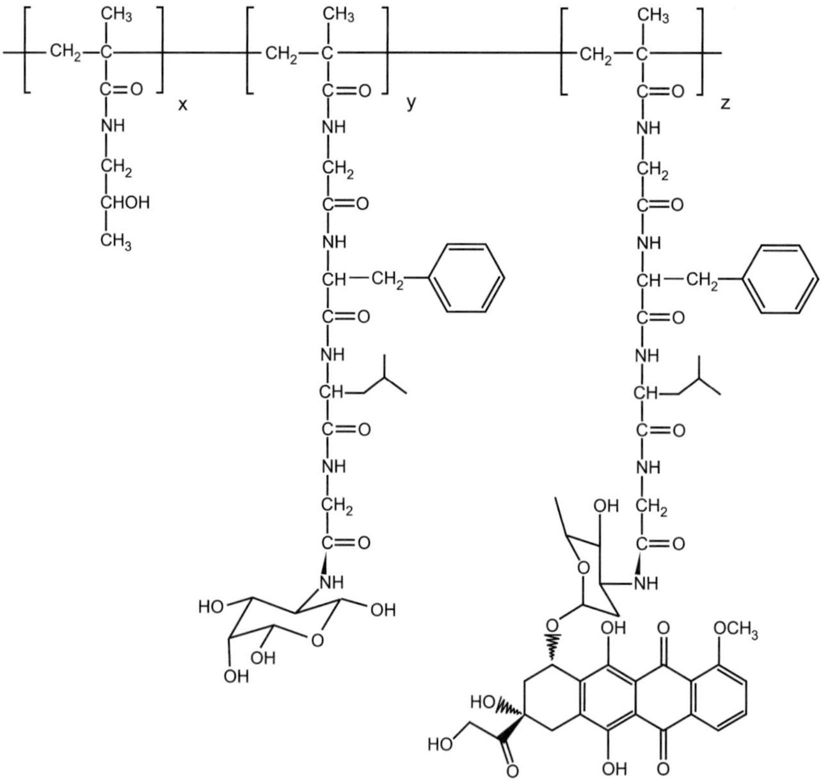

Fig. 5 HPMA copolymer-doxorubicin-galactosamine (PK2, FCE28069)

properties of protein pharmaceuticals, polymer-drug conjugation was seen as a means to improve the cell specificity of low-molecular-weight drugs. Supplementary features are needed to design an effective polymer-drug conjugate. These include: a bioresponsive polymer-drug linker that is stable during conjugate transport and able to release drug at an optimum rate on arrival at the target site; adequate drug-carrying capacity in relation to the potency of the drug being carried; and the ability to target the diseased cell or tissue by an active (receptor-ligand) or a passive (pathophysiological) mechanism. As the drugs carried often exert their effects via an intracellular pharmacological receptor, it is essential that they eventually access the correct intracellular compartment [12, 86, 87]. Routinely used cytotoxic chemotherapy distributes randomly in the body, and this feature, which is frequently combined with poor tumour selectivity in the mechanism of action, results in a relatively low therapeutic index. Those common solid tumours (breast, prostate, lung and colon cancer) that are the major cause of cancer mortality are particularly difficult to treat, hence the global quest for improved tumour targeting. Many researchers are trying to design improved low-molecular-weight prodrugs [88]. However, covalent attachment of chemotherapy to a polymeric carrier is particularly attractive, as the increased molecular weight produces a radical change in the pharmacokinetics at both the whole body and cellular levels [89] (Fig. 2). Initially, it was believed that receptor-mediated targeting would be a prerequisite for tumour selectivity, and conjugates have been synthesised to contain a plethora of ligands, including antitumour antibodies and peptides. So far, no tumour-specific conjugate has progressed into clinical development. The realization that the prolonged plasma circulation of polymer-conjugated drug itself led to significant passive tumour targeting [21, 90, 91] by the "enhanced permeability and retention" (EPR) effect (Fig. 2) [21] did, however, pave the way for the continued clinical development of simpler polymer conjugates that contain only covalently bound drug but no targeting ligand (Table 2). Careful tailoring of polymer-drug linkers is essential to the creation of a polymeric prodrug that is inert during transport but allows drug liberation at an appropriate rate intratumourally.

6.1
HPMA copolymer-Gly-Phe-Leu-Gly-doxorubicin (PK1, FCE28068)

HPMA homopolymer was originally developed by Kopecek and colleagues as a plasma expander [15]. Collaborative research with Duncan and colleagues in the early 1980s produced two HPMA copolymer-doxorubicin conjugates [15] that subsequently progressed into Phase I/II evaluation. PK1 was synthesised by aminolysis of an HPMA copolymer-Gly-(D,L)Phe-Leu-Gly-ONp precursor (HPMA:MA-peptide-ONp 95:5) with doxorubicin (Fig. 4) [40–42]. The conjugate used in the Phase I clinical trial had a molecular weight of $\sim 30\,000$ Da; a doxorubicin content ~ 8.5 wt % and a free doxorubicin content < 2% of the total doxorubicin. This tetrapeptide Gly-Phe-Leu-Gly linker is stable in the circulation [92], but is cleaved by the lysosomal thiol-dependent proteases, particularly cathepsin B [93, 94] following endocytic uptake of conjugate from the tumour interstitium (Fig. 2). This approach resulted in the concentration of approximately 70 times more doxorubicin in mouse melanoma tumours than in normal tissues. Importantly, this approach also increased the maximum tolerated dose of the polymer-drug by up to ten times that of the free drug.

Preclinical studies showed remarkable antitumour activity of PK1 coupled with reduced toxicity. These results led to the move to clinical evaluation [20, 95]. During Phase I PK1 was administered as a short infusion every three weeks [96]. Dose escalation progressed cautiously as neither polyHPMA nor any HPMA copolymers had previously been administered to humans. A starting dose of 20 mg/m^2 (doxorubicin-equivalent) was chosen (1/10th the LD$_{10}$ in mouse preclinical studies [97]) and escalation progressed to a maximum tolerated dose (MTD) of 320 mg/m^2 (doxorubicin-equivalent) [96]. No polymer-related toxicity (or immunogenicity) was observed. The dose-limiting toxicity (DLT) was typical of the anthracyclines and included febrile neutropenia and mucositis. Alopecia was not seen until doses > 180 mg/^{2} and nausea was mild without the need for antiemetics until doses of $\geq$ 240 mg/^{2}.

PK1 administration produced two partial and two minor responses in the cohort of 36 patients enrolled. These were in non-small-cell lung cancer (NSCLC), colorectal cancer and anthracycline-resistant breast cancer at 80 mg/m^2 (doxorubicin-equivalent) and anthracycline naive breast.

Preclinical pharmacokinetics showed very little free doxorubicin in plasma after PK1 administration to mice [98]. In Phase I trials PK1 was administered as a short infusion every three weeks, and the maximum tolerated dose was 320 mg/m^2 (doxorubicin-equivalent) [96]. This is a four- to five-fold increase compared with the normal safe dose of free drug. The dose-limiting toxicities seen were typical of the anthracyclines, and included febrile neutropenia and mucositis. Despite cumulative doses of up to 1680 mg/m^2 (doxorubicin-equivalent), no cardiotoxicity – a side-effect that is typical of anthracyclines – was observed. Antitumour activity seen in patients considered to be chemotherapy resistant/refractory and at lower doxorubicin doses (80–180 mg/m^2) was consistent with EPR-mediated targeting, although gamma camera imaging conducted as part of this study had poor resolution and failed to show clear evidence of selective tumour localization in all patients. Clinical pharmacokinetics assessed by HPLC and gamma camera imaging using a ^{131}I-labelled analogue confirmed prolonged plasma half-life when doxorubicin was administered in conjugate form ($t_{1/2}\alpha = 1.8$ h and $t_{1/2}\beta = 93$ h), absence of liver accumulation and rapid renal elimination (50–75% over 24 h) [96]. Polymer-bound doxorubicin detected in plasma was always higher (> 1000) than levels of free doxorubicin. Uptake was seen in known tumour sites in 6 of the 21 patients studied. Tumour levels of radioactivity were 2.2% dose at 2–3 h, 1.3% dose at 24 h and 0.5% dose after eight days. Neither dose-dependency in pharmacokinetics, nor contributory clinical factors causing changes in the clearance of polymer-bound or free drug were observed [96, 99]. Most importantly, this pivotal clinical study confirmed that cumulative doses of HPMA copolymer > 20 g/m^2 could be administered without signs of immunogenicity or polymer-related toxicity. The fact that the rodent models established to document the preclinical pharmacokinetics and toxicology of PK1 [15] correlated well with the subsequent clinical observations [96] validated the approach as a useful preclinical predictive tool.

A PK1 dose of 280 mg/m^2 was recommended as the Phase II dose. PK1 is currently undergoing phase II trials for breast, non-small-cell lung and colon cancers. Although full publication of the Phase II results are awaited it has been reported that PK1 showed no activity in a colorectal cancer, but activity has been seen in breast and NSCLC.

6.2
HPMA copolymer-Gly-Phe-Leu-Gly-doxorubicin-galactosamine (PK2, FCE28069)

PK2 is the only polymer-drug conjugate bearing a targeting ligand to be tested clinically. It was designed as an asialoglycoprotein (ASGP) biomimetic with the aim of targeting the hepatocyte and hepatocellular carcinoma ASGP receptor for the treatment of liver cancer. It should be emphasized that both normal hepatocytes and hepatoma bear the ASGP receptor. In this case the polymer precursor is synthesised using a monomer feed ratio of HPMA : MA-peptide-ONp adjusted to 90 : 10 [100–102] to provide the additional side-chains needed for conjugation of both doxorubicin and galactosamine. PK2 has a MW ~ 25 000 g/mol, a doxorubicin content of ~ 7.5 wt %, and a galactosamine content of 1.5–2.5 mol % (Fig. 5). Free doxorubicin content < 2% total doxorubicin.

This galactosamine-containing conjugate was the first synthetic biomimetic (of an asialoglycoprotein) polymer. It is also the only "targeted" polymer-drug conjugate so far to enter GCP clinical trial. Physiologically, loss of terminal sialic acid residues from ageing glycoproteins exposes galactose signalling plasma clearance. Endocytic uptake by hepatocytes after interaction with the asialoglycoprotein receptor (ASGPR) leads to lysosomal trafficking and their subsequent degradation [103]. PK2 was designed with the aim of improving treatment of primary hepatocellular carcinoma and metastatic liver disease. The latter would only be feasible if released drug diffused into liver metastases and could act via the "bystander effect". PK1 is not inherently hepatotropic, but preclinical studies confirmed that addition of galactosamine to HPMA copolymers promoted significant hepatocyte targeting after i.v. injection (~ 80% of dose) [104]. The magnitude of targeting (% dose) was markedly dose-dependent due to ASGP receptor saturation [105].

Phase I/II trials were conducted in 31 patients of which 23 had primary hepatoma [85] and the compound was given by intravenous infusion every three weeks, initially at an infusion rate of 4.16 ml/min (2 mg/ml doxorubicin-equivalent), but due to pain during infusion this was reduced to an infusion rate of 2 ml/min with a 1.0 mg/ml solution. Six patients were given FCE28069 by 24 h infusion to see if this improved targeting efficiency. The starting dose was again 20 mg/m^2 (doxorubicin-equivalent) and the FCE28069 MTD was 160 mg/m^2. The dose-limiting toxicity was typical of anthracyclines, principally myelosuppression and mucositis [85]. Interestingly the PK2 MTD was significantly lower than seen for PK1 and the authors speculated that this might be due to the presence of extrahepatic galactose receptors. This is possible, but the pharmacokinetic data were not indicative of normal tissue targeting, other than the hint of lung localisation at early times in the gamma camera images. It is more likely that the reduced solubility of FCE28069 contributed to this change in toxicity profile. The reactions seen on infusion would be consistent with this hypothesis, as would transient lung localisation of polymer aggregates.

Of the 23 patients entered, two had a measurable partial response, a third patient showed reduction in tumour volume, and there were 11 others with stable disease [85]. Gamma camera imaging confirmed galactose-mediated liver targeting to 15–20% dose at 24 h [85, 106]. Although most of the conjugate was localized to normal liver hepatocytes (16.9% vs 3.2% dose in the hepatic tumour), it was estimated that the doxorubicin concentration in hepatoma tissue would still be 12–50-fold higher than could be achieved by administration of free drug.

The extent of FCE28069 liver localisation seen clinically was lower than observed in preclinical studies [104]. This could be due to a lower number of ASGP receptors present on human hepatocytes compared to those in the rat. However, it is noteworthy that the FCE28069 conjugates used in the clinic had lower galactose (1.5–2.0 mol %) content than suggested as optimal for liver targeting in preclinical studies ($\sim$ 4 mol %) [104]. It is well known that the asialoglycoprotein receptor requires a multivalent ligand so a high galactose density along the polymer chain is crucial if maximum liver targeting is to be achieved. Lack of increased liver uptake after a 24 h infusion of FCE28069 was suggested as indicative of lack of receptor saturation. However, comparisons of liver levels were made at 24 h and this may not be the best time point for comparative studies since hepatocyte targeting is usually rapid ($\sim$ 5 min), and at longer time points liver levels are complicated by elimination due to hepato-biliary transfer and excretion.

6.3
HPMA copolymer-antibody-doxorubicin conjugates

Rihova, Ulbrich and colleagues have reported preliminary clinical experiments in six patients with refractory disease (angiosarcoma, and breast carcinoma) [107, 108]. HPMA coplymer-Gly-Phe-Leu-Gly-doxorubicin (or epirubicin)-human immunoglobulin (HuIg) has been synthesised on a case-by-case basis for patient treatment. The HuIg used was either autologous IgG from sera by precipitation with 40% ammonium sulfate, or was a commercially available allogenic human γ-globulin. The primary aim of these preliminary studies was to evaluate the immunomodulatory effects of these conjugates. Disease progression was monitored and a large number of biochemical and immunological parameters were assessed. Despite difficulty assessing the data obtained objectively, it is interesting to note that in some patients antitumour effects were seen. The conjugate did not seem to induce anti-Ig antibodies. Increased levels of $CD16^+56$ and $CD4^+$ cells in peripheral blood and activation of NK and LAK cells supported the suggestion that HPMA copolymer-doxorubicin conjugates can be immunostimulatory.

6.4
HPMA copolymer-paclitaxel (PNU166945)

The poor water solubility of paclitaxel combined with hypersensitivity reactions associated with the standard ethanol and Cremophor formulation made it a good candidate for polymer conjugation. An HPMA copolymer conjugate of paclitaxel was developed by Pharmacia, with the aim of improving drug solubility and subsequent controlled release of paclitaxel thereafter. A glycine derivative of paclitaxel was synthesised via the 2' position of paclitaxel and this was attached to an HPMA copolymer precursor containing -Gly-Phe-Leu- peptide side-chains. In this case the drug is linked via the same tetrapeptide linkage used to create PK1 and PK2 with an additional terminal ester bond. The resulting conjugate (PNU166945) (Fig. 6) was more soluble than paclitaxel (> 2 mg/ml conjugate compared to 0.0001 mg/ml paclitaxel) and had a low drug content of $\sim$ 5 wt % [109]. Theoretically paclitaxel or peptidyl derivatives will be released from the polymer by hydrolytic or enzymatic (esterases) degradation of the ester bond, or proteolytic cleavage of the peptidyl linker.

In Phase I study, PNU166945 was administered by a 1 h infusion every three weeks [109]. The starting dose of 80 mg/m^2 (paclitaxel-equivalent) was one-third of its MTD in dogs. The highest PNU166945 dose administered was 196 mg/m^2 (paclitaxel-equivalent), although no DLTs were seen at this level. The toxicity observed was consistent with commonly observed paclitaxel toxicities including flu-like symptoms, mild nausea and vomiting, mild haematological toxicity and neuropathy. Neurotoxicity grade 2 occurred in two patients at a dose of 140 mg/m^2 (although grade 1 was pre-existing on their entry) and one patient at 196 mg/m^2 had grade 3 neuropathy after the fourth cycle. Interestingly, alopecia was absent throughout. Studies were curtailed due to concerns of potential clinical neurotoxicity following observations in preclinical animal studies [109]. Even in this small patient cohort, antitumour activity was observed. A paclitaxel-refractory breast cancer patient showed remission of skin metastases after two courses at 100 mg/m^2

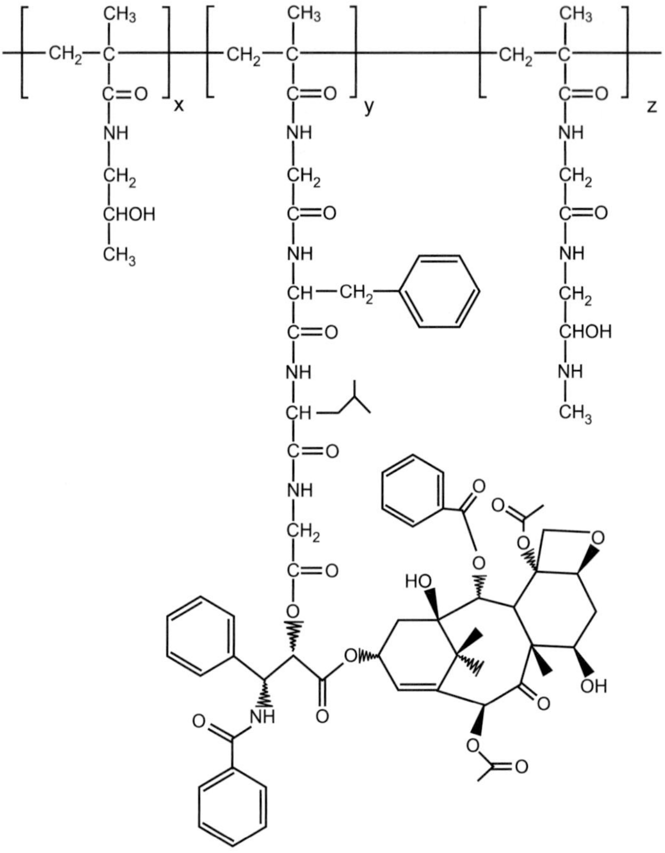

Fig. 6 HPMA copolymer-paclitaxel (PNU166945)

(paclitaxel-equivalent). Two other patients had stable disease at a dose of 140 mg/m^2. Plasma pharmacokinetics measured over 48 h using HPLC was linear with dose for both PNU166945 and the released paclitaxel. The conjugate had a $t_{1/2} \sim 6.5$ h and its volume of distribution indicated plasma circulation. Free paclitaxel released from the conjugate had a $t_{1/2} \sim 1.2$ h and free drug levels were low, ~ 1% of the paclitaxel present in plasma as conjugate. Antitumour activity was seen at a relatively low paclitaxel dose (100 mg/m^2). Conclusion of the Phase I to identify the Phase II dose and possible schedule optimisation would have been interesting.

6.5
HPMA copolymer-camptothecin (MAG-CPT; PNU 166148)

Camptothecin (CPT) from an Oriental tree, *Camptotheca acuminata*, is an inhibitor of topoisomerase I and has exhibited a promising antitumour activity in various experimental tumours. However, CPT is extremely water insoluble, and this feature has severely restricted its clinical application [110]. Recently, novel CPT analogues with an improved water solubility, CPT-11 and topotecan, have emerged as a new class of antitumour agents [111]. CPT-11 has been available in Japan since 1994, where it is approved for the management of colorectal, cervical, ovarian, uterine, gastric, breast, and lung cancer. In the United States, CPT-11 is presently used for the treatment of advanced colorectal adenocarcinoma. Camptothecin's poor solubility and nonspecific toxicity has made it an attractive candidate for polymer conjugation [113]. The only option for drug conjugation was an ester linkage. The HPMA copolymer-camptothecin conjugates (so-called MAG-CPT derivatives) were synthesised from an HPMA copolymer precursor composed of HPMA:methacryloyl-glycine (MA-Gly)-ONp 95:5 or 90:10 (Fig. 7); hence the MAG acronym used to describe this copolymer. Camptothecin was first modified at the C-20 α-hydroxy group to give a peptidyl prodrug (such as Gly-camptothecin) and then bound to the polymer intermediate. The resul-

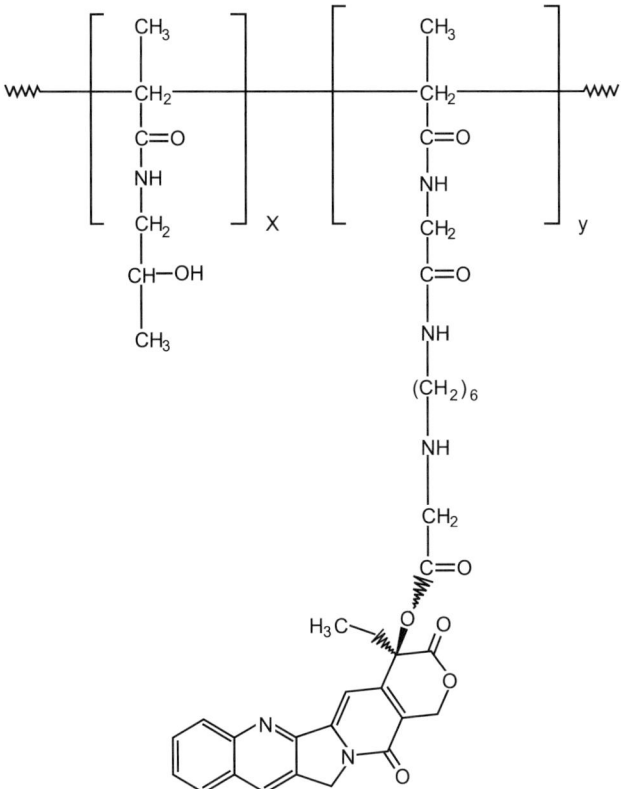

Fig. 7 HPMA copolymer-camptothecin (MAG-CPT; PNU 166148)

tant conjugates had a MW of 20 000–30 000 Da depending on their side-chain content, and a camptothecin loading of 5–10 wt %. Although conjugates containing a library of different peptidyl linkers were examined preclinically, the conjugate PNU166148 (MAG-CPT) containing the Gly-C_6-Gly-linkage was selected for Phase I clinical studies. Camptothecin is released from this conjugate by either chemical or esterase-mediated hydrolysis.

Two dosing schedules were studied during Phase I evaluation of MAG-CPT. An i.v. infusion over 30 min every 28 days [114], and as an alternative, daily treatment ($\times 3$) repeated every four weeks [115]. In the first study [114] 62 patients were entered starting at a dose of 30 mg/m^2 (camptothecin-equivalent). Dose escalation progressed to an MTD of 240 mg/m^2 with 200 mg/m^2 the recommended dose for further studies. At 240 mg/m^2 the DLTs included grade 4 neutropenia and thrombocytopenia, and grade 3 diarrhoea. Severe and unpredictable cystitis was also seen.

In the recently reported Phase I study [114], MAG-CPT was administered as a 30 min infusion on three consecutive days every four weeks. The starting dose was 17 mg/m^2/day and this was escalated to 130 mg/m^2/day; total dose per cycle = 390 mg/m^2. Haematological toxicity was rare, but cumulative bladder toxicity was dose-limiting at doses of 68 mg/m^2 or greater. This could only be resolved by withdrawal of treatment. Of the 16 patients entered in this trial, 11 were evaluable for clinical responses after two courses. These two Phase I studies were the first involving HPMA copolymer conjugates in which no objective clinical responses were seen. However, one patient with renal cell carcinoma had tumour shrinkage and a colon patient had stable disease for 62 days.

Using HPLC analysis no dose-dependency of plasma clearance of either MAG-CPT or the released drug was seen [114, 116]. Plasma levels of free camptothecin were 100 times lower than conjugated drug and there was no significant difference in terminal half-lives $\sim$ 8–10 days of free and polymer-bound drug. Camptothecin (measured as total drug) was still appearing in urine at four weeks with $\sim$ 69% dose excreted in urine after four days. Patients suffering from the worst renal toxicity had a relatively higher plasma AUC of MAG-CPT, suggesting that it was due to impaired renal function.

Bladder toxicity of MAG-CPT had not been anticipated. However early studies with camptothecins did highlight toxicities including vomiting, diarrhoea and chemical or haemorrhagical cystitis. The acidic environment in the bladder causes the formation of the insoluble lactone form of camptothecin. The bladder toxicity of MAG-CPT has been attributed to variable conversion of the inactive open ring form to the active closed ring form of the drug. However the biodistribution of the compound would also have contributed. HPMA copolymer molecular weight has been specifically optimised to ensure effective renal elimination. This has been clearly visualised by gamma camera imaging of other HPMA copolymer conjugates. There is always a likelihood of kidney or bladder toxicity if the polymer-drug linker degrades in urine to deliver high local doses of any cytotoxic agent. As urinary excretion of the MAG-CPT is very high (66% at 24 h) and the conjugate was still detectable in urine after four weeks [114, 116], bladder toxicity was perhaps not surprising.

By contrast, Phase I results with HPMA copolymer-paclitaxel and HPMA copolymer-camptothecin were disappointing, and underline the need for careful optimisation of the polymer-drug linker to ensure stability during transit. Both conjugates contain relatively low drug loading (< 10 wt % compared to 37 wt % paclitaxel in CT-2103). Rapid hydrolysis of the polymer-drug ester linkage could explain why MAG-CPT displayed dose-limiting cumulative bladder toxicity – probably due to drug liberation during renal elimination – and HPMA copolymer-paclitaxel displayed neurotoxicity, which is typical for free paclitaxel [114].

6.6
HPMA copolymer-platinate (AP5280)

A number of HPMA copolymer-platinates synthesised as "cisplatin" or "carboplatin" mimetics have been recently described [91, 117, 118] (Fig. 8). These conjugates were prepared from HPMA copolymer precursors containing either -Gly-Gly-ONp or Gly-Phe-Leu-Gly-ONp side-chains (5 or 10 mol %). The side chains were modified by hydrolysis (– COOH), or aminolysis with ethylenediamine (en), aminomalonate or aminoaspartate to provide the terminal ligands for platination. A HPMA copolymer Gly-Phe-Leu-Gly-en-Pt required lysosomal activation to release active platinum species, confirmed by the observation that conjugates containing the nondegradable linker -Gly-Gly-en-Pt were completely inactive in vivo. Whereas the – COO-Pt released platinum species much too rapidly for useful delivery, the malonate derivative showed a slower, more useful rate of hydrolysis [118]. Using platinum NMR it has been shown that the malonato ring rearranges with time to the more thermodynamically favourable structure (AP5280) shown in Fig. 8. The pharmaceutical product AP5280 contains approximately 8.5 wt % of Pt and has a molecular weight of approximately 25 000 and $MW/Mn = 1.7$.

Phase I studies conducted in Europe were recently described [117]. Patients with solid tumours received AP5280 as a 1-h i.v. infusion every 21 days. Twenty-nine patients were treated at eight dose levels (90–4500 mg Pt/m^2). The dose-limiting toxicity was Common Toxicity Criteria grade 3 vomiting and was experienced at 4500 mg Pt/m^2 in two of six patients. The maximum tolerated dose on this schedule was therefore 4500 mg Pt/m^2, and the recommended dose for a Phase II study is 3300 mg Pt/m^2 once every three weeks. Renal toxicity, neurotoxicity and myelosuppression, toxicities typically observed with cisplatin and carboplatin, were minimal for AP5280. The area under the curve of total Pt increased with increasing AP5280 dose. Plasma clearance of total Pt was 644 ± 266 ml/h, and the terminal half-life was 116 ± 46.2 h. After AP5280 administration, Pt-guanine-guanine DNA adduct con-

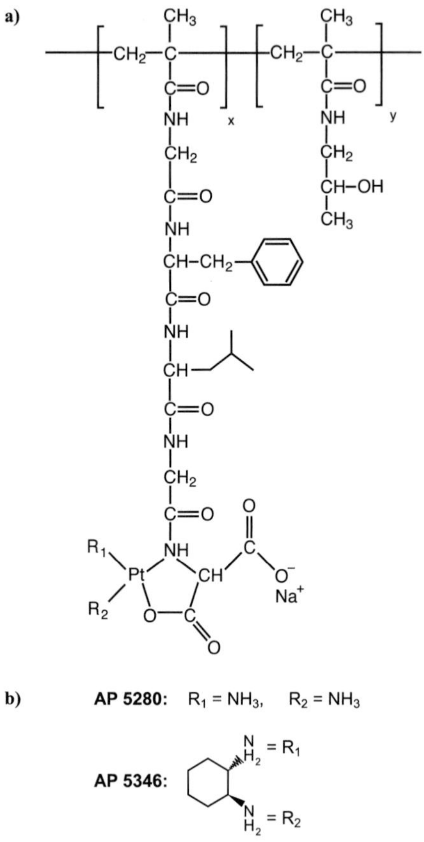

Fig. 8 a HPMA copolymer-platinate (AP5280) **b** HPMA copolymer-platinate (AP5346)

centrations in WBCs ranged from 70 to 1848 μmol/μg DNA, concentrations that were substantially lower than the concentrations measured after administration of therapeutic doses of cisplatin.

6.7
HPMA copolymer-DACH platinate (AP5346)

A second lead HPMA copolymer-platinate (AP5346) has been identified with a similar Gly-Phe-Leu-Gly-aminomalonate side chain, but in this case terminating in a 1,2-diaminocyclohexyl (DACH) palatinate (Fig. 8). In Phase I clinical trial, it was administered as intravenous infusion of 80–1280 mg Pt/m^2 once a week in 28-day cycle to patients with a broad cross-section of tumour types. Out of the 12 evaluable patients, one demonstrated a partial response. Dose limiting toxicity was neutropenia but also nausea, vomiting, asthenia, fatigue and diarrhoea were observed (as presented in October 2004 at the 16th

EORTC-NCI-AACR Symposium in Geneva). An Investigational New Drug application was filed with the FDA in December 2004.

6.8
Poly-*L*-glutamic(PG)-paclitaxel (CT-2103, XYOTAX)

Poly-*L*-glutamic acid (PG) was effectively used in the 1990s by Wallace and colleagues [119] and a PG-paclitaxel conjugate (acquired by Cell Therapeutics Inc.), CT-2103, has been advancing successfully and rapidly through an early clinical development programme. CT-2103 contains 37 wt % paclitaxel linked through the 2' position – that is, via an ester bond – to the γ-carboxylic acid of PGA (MW $\sim$ 40 000 g/mol) (Fig. 9). Moreover, it is 80 000 times more soluble than paclitaxel. This conjugate has the advantage of a biodegradable PG polymer backbone, and it is cleaved by cathepsin B to liberate diglutamyl-paclitaxel [120]. Remarkable antitumour activity was seen in a variety of in vivo tumour models which, together with evidence of tumour targeting by the EPR effect, paved the way for clinical testing [119, 121, 122]. Interestingly, it has been shown in preclinical studies combining conjugate administration with radiation treatment that tumour targeting of PG-paclitaxel by the EPR-effect is significantly increased, leading to enhanced antitumour activity [123]. This has important implications for possible clinical development of this and other polymer-drug conjugates. CT-2103 is currently undergoing an extensive Phase I/II programme in Europe and the USA [120, 124–126]. In one Phase I study, CT-2103 was administered as a single agent i.v. over 30 min every three weeks. The starting dose was $\sim$ 11 mg/m^2 (paclitaxel-equivalent) and dose escalation progressed to a MTD of 266 mg/m^2. In another Phase I study, a fixed dose of cisplatin (75 mg/m^2) or carboplatin was given with escalating doses of CT-2103 every 21 days. CT-2103 was administered first by a 10 min i.v. infusion followed by platinate by i.v. infusion. In these studies, CT-2103 has shown manageable toxicity and a significant number of

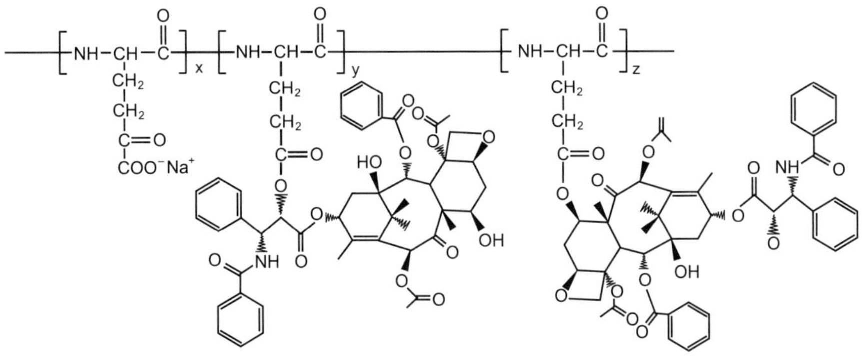

Fig. 9 Poly-*L*-glutamic-paclitaxel (CT-2103, XYOTAX)

patients displayed partial responses or stable disease (mesothelioma, renal cell carcinoma, NSCLC and a paclitaxel-resistant ovarian cancer patient). In the Phase II programme, CT-2103 is being evaluated against recurrent colorectal cancer, recurrent ovarian, fallopian tube or peritoneal cancer, and NSCLC. Conjugate is administered at doses of 175 mg/m^2 or 210 mg/m^2 (Pt-equivalent) and in certain trials is given in combination with cisplatin or carboplatin. Early results show antitumour activity and minimal toxicity. Except for some hypersensitivity reactions, no serious drug-related events have been reported. In addition, Phase III trial has recently started in which CT-2103 is given in combination with carboplatin to ovarian cancer patients. Clinical pharmacokinetics data show that CT-2103 is stable in plasma; data are consistent with prolonged tumour exposure and reduced systemic exposure to active drug. Based on the promising results in phase I/II studies, three phase III trials of CT-2103 were initiated in advanced non-small-cell lung cancer (NSCLC) [127]. These Selective Targeting for Efficacy in Lung Cancer, Lower Adverse Reaction (STELLAR) trials represent the largest randomized phase III programs in patients with NSCLC and a poor performance status.

Preliminary reports from the STELLAR 3 and STELLAR 4 phase III clinical studies of XYOTAX for the first-line treatment of non-small-cell lung cancer (NSCLC) patients with poor performance status 2 (PS2) were recently presented at the 2005 Annual Meeting of the American Society of Clinical Oncology (ASCO). STELLAR 3 and STELLAR 4 were designed to determine if XYOTAX could improve survival while reducing serious side-effects when compared to standard single agents, gemcitabine or vinorelbine (STELLAR 4), or when used with a second chemotherapeutic agent, carboplatin (STELLAR 3). For platinum-based doublet therapy, XYOTAX provided an easier administration, was better tolerated, and a more convenient first-line treatment for PS2 patients over paclitaxel. However, the combination of XYOTAX/carboplatin yielded similar rates of overall survival in the first-line treatment of PS2 patients with NSCLC to the current standard paclitaxel/carboplatin. As a single agent, XYOTAX offered a better tolerated, less toxic treatment option than chemotherapeutic agents currently available, and showed a 40% improvement in overall survival over the approved agent vinorelbine (40%). At one year, 26% of XYOTAX recipients were alive compared to 7% of vinorelbine patients, and at two years, 15% of XYOTAX patients were alive compared to none of the vinorelbine and 12% of gemcitabine recipients. An overall significant reduction in on-study deaths compared to both gemcitabine and vinorelbine, a significant increase in the number of patients able to complete the full six doses of therapy compared to vinorelbine, and a significant reduction in nausea/vomiting and in potential life threatening side-effects (grade 3/4) associated with gemcitabine and vinorelbine were also observed. While additional results for both STELLAR trials are still awaited, these initial results show promise for XYOTAX as a single agent in the first-line treatment setting for PS2 NSCLC patients.

Preliminary data from a study of XYOTAX in combination with carboplatin produced a major tumor response in 98% of first-line ovarian cancer patients (Presented at ASCO, 2005). The phase II study was for first-line induction and single-agent maintenance therapy of advanced stage III/IV ovarian cancer. XYOTAX was administered over a 10-minute infusion at doses of 210 mg/m^2 in 20 patients and 175 mg/m^2 in 62 patients followed by carboplatin (AUC = 6) on day one of each 21-day cycle for up to six cycles. After six cycles, patients with stable disease or better continued on to receive single-agent XYOTAX at a dose of 175 mg/m^2 for up to 12 four-week cycles. Of the 82 patients studied, 80 patients, or 98%, achieved a major tumor response (CR+PR based on a reduction in CA-125 levels) during the induction phase of the therapy, including 85% with complete response and 12% with partial response. Some grade 3/4 side-effects at the 175 mg/m^2 dose in combination with carboplatin included neuropathy, nausea, vomiting, febrile neutropenia, thrombocytopenia and anemia. Data presented at ASCO include only the induction phase of the trial. The maintenance portion of the trial is ongoing. CTI and the Gynecologic Oncology Group (GOG) are presently evaluating XYOTAX (135 mg/m^2) as monthly maintenance in a phase III clinical trial in ovarian cancer patients who have achieved a complete response following standard first-line chemotherapy.

Separately, Cell Therapeutics reported initial data from a phase I study of weekly XYOTAX given in combination with radiation for patients with esophageal and gastric cancer. Of the 11 patients with locoregional disease that could be evaluated for tumor response, four patients achieved a complete disappearance of their tumor and five patients achieved a 50% or greater shrinkage of their cancer.

6.9
Poly-L-glutamic(PG)-camptothecin (CT-2106)

Camptothecins, topoisomerase I inhibitors, are an important and rapidly growing class of anticancer drugs. However, like taxanes, their full clinical benefit is limited by poor solubility and significant toxicity. Oral analogs such as topotecan and irinotecan are soluble, but are less effective in combating tumours. Camptothecins are important drugs in the treatment of advanced colon, lung, and ovarian cancers. Cell Therapeutics Inc. is also developing a camptothecin conjugate using the same polyglutamate polymeric carrier, CT-2106 [128]. Conjugates containing different linkers including -Gly, -glycolic acid, -γ-Glu and -β-Ala have been described [129–131]. These were synthesised using polymers of molecular weight 33 000–74 000 Da. Linking a camptothecin to polyglutamate polymer renders it water-soluble [130], and preclinical studies suggest that it permits up to 400% more drug to be administered without an increase in toxicity. CT-2106 as a single agent and/or in combination with 5FU showed significantly enhanced antitumour activity in several animal tumour models [131]. The lead conjugate PG-Gly-CPT (CT-2106) containing 33–35 wt % camptothecin and of molecular weight of 55 000 Da entered Phase I trials in 2002. Phase I/II clinical trials of CT-2106 in patients with advanced cancers were initiated in 2004.

6.10
PEG-camptothecin (PROTHECAN)

Following its successful application to protein conjugation, PEG has also been used to create drug conjugates [132, 133]. The safety profile of PEG is well documented so its use as a potential drug carrier has been obvious for > 20 years and many conjugates have been described in the literature [44, 133]. Whilst the HPMA copolymer conjugates and PG conjugates described above all contain multiple pendant functional groups for drug attachment, PEG contains only two terminal – OH groups suitable for conjugation. This severely limits drug-carrying capacity to two drug molecules per PEG chain unless more sophisticated chemistry is used to amplify the number of terminal binding sites. However, the PEG polymer chain can be reproducibly synthesised to give molecules of uniform molecular weight (polydispersity, MW/MN ~ 1.0), which results in a homogenous product. Enzon Inc. selected a PEG-camptothecin conjugate (PEG-CPT; PROTHECAN®, Pegamotecan) for first Phase I pharmacokinetic and safety trials [132] (Fig. 10). Camptothecin is linked to PEG at the C – 20 – OH position thus favouring the desired lactone ring configuration. The ratio of PEG-CPT to active drug is reported to be 60:1, indicating a drug content of 1.7 wt %. This is a rather low loading and illustrates the limitation of PEG as a drug carrier, namely that drug can only be bound via the two reactive termini. In Phase I study, PEG-CPT was administered every three weeks at doses of 600–4800 mg/m² (conjugate) (estimated to represent $\sim$ 10–82 mg/m² camptothecin-equivalent). It had a maximum tolerated dose of 200 mg/m² (camptothecin-equivalent) [120]. DLT was neutropenia and thrombocytopenia, observed at the highest dose level. Preliminary pharmacokinetic studies suggested that PEG-CPT produced prolonged circulating levels of camptothecin ($t_{1/2} > 72$ h). Phase II trial in patients with small cell lung cancer (SCLC) was initiated.

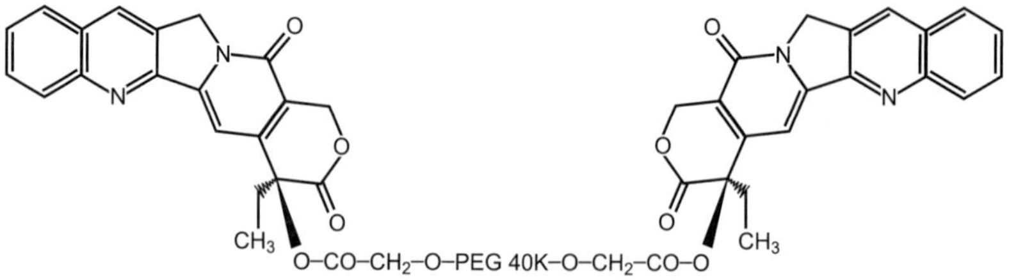

Fig. 10 PEG-camptothecin (Prothecan)

6.11
PEG-paclitaxel

In May 2001, Enzon Inc reported the start of a Phase I clinical trial using a PEG-paclitaxel conjugate. The protocol has been designed to determine the safety, tolerability and pharmacology of PEG-paclitaxel in patients with advanced solid tumours and lymphomas. Although PEG-paclitaxel conjugates have been reported in preclinical studies [115, 134], there is not yet information on the chemistry of this particular conjugate or its clinical progress to date.

6.12
Dextran-doxorubicin (AD-70, DOX-OXD)

Polysaccharides have long been a popular choice for synthesis of polymer-drug conjugates [135, 136]. Dextran (mainly α-1,6-polyglucose with some α-1,4 branching) has been particularly popular owing to its clinical approval for use as a plasma expander. A dextran-doxorubicin conjugate (AD-70) was tested clinically. A dextran polymer with a molecular weight of $\sim 70\,000$ Da was used to prepare the conjugate. Drug conjugation seemed to be by Schiff base formation using oxidised dextran also modified with glycine as a pendant group for reaction with the anthracycline [137]. The rationale of this conjugation approach was to utilise hypoxic conditions in the tumour to promote drug liberation.

In a Phase I trial involving 13 patients, AD-70 was administered every 21–28 days by a 30 min infusion. A starting dose of 40 mg/m^2 (doxorubicin-equivalent) was chosen as 1/10 the mouse LD$_{10}$. As unexpected toxicities occurred, the dose was reduced to the range 12.5–40 mg/m^2 [137]. At the lowest level there was minimal toxicity and 12.5 mg/m^2 was suggested as a Phase II dose. The conjugate MTD was 40 mg/m^2 (doxorubicin-equivalent). Thrombocytopenia and severe hepatotoxicty were the DLTs. Hepatotoxicity lasted for several weeks suggesting liver localisation of the conjugate with slow release of doxorubicin over time thereafter. Toxicity was attributed to uptake of the polysaccharide by the reticuloendothelial cells in the liver. This would result from use of dextran (polyglucose) as a carrier (there is a macrophage receptor for glucose) and/or the fact that doxorubicin was conjugated to oxidised dextran via a Schiff base – residual aldehydes would surely be present after drug conjugation.

6.13
Polymeric micelles

Block copolymers spontaneously assemble into nanoscaled polymeric micelles. Self-assembling block copolymer micelles have long been explored as drug carriers. A pluronic block copolymer micelle incorporating doxorubicin and able to circumvent p-glycoprotein-mediated resistance has recently shown promising results in Phase II clinical evaluation [138]. Like other more traditional micellar formulations, the drug was noncovalently entrapped in this case. By contrast, Kataoka and colleagues have designed self-assembling polymeric micelles (NK911; 42 nm in diameter) using block copolymers of PEG (MW $\sim$ 5000 g mol^{-1})-poly(aspartic acid) that also include a fraction of doxorubicin that is covalently bound to the polymer ($\sim$ 45%), as well as free drug [139, 140]. This is, therefore, truly a polymer therapeutic as defined above. NK911 accumulates preferentially in tumour tissue by the EPR effect, leading to a three- to four-fold improvement in targeting [140]. However, in this case, the covalently bound drug is inactive, and it is the free drug slowly escaping over 8–24 h that destroys tumour cells. Polymeric micelles provide a promising formulation for cancers with limited vasculature [141].

6.14
Brain tumour implants – local delivery of chemotherapy

Cancer drugs can cause enormous toxicity. By placing the drug next to the tumour environment in a polymer drug delivery agent following surgery, both the safety and efficacy of cancer chemotherapy can be improved. Higher local drug concentrations can be achieved, and the systemic toxicity usually associated with standard drug treatments can be minimized. Novel polymers such as polyanhydrides have been designed and used for local delivery of chemotherapeutics [142]. In brain cancer, polyanhydride polymers shaped like wafers have been used to locally deliver drugs such as carmustine (1,3-bis (2-chloroethyl)-1-nitrosourea, BCNU) [143] following surgery. After removal of as much of the tumour as is possible, small polymer drug wafers are placed at the surface of the brain within the tumour resection cavity. The wafers release the entrapped drug in a controlled fashion, locally delivering it for approximately three weeks to destroy any remaining tumour. Side-effects associated with systemic delivery are minimized. Although these wafers do not fully fit under the description of "polymer therapeutics" as the chemotherapeutics are not covalently bound but are rather entrapped, they represent one of the most powerful and successful clinical uses of polymers in cancer treatments to date. One clinical trial showed that after two years, 31% of the patients treated were alive whereas only 6% of patients receiving standard brain tumour therapies survived [144]. In 1996, the US Food and Drug Administration (FDA) approved brain tumour implants for patients with recurrent glioblastoma, the first new brain cancer therapy approved in over 20 years. In 2003, the FDA approval was extended to include initial surgery for malignant glioma based on two additional randomized prospective studies that demonstrated improved survival and safety [145]. Studies have also reported benefits for experimental brain metastases [146] and invasive pituitary adenomas [147]. Local delivery of chemotherapeutics from longlasting implantable lipid formulations to spinal fluid has also been used clinically to treat carcinomatous meningitis [148].

7
Other compounds in preclinical stage

7.1
DE-310

A polysaccharide-camptothecin (DE-310) conjugate containing the new camptothecin analogue DX-8951f is currently in Phase I trial (Fig. 11). The drug is covalently bound to the carrier via a Gly-Gly-Phe-Gly peptide linker [149–151]. Duncan and colleagues also developed a dextrin-doxorubicin (α-1,4-polyglucose) conjugate [152] which is currently in preclinical development (ML Laboratories Inc.).

7.2
Carboxymethyl dextran-CPT analogue (T-2513) conjugate (T-0128)

The main clinical adverse effects of CPTs are myelosuppression and gastrointestinal toxicity, especially severe diarrhoea. These side-effects are closely related to its pharmacokinetic properties [153, 154]. To alter its pharmacokinetic behavior in order to decrease its toxicity and increase the therapeutic efficacy of CPT, a new macromolecular prodrug, denoted T-0128, was synthesised (Fig. 12). This prodrug is a novel CPT analogue (T-2513)-carboxymethyl (CM) dextran conjugate via a triglycine spacer, with a molecular weight of M_r 130 000 [155–158]. The in vivo antitumour study against Walker-256 carcinoma demonstrated that T-0128 was ten times as active as T-2513. Additionally, comparative efficacy studies of T-0128, T-2513, CPT-11, and topotecan were performed using a panel of human tumour xenografts in nude mice, showing the advantage of T-0128 over these CPTs [158]. A single i.v. injection of T-0128 at 6 mg/kg (based on the amount of T-2513 bound to CM dextran) induced complete regression of MX-1 mammary carcinoma. T-0128 at 10 mg/kg weekly for three weeks (one-tenth of its MTD) cured LX-1 lung carcinoma, St-4 gastric and HT-29 colorectal tumour xenografts that are highly refractory to CPTs. These demonstrate the broad range of therapeutic doses achieved with T-0128. Pharmacokinetic studies showed that after i.v. administration of T-0128, the conjugate continued to circulate at a high concentration for an extended period, resulting in tumour accumulation in Walker-256 tumour model. The significant increases in the amount and exposure time of released T-2513 in the tumour explain the enhanced efficacy of T-0128 well.

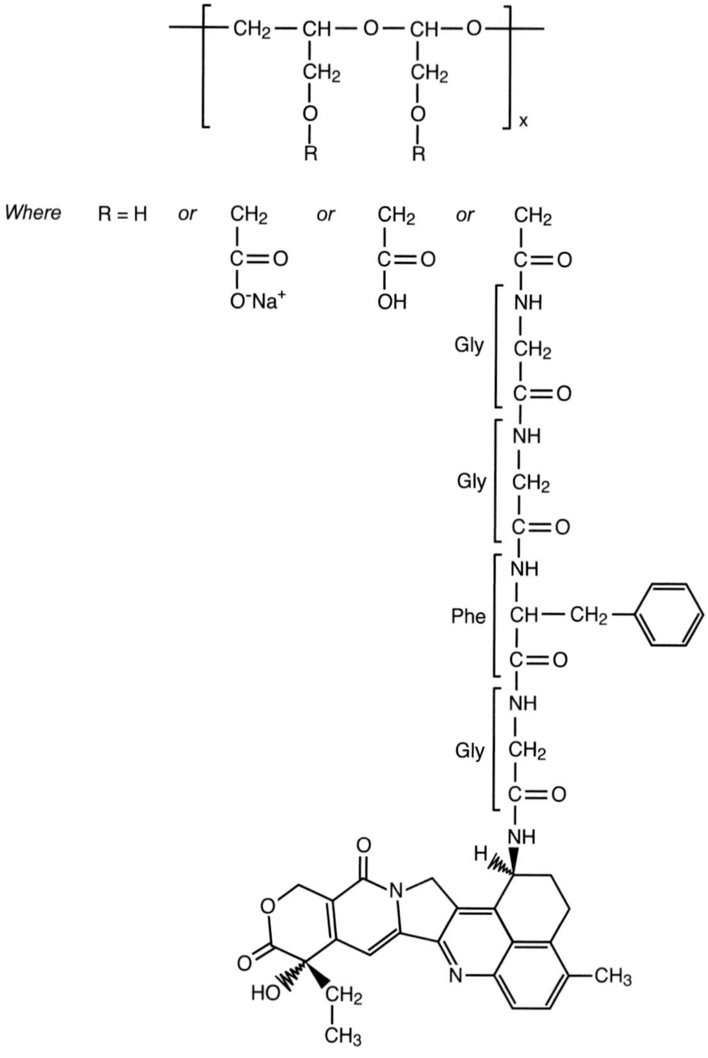

Fig. 11 Polysaccharide-camptothecin (DE-310)

7.3
Polyacetal-diethylstilboestrol

An attractive approach was developed by Vicent et al. containing the bioactive agent as an integral part of the polymer backbone. Polyacetals were synthesised incorporating a drug with bishydroxyl functionality into the polymer backbone. Degradation of the polymer backbone in the acidic environment of the lysosome or the extracellular fluid of some tumours would

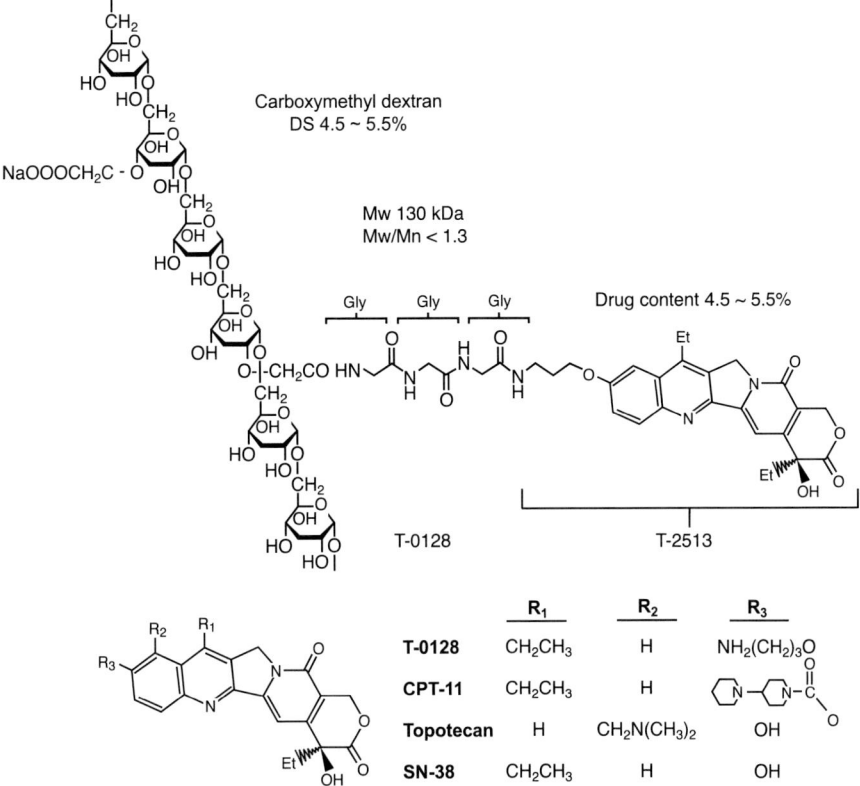

Fig. 12 Camptothecin analogue-carboxymethyl dextran (T-0128)

then trigger drug release, eliminating the need for a biodegradable linker. A tert-polymerisation approach was used to incorporate nonsteroidal oestrogen diethylstilboestrol (DES) into the main chain of water-soluble polyacetals synthesised using as comonomer PEG of MW 2900 or 3400 g/mol [159]. When PEG2900 was used the resultant polymer had a MW of 18 900 g/mol, a MW/MN of 1.9 and a DES loading 4.3 wt %. With PEG3400 the polymer MW was 43 000 g/mol, MW/MN = 1.8 and it had a DES loading 4.7 wt %. DES-polyacetal displayed greater cytotoxicity than DES against human and murine tumour cell lines (IC_{50} = 48 and 420 µg/ml against MCF-7 human breast cancer cells and IC_{50} = 97 and 560 µg/ml against B16F10 murine melanoma cells, respectively). These polymers showed no significant haemolysis at concentrations up to 20 mg/ml, confirming their suitability for further in vivo evaluation. An enhanced rate of hydrolytic degradation of the polymer backbone was seen at pH 5.5, (65% trans-DES released in 96 h), compared to pH 7.4 (4% trans-DES released in 96 h). These bioresponsive DES-polyacetals tert-polymers are the first water-soluble anticancer polymeric drugs designed for acidic pH-triggered release of a drug incorporated into the polymer main chain. Their in vitro characteristics suggest that further in vivo evaluation is warranted.

7.4
HPMA copolymer-1,5-diazaanthraquinone

1,5-Diazaanthraquinones (DAQs) are promising anticancer drugs, but their clinical potential is limited due to poor solubility. A library of HPMA copolymer conjugates containing a novel amino-functionalised 1,5-diazaanthraquinone derivative (amino-DAQ) have been synthesized [160]. Conjugation to HPMA copolymers improved amino-DAQ aqueous solubility by 7-fold. The HPMA copolymer-amino-DAQ conjugates were slightly less haemolytic than the parent compound. When conjugates were incubated with isolated rat liver lysosomal enzymes (tritosomes), the rate of amino-DAQ release was influenced by both drug loading and the composition of the peptidyl side chain used to link the drug to the carrier. The higher the drug loading, the lower the rate of drug release. Whereas the GG linker did not release amino-DAQ, up to 26% of the amino-DAQ was released from a GFLG linker over 24 h. HPMA copolymer-amino-DAQ conjugates showed much lower in vitro cytotoxicity than the free drug against B16F10 murine melanoma and MCF-7 breast cancer cell lines. Nonetheless, the observed lysosomal activation of the HPMA copolymer-Gly-Phe-Leu-Gly-amino-DAQ conjugates suggests that evaluation of the antitumour potential in vivo is warranted.

8
Targeting tumour vasculature

An exciting alternative to cytotoxic chemotherapy is the destruction of angiogenic vasculature itself. The number of antiangiogenesis therapeutic drugs in clinical trials is steadily growing. Many of the first-generation antiangiogenic proteins in clinical testing are delivered systemically and, for the most part, target active endothelium, such as that feeding a solid tumour, as opposed to the quiescent endothelium that supports normal, healthy tissues. These agents consist of small molecules, proteins, and antibodies, and their efficacy could potentially be improved by drug delivery systems. If these angiogenesis inhibitors could be selectively targeted to the metabolically active endothelium of tumours, a much higher therapeutic index could be achieved.

One can envision different approaches to using polymer therapeutics to target the tumour vasculature. The first consists of using antiangiogenesis therapeutic drugs, which inhibit endothelial cell proliferation, and to conjugate them to a polymer as a polymer therapeutic, in order to direct them to the tumour compartment actively or passively. Although a large number of antiangiogenic agents are already in clinical development [161], this goal was recently achieved with the development of a water-soluble polymer conjugate used to deliver an antiangiogenic agent, HPMA copolymer-TNP470 conjugate (Caplostatin) [10]. In addition to this compound, the PEG backbone has been conjugated to IFN-α-2b as we previously described (Sect. 6.4), and when used in combination with thalidomide [162] or docetaxel [58], they have yielded very encouraging results. PEGylation of xanthine oxidase (PEG-XO) [163] reduced the affinity of the native protein for all endothelium while inducing tumour accumulation of the protein by the EPR effect, leading to significant suppression of tumour growth following administration of hypoxanthine (described in detail in Sect. 9.3.1).

8.1
Use of targeting moieties to deliver drugs to the tumour vasculature

An alternative approach for targeting the tumour vasculature with a polymer therapeutic consists of functionalizing the cytotoxic drug-polymer conjugate with a targeting moiety that specifically recognizes the tumour endothelial cell. Directing targeting moieties to the tumour endothelium rather than to the tumour cells offers multiple advantages. Tumour endothelium is significantly more accessible than the tumour cells themselves. Because each tumour endothelial cell can support the growth of up to 100 tumour cells, killing the tumour endothelial cell should be more effective than killing each individual tumour cell. Because of the intrinsic genetic instability of tumour cells, their patterns of tumour marker expression are often heterogenous and change with time [164] (Table 4).

8.1.1
Targeting tumour vessels using markers of angiogenesis

Tumour endothelium differs from normal endothelium, and few markers of angiogenesis, located either on the surface of endothelial cells or in the modified subendothelial extracellular matrix, have been described and characterized (reviewed in [164]). Tumour endothelial specific markers are useful for targeting with drug delivery systems. Some of these include VEGFR2 (vascular endothelial growth factor receptor II), integrin $\alpha_v\beta3$ and $\alpha_v\beta5$ [165–167], ED-B domain of fibronectin [168], the large isoforms of tenascin C [169], phosphatidyl serine phospholipids [170], and endoglin [171]. VEGFR2 is the receptor for VEGF, one of the principal mediators of vascularization in solid tumours. The humanized neutralized antibody to VEGF (Avastin, Genentech) has been recently approved for colorectal cancer. Antibodies to VEGFR2, such as 2C3, have antitumour activity against tumour xenografts in mice [172]. One of the best targeting moieties is the human antibody fragment L19 in single-chain Fv antibody fragment configuration "scFv" recognizing the

Table 4 Targeted delivery systems to tumour vasculature under development

Compound	Name	Mouse models	Refs.
Polymer-angiogenesis inhibitor conjugate			
HPMA copolymer-TNP-470	HPMA-TNP-470	Melanoma, Lewis Lung carcinoma	[10]
Antiendothelial immunoconjugates			
Anti-ED-B domain of FN (L19)-tissue factor	L19-TF	Teratocarcinoma Colon carcinoma	[249]
L19-interleukin 12	L19-IL12	Teratocarcinoma Colon carcinoma	[250]
L19-interleukin 2	L19-IL2	Teratocarcinoma	[251]
L19-astatine 211	L19-^{211}At	Teratocarcinoma, sarcoma	[175]
Anti-endoglin-deglycosylated Ricin A (dgRA)	Y4-2F1-dgRA	Breast carcinoma	[252]
Anti-endoglin-dgRA	P3-2G8-dgRA	Breast carcinoma	[252]
Antiendothelial fusion proteins			
Endostatin-human prolactin (hPRL) antagonist	Endostatin-G129R	Breast carcinoma	[253]
Anginex-albumin (EPR effect)	Anginex-albumin	Ovarian carcinoma	[254]
Antiendothelial peptide conjugates			
NGRpeptide (targeting CD13)-Doxorubicin (Dox)	NGRpep-Dox	Breast cancer	[193], [255]
RGDpeptide (targeting $\alpha_v\beta 3$)-Dox	RGDpep-Dox	Breast cancer	[255]
RGDpeptide (targeting $\alpha_v\beta 3$)-Human monoclonal antibody	RGDpep-HuMAb	Melanoma	[256]
Antiendothelial liposomes			
APRPGpeptide (homing angiogenic peptide)-liposome-DPP-CNDAC (antitumour nucleoside)	APRPGpep-Lip-DPP-CNDAC	Melanoma	[203], [257], [258]
APRPGpeptide-liposome-Dox	APRPGpep-Lip-Dox	Melanoma	[203], [257], [258]
Ab fragment to ED-B domain of FN-liposomes-2′-deoxy-5-fluorouridylyl-N^4-octadecyl-1-β-D-arabinofuranosylcytosine	ScFv-5FdU-NOAC-liposomes	Teratocarcinoma	[259]

ED-B domain of fibronectin [173], which has been used to target tumour and nontumour angiogenesis in animal models [174–177] and in cancer patients [178]. Radiolabeled monoclonal antibodies for large tenascin C have been investigated clinically for both diagnostic and immunotherapeutic applications [179–182]. A more specific antibody that recognizes an extradomain C within the large isoform specific for aggressive tumours and undetectable in normal tissues has been recently described [183].

In addition to these proteins, novel candidate markers preferentially expressed in tumour endothelium have been generated by proteomic [184–186] and transcriptomic techniques [3, 187, 188]. Biopanning of phage display libraries [189, 190] has revealed peptide motifs [190, 191] and antibodies [192] that specifically home in to tumour vasculature and are useful targeting moieties. Some surface proteins in tumour endothelium recognized by peptides motifs, such as aminopeptidase N (for NGR-containing peptides) [193] and integrin $\alpha_v\beta 3$ (RGD-containing peptides) [194], have been validated, and the protein expression of a few of the novel tumour endothelial markers (TEMs) identified by serial analysis of gene expression [3, 187] has also been confirmed [195, 196]. Although comprehensive immunohistochemistry analysis using specific monoclonal antibodies as well as biodistribution studies in animal models of angiogenesis-related diseases is still necessary to confirm the specificity of most novel markers [164], it is foreseeable that many could serve as targeting moieties for protein therapeutics.

8.1.2
Drug targeting to angiogenic vessels using peptide motifs

In the last decade, the use of peptide-targeting moieties to deliver drugs to the tumour vasculature has increased exponentially. RGD-containing peptides are the most well characterized; RGD-containing peptides or biomimetic ligands recognizing integrin $\alpha_v\beta 3$ have been effectively used as targeting moieties to deliver drugs to the tumour compartment [34, 197] and for radioimaging of tumour animal models [198, 199].

Pasqualini and Ruoslahti reported a novel in vivo phage display that distinguished between active proliferating microvascular EC in a tumour and quiescent nonproliferating EC elsewhere in the vasculature [200]. This methodology permitted angiogenesis-related targeting of tumour blood vessels. Moreover, they demonstrated that a small peptide could be specifically targeted to tumour vasculature inhibiting angiogenesis, tumour growth and invasion [201, 202]. This approach has been pursued by several investigators, and a number of peptide motifs that home in to all tumour endothelium or to organ-specific tumour vasculature have been identified [203]. The vast majority of the proteins/sugars recognizing the peptide sequences are unknown, so it is still uncertain as to how specific they will be with respect to nonendothelial cells. Because the peptides are delivered by intravenous administration, binding to potential nonvascular cells should be reduced. To date, many peptide sequences identified by phage display have been successfully used in mice as targeting vehicles for drugs such as toxins (Fig. 1A) and liposomes (Fig. 1E) to the tumour vasculature (Table 4). Recently, Arap, Pasqualini and their colleagues reported in vivo screening of a peptide library in a patient for the first time [204]. Specific peptide motifs were found to home to the endothelium of different organs, proving that the vasculature of each organ is unique. If such a molecular map of the human vasculature is eventually achieved and the results are taken together with the recently identified genes that encode endothelial markers overexpressed during tumour angiogenesis [3], a novel pharmacologic approach to angiogenesis-dependent diseases can be envisioned. Some of the many potential clinical applications of this elegant technology were reviewed previously [201, 205].

Few examples of polymer conjugates using peptides as endothelial targeting moieties have been reported. These are the HPMA copolymer-doxorubicin-RGD conjugates [206], the technetium 99 m-labeled-HPMA-copolymer carrying doubly cyclized RGD motifs (HPMA-copolymer-RGD4C conjugate) [207] and the PEGylated cyclic RGD radiotracers (64-Cu-DOTA-PEG-RGD and ^{125}I-RGD-mPEG) [208, 209]. In addition, in vitro studies of PEG-liposomes coupled to cyclic RGD or ATWLPPR peptides that target tumour vasculature have been performed [210]. These polymers provide a foundation for targeted polymer delivery of drugs to the tumour vasculature of solid tumours.

8.2
HPMA copolymer-TNP-470 (caplostatin)

The tumour endothelium has proven to be an exciting target for anticancer drugs whose goal is to stop the angiogenesis required for tumour growth and progression. TNP-470, a low molecular weight analogue of fumagillin, was first shown to be antiangiogenic in 1990 by Ingber and colleagues [31]. More recently, when tested in clinical trials against a variety of tumours, TNP-470 treatment showed promising antitumour activity when used alone or in combination with conventional chemotherapy [211, 212]. However, the promise of this drug was significantly limited by neurotoxicity that occurred at the optimal anticancer dose [213, 214]. Using an approach that combines a drug-polymer cojugate with targeted delivery to the neovasculature, Satchi-Fainaro and coworkers were able to achieve enhanced and prolonged activity of TNP-470 in a variety of in vivo models [10]. We designed and synthesised a water-soluble conjugate of N-(2-hydroxypropyl)methacrylamide (HPMA) copolymer, a Gly-Phe-Leu-Gly-ethylenediamine linker, and TNP-470 (Fig. 13). HPMA copolymer-Gly-Phe-Leu-Gly-en-TNP-470 conjugate selectively accumulated in the tumour microvasculature due to the passive targeting phenomenon first described by Matsumura and Maeda, the EPR

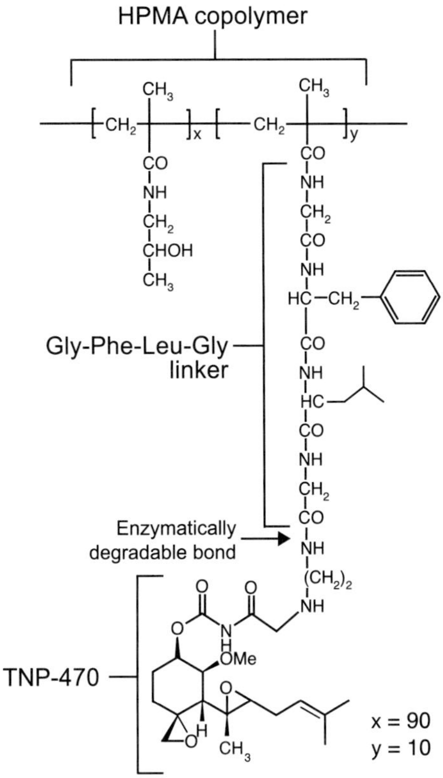

Fig. 13 HPMA copolymer-TNP-470

effect [21]. The HPMA copolymer-TNP-470 conjugate potently inhibited tumour angiogenesis and subsequent tumour growth in both in vivo tumour models (A2058 human melanoma, U87 human glioblastoma, COLO-205 human colon carcinoma, PC3 human prostate carcinoma and Lewis Lung carcinoma) and a hepatectomy model. In addition, this conjugate did not cross the blood–brain barrier and did not induce neurotoxicity as did the unconjugated TNP-470. This very promising approach is worthy of further development. In fact we hope that combination of HPMA copolymer conjugates containing RGD motifs to promote integrin $\alpha_V\beta_3$ receptor-mediated targeting [207] with antiendothelial chemotherapy might enhance the effects seen further. In situations where the tumour is well vascularised, but vasculature permeability is poor, this strategy might be essential.

Approaches such as these hold significant promise for the development of new targeted antiangiogenic therapies as well as for the optimization of existing antiangiogenic drugs. Current interest is focusing on an even earlier stage in tumour progression, the point at which a dormant, avascular tumour acquires the ability to grow and metastasize by "switching on" angiogenesis [215]. As sensitive biomarkers and imaging systems capable of detecting the nascent microvasculature are developed, it is possible to imagine using a potent angiogenesis inhibitor that targets the first generation of angiogenic vessels developing in a tiny tumour lesion that is in the process of acquiring the angiogenic phenotype. Such an agent might be capable of maintaining the dormancy of that lesion indefinitely [216].

8.3
Delivery schedules and vehicles to target angiogenesis

In the last several decades, the development and use of controlled release polymers has enabled the design of bioassays for the in vivo identification and testing of angiogenesis inhibitors. This, in turn, has led to the introduction of a significant group of new cancer therapeutics, which target the new capillary growth that invades and nurtures developing tumours. This antiangiogenic strategy to treat human cancer, pioneered by Judah Folkman, was recently validated in a large randomized clinical trial. In this trial, reported by Hurwitz and coworkers, bevacizumab (Avastin), a vascular endothelial growth factor (VEGF)/vascular permeability factor (VPF) antagonist, was administered along with conventional chemotherapy and significantly improved the survival of patients with metastatic colorectal cancer. Bevacizumab, a recombinant humanized anti-VEGF monoclonal antibody, was given in combination with the standard chemotherapy of irinotecan, fluorouracil, and [calcium folinate] leucovorin (IFL) to a cohort of over 800 patients. Patients who received the combination therapy showed a median survival of 20.3 months compared to 15.6 months for the placebo group. Studies aimed at achieving maximum therapeutic efficacy of some antiangiogenic drugs have led to the development of what is now referred to as an "antiangiogenic scheduling" of conventional chemotherapies [217]. This regimen is able to circumvent the drug resistance induced by these same anticancer drugs delivered on the traditional chemotherapeutic schedule, as the main target is now the tumour endothelial cell. The first example of this phenomenon was seen in studies with the chemotherapeutic agent cyclophosphamide, when it was delivered on a low-dose, high-frequency schedule as opposed to standard bolus administration [217]. This schedule resulted in effective control of tumour growth, with a concomitant lack of drug resistance in a number of tumour models. This delivery schedule, also termed "low-dose metronomic chemotherapy" is currently being tested in clinical trials. The unique vasculature of tumours, characterised by increased permeability and their complex 3-D architecture, has recently been exploited as an approach to delivering tumour-suppressing drugs with increased efficiency. By manipulating the physicochemical properties of liposomes, they can be made responsive to the specific physiologic features of a tumour (such as low pH), so that the liposomes release their carried drug selectively in tumour tissue. Liposome-mediated delivery of anticancer drugs has improved significantly with a concomitant increased accumulation of drug in tumour vessels.

8.4
Related technologies: PEGylated-liposomes to target tumour vasculature

Long-circulating liposomes, such as PEGylated liposomes (sometimes referred to as Stealth [218]), are also accumulated at the tumour site by the EPR effect [219]. This accumulation is followed by the gradual release of drug in situ and its subsequent diffusion to the intracellular tumour compartment. As in the case of the polymeric wafers for brain chemotherapy, these PEGylated-liposomes do not fully fit under the description of "polymer therapeutics", as the chemotherapeutics are not covalently bound but are rather entrapped. However, they represent a beautiful example of hybrid technologies. Tumour biopsy following the administration of the liposomal anthracyclines Doxil and DaunoXome confirmed EPR mediated tumour targeting in the clinical setting.

8.4.1
PEGylated-liposome-Raf mutant

A further step in liposomal drug delivery to tumour angiogenesis is to devise an active targeting strategy by coupling ligands to the liposome surface that will recognize specific receptors of the tumour endothelial cell (Fig. 1E). By delivering a large drug payload directly into the endothelial cell, an increased therapeutic effect is expected. As an example of this approach, a polymerised cationic liposome has been linked to an endothelial targeting ligand recognizing $\alpha_v\beta_3$ and used to deliver a mutant Raf gene. Systemic injection of the conjugate into mice resulted in apoptosis of the tumour-associated endothelium, leading to tumour cell apoptosis and regression of established primary and metastatic tumours [34].

8.4.2
Liposomal-PEG-Ala-Pro-Arg-Pro-Gly (DSPE-PEG-APRPG)

Another example of targeted PEGylated-liposomal delivery to the tumour vasculature is liposomal-PEG-Ala-Pro-Arg-Pro-Gly (DSPE-PEG-APRPG) [220]. In this compound, the peptide Ala-Pro-Arg-Pro-Gly (APRPG) [203], identified from phage-display libraries as a target moiety for tumour endothelium, is conjugated to hydrophobized polyethylene glycol (distearoylphosphatidylethanolamine [DSPE]-PEG) [220]. Liposomes containing DSPE-PEG-APRPG specifically bound to human umbilical vein endothelial cells stimulated by vascular endothelial growth factor in vitro, and showed enhanced accumulation in tumours in vivo when tested in colon 26 NL-17 carcinoma-bearing mice. The results indicate that PEG conjugate endowed liposomes with long-circulating character through avoidance of RES trapping. Furthermore, PEG-APRPG conjugate liposome accumulated in tumour more than PEG liposome at 24 h after the injection, and adriamycin-encapsulated liposomes modified with APRPG-PEG caused more efficient tumour growth suppression than adriamycin-encapsulated liposomes modified with PEG alone (Fig. 1E). Thus, the improved delivery of liposomes to the tumour tissues was both due to enhanced passive targeting through the EPR effect (via PEGylation), as well as active targeting to the tumour angiogenic vasculature (via conjugation of the targeting moiety APRPG). This study and the PEGylated-liposome-Raf mutant study described above demonstrate that long-circulating liposomes encapsulating anticancer drugs and targeting the tumour neovasculature can be effectively used to eradicate cancerous cells by damaging angiogenic endothelial cells.

9
Combination of polymer therapeutics

A prerequisite for pharmacological activity of constructs delivered via the lysosomotropic or endosomotropic routes (Fig. 2) is cellular internalization and high levels of activating enzyme. Slow cellular uptake, transient cessation of endocytosis or too low levels of the activating enzyme could, in theory, render cells resistant. To thwart this potential problem, polymer-based approaches containing cancer chemotherapy, or alternatively antiangiogenic agents, are now being designed for extracellular drug delivery (Figs. 14A and 15A).

9.1
Polymer directed enzyme prodrug therapy (PDEPT)

PDEPT is a two-step approach combining a polymeric prodrug and polymer-enzyme conjugate, thereby generating cytotoxic drug selectively in the tumour interstitium (Fig. 14A). HPMA copolymer-cathepsin B combined with HPMA copolymer-Gly-Phe-Leu-Gly-doxorubicin (Fig. 14B), and an HPMA copolymer-β-lactamase conjugate and an HPMA copolymer-Gly-Gly-cephalosporin-doxorubicin combination (Fig. 14C), have shown in vivo proof of concept [45, 221, 222]. To achieve proof of concept HPMA copolymer-Gly-Phe-Leu-Gly-doxorubicin (PK1) was selected as the model prodrug, and HPMA copolymer-cathepsin B as the activating enzyme conjugate (Fig. 14B). In vivo, ^{125}I-labelled HPMA copolymer-cathepsin B and PK1 showed tumour targeting by the EPR effect in a sc B16F10 model. Moreover, when

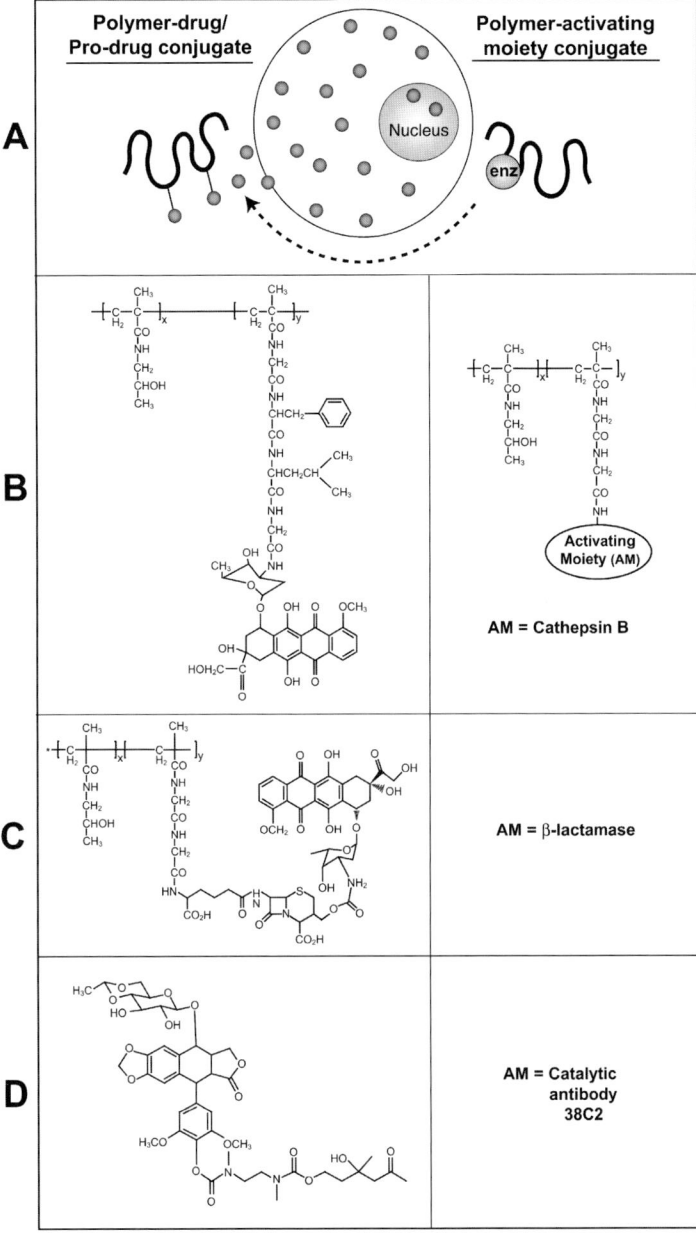

Fig. 14 A Polymer directed enzyme prodrug therapy (PDEPT) is a two-step approach that relies on activation of a polymer-drug conjugate by a complementary polymer-enzyme conjugate; **B** PK1 activated HPMA copolymer-Gly-Gly-cathepsin B; **C** HPMA copolymer-Gly-Gly-cephalosporin-doxorubicin activated by HPMA copolymer-Gly-Gly-β-lactamase; **D** Etoposide prodrug (candidate for polymer conjugation) activated by HPMA copolymer-catalytic antibody 38C2

PK1 (10 mg/kg doxorubicin-equivalent) i.v. injection was followed after 5 h by the cathepsin B conjugate, there was a rapid increase in doxorubicin release within the tumour tissue (3.6-fold increase in the AUC compared to that seen for PK1 alone). This confirmed the ability of triggered drug release in the extracellular space [221]. Subsequently, a nonmammalian enzyme combination, HPMA-copolymer-Gly-Gly-cephalosporin-doxorubicin (HPMA copolymer-C-Dox) and HPMA copolymer-Gly-Gly-β-lactamase were synthesized (Fig. 14C) [45]. Again the two-step administration led to release of free doxorubicin in vivo, and this PDEPT combination caused a significant decrease in tumour growth, in a B16F10 tumour model that was nonresponsive to free doxorubicin and HPMA copolymer-C-Dox. As the PDEPT combination displayed no general toxicity at the doses used and did not lead to an increase in free doxorubicin concentration in normal tissues, further development of this concept is warranted [222].

The use of HPMA copolymer conjugates of catalytic antibodies for prodrug activation is an imaginative step further along this road [222] (Fig. 14D). We reported the preparation of a novel catalytic antibody-polymer conjugate for selective prodrug activation [222]. Antibody 38C2 catalyses a sequence of retro-aldol retro-Michael cleavage reactions, using substrates that are not recognized by human enzymes. Therefore, nonspecific prodrug activation should be minimised. Furthermore, the antibody has recently demonstrated its efficiency in activating several prodrugs in vitro and in vivo [223, 224]. HPMA copolymer was conjugated to catalytic antibody 38C2 through an amide bond formation between the ε-amino group of the lysine residue from the antibody molecule and a p-nitrophenyl ester of the polymer. The conjugate was purified over a size exclusion column using FPLC, similar to the way it was done for HPMA copolymer-Gly-Gly-β-lactamase [45]. The resulting conjugate retained most of its catalytic activity (75–81%) in comparison to the free antibody. Furthermore, the conjugate inhibited tumour cell growth in vitro by activating an etoposide prodrug (Fig. 14D). Cell growth inhibition in the presence of the prodrug and the conjugate was almost identical to inhibition by the free antibody and the prodrug. This is the first time that a catalytic antibody was conjugated to a passive targeting moiety while retaining its catalytic ability to activate a prodrug. HPMA copolymer-catalytic antibody 38C2 can be used for selective activation of prodrug in the PDEPT approach by replacing the enzyme component with catalytic antibody 38C2 and conjugating the etoposide prodrug to a polymer.

9.2
Polymer-enzyme liposome therapy (PELT)

Similarly, HPMA copolymer-phospholipase C (PLC) conjugates can accelerate drug release from liposomes (Fig. 15B); this "polymer-enzyme liposome therapy" (PELT) [13] could present another opportunity for combi-

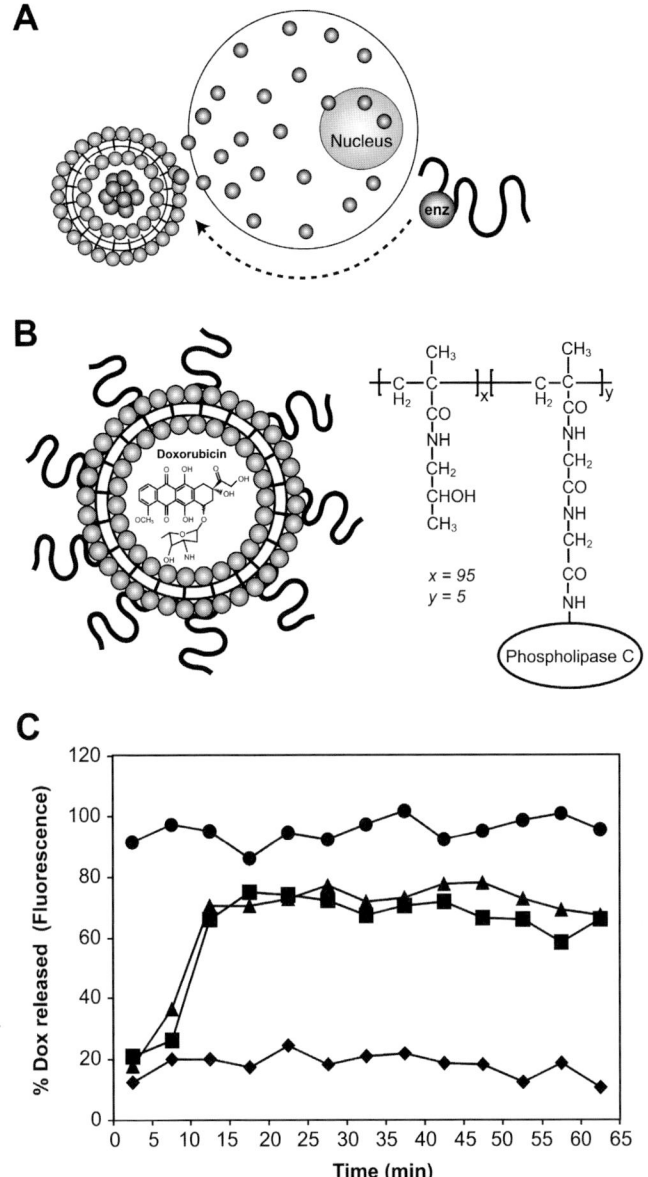

Fig. 15 A Polymer-enzyme liposome therapy (PELT) relies on the liberation of drug from liposomes by the action of a polymer-phospholipase conjugate. **B** PEGylated liposomal doxorubicin (Doxil) activated by HPMA copolymer-Gly-Gly-phospholipase C conjugate. **C** Comparison of enzymatic activity of free phospholipase C and HPMA copolymer Gly-Gly-phospholipase C in vitro. Release of doxorubicin from Daunoxomes in the presence of phospholipase C (■), HPMA copolymer-Gly-Gly-phospholipase C (▲) or Triton X-100 (1%) (•) and in the absence of enzyme as a control (♦) (R. Satchi and R. Duncan, unpublished results)

nation chemotherapy. Incubation of HPMA copolymer-PLC with liposomal-doxorubicin (DaunoXome) in buffered solutions led to complete release of free doxorubicin (Fig. 15C). HPMA copolymer-PLC conjugate retained enzymatic activity as the free PLC (Fig. 15C). Nonspecific release of doxorubicin from the liposomes was not observed in buffered solutions in the absence of PLC over 24 h (R. Satchi and R. Duncan, unpublished results).

9.3
Combination of polymer therapeutics inducing oxidative stress

Reactive oxygen species (ROS) are highly cytotoxic. Many antitumour drugs and proteins exhibit their antitumour effects based on their ability to generate ROS. However, systemic distribution of these drugs causes many undesirable effects [225]. Maeda and coworkers have reported three polymer therapeutics designed to generate oxidative stress within the tumour tissue: PEGylated xanthine oxidase (PEG-XO) [163], PEGylated D-amino acid oxidase (PEG-DAO) [226], and PEGylated zinc protoporphyrin (PEG-ZnPP) [227–230], all of which show tumour-selective targeting via the EPR effect. Individually, each polymer therapeutic has potent antitumour activity, with tumour growth suppressed even after discontinuation of the treatment. The combination of PEG-DAO and PEG-ZnPP has been shown to be a powerful therapy, with tumour regressions observed in 3/8 treated mice.

9.3.1
PEG-XO

Native xanthine oxidase, which generates superoxide anion (O_2^-) and hydrogen peroxide (H_2O_2), has potent antitumour effects [231, 232]. Yet, it normally binds to all endothelial cells via its ε-amino group in lysine residues, causing accumulation in normal tissues and systemic toxicity. In an elegant study, Sawa and coworkers used PEGylation of xanthine oxidase to its lysine residues in order to achieve both (a) hindrance of the high affinity interaction of native XO for all vascular cells, and (b) accumulation of the protein within the tumour environment via the EPR effect [163]. PEG conjugation significantly increased both blood and tumour accumulation of PEG-XO compared with that of native XO, and caused significant suppression of tumour growth of mouse sarcoma S-180 cells following administration of hypoxanthine, which was not observed with native XO.

9.3.2
PEG-DAO

PEGylated D-amino acid oxidase (PEG-DAO) is another oxidative chemotherapeutic agent that generates hydrogen peroxide in the tumour environment after administration of its substrate D-proline [226]. Because D-amino acids do not exist in mammalian systems to a significant level, the activity of the enzyme can be regulated by exogenous administration of its substrate. PEGylation of DAO increased the protein's short half-life in vivo, as the molecular size of the native protein (M_r 39 000) is slightly smaller than the renal excretion threshold ($M_r \sim 50 000$), and also targeted the protein to the tumour environment via the EPR effect. Tumour growth of mouse sarcoma S-180 cells was significantly suppressed in mice administered PEG-DAO plus D-proline. Growth suppression continued to at least 27 days after tumour implantation, which was 15 days after the last treatment with PEG-DAO and D-proline. In contrast, no significant antitumour effect was observed in mice treated with native DAO plus D-proline.

9.3.3
PEG-ZnPP

Another skillful approach to targeting the tumour compartment by oxidative stress is to make the tumour more vulnerable to oxidative insults. This was achieved when PEG was conjugated to zinc protoporphyrin (ZnPP), a heme oxygenase (HO) inhibitor [227–230, 233]. Tumour and normal cells are protected from oxidative damage by the stress inducible heme oxygenase (HO – 1), which is induced in cells after insults such as hypoxia [234], ROS [235], and UV irradiation [235]. HO – 1 catalyses the initial and rate-limiting step of heme degradation; oxidative cleavage of the heme leads to formation of the antiapoptotic molecule carbon monoxide (CO), free iron, and biliverdin, the latter of which is subsequently reduced to the potent antioxidant bilirubin [236, 237]. HO – 1 exerts its protective role by multiple mechanisms [228], including decreasing pro-oxidant levels (heme) [238], and increasing antioxidant (bilirubin) levels [239]. Zinc protoporphyrin (ZnPP) is a specific HO inhibitor. PEG-ZnPP, a water-soluble derivative of ZnPP, exerts cytotoxic effects by itself [227]. Because it makes cells more vulnerable to toxic insults, PEG-ZnPP also effectively potentiates the toxicity induced by peroxides and anticancer agents both in vitro and in vivo [228]. When PEG-ZnPP was used to treat mouse sarcoma S-180 tumours in mice, potent antitumour effects were observed. PEG-ZnPP treatment produced tumour-selective suppression of HO activity as well as induction of apoptosis, presumably through increased oxidative stress [227–230].

9.3.4
PEG-DAC and PEG-ZnPP combination

Building on these findings, Fang and coworkers reported a sophisticated example for targeting the tumour compartment by oxidative damage using the combined actions of PEG-DAC and PEG-ZnPP [228]. PEG-DAC selectively delivers oxidative damage to the tumours and PEG-ZnPP makes them more susceptible to these oxidative insults. In this example, tumour growth was almost completely suppressed when the mice were pretreated with PEG-ZnPP followed by PEG-DAO plus D-proline [229]. Importantly, lower doses of PEG-DAO and PEG-ZnPP were used in these experiments than in the reports described above. Continuous suppression of tumour growth was observed even 22 days after the last treatment with PEG-DAO and D-proline, and complete regression of tumour growth was observed in 3/8 mice. This study demonstrates that polymer therapeutics are promising effective mediators of oxidative therapy, and warrants further investigation for their clinical application.

10
Conclusions

In general, an ideal delivery system is one that can enable the conjugation of a targeting moiety and an active entity in a simple chemical platform. In the near future the various technologies for drug delivery will be combined to target pathological angiogenesis with minimal side-effects (schematically shown in Fig. 1D). Paul Ehrlich's magic bullet dream (1906) may not have materialised yet, but combination therapy targeting tumour vasculature with drug delivery systems of cytotoxic chemotherapeutic drugs will probably be the ideal attack on both compartments of the tumour; the endothelial and the tumour cell.

Acknowledgements We would like to thank Dr Judah Folkman for useful discussions. We thank Kristin Gullage for her artwork.

References

1. Gregoriadis G (1989) In: Roednick F, Kroon A (eds) Drug carrier systems, Horizons in biochemistry and biophysics, vol 9. Wiley, New York, p 1
2. Allen TM (2002) Nat Rev Cancer 2:750
3. St Croix B, Rago C, Velculescu V, Traverso G, Romans KE, Montgomery E, Lal A, Riggins GJ, Lengauer C, Vogelstein B, Kinzler KW (2000) Science 289:1197
4. Moses MA, Brem H, Langer R (2003) Cancer Cell 4:337
5. Bolla M, Collette L, Blank L, Warde P, Dubois JB, Mirimanoff RO, Storme G, Bernier J, Kuten A, Sternberg C, Mattelaer J, Lopez Torecilla J, Pfeffer JR, Lino Cutajar C, Zurlo A, Pierart M (2002) Lancet 360:103
6. Tsukagoshi S (2002) Gan To Kagaku Ryoho 29:1675
7. Heyns CF, Simonin MP, Grosgurin P, Schall R, Porchet HC (2003) BJU Int 92:226
8. Langer R (1998) Nature 392:5
9. Brem H, Ewend MG, Piantadosi S, Greenhoot J, Burger PC, Sisti M (1995) J Neurooncol 26:111

10. Satchi-Fainaro R, Puder M, Davies JW, Tran HT, Sampson DA, Greene AK, Corfas G, Folkman J (2004) Nat Med 10:255
11. Satchi-Fainaro R, Birsner AE, Short SM, Butterfield C, Folkman J (2005) 32nd Annual Meeting of the Controlled Release Society, 18–22 June 2005, Miami, FL
12. Duncan R (2003) Nat Rev Drug Discov 2:347
13. Duncan R, Gac-Breton S, Keane R, Musila R, Sat YN, Satchi R, Searle F (2001) J Control Release 74:135
14. Vandermeulen GW, Klok HA (2004) Macromol Biosci 4:383
15. Duncan R (2004) In: Budman D, Calvert H, Rowinsky E (eds) Handbook of anticancer drug development. Lippincott, Williams & Wilkins, Baltimore, MD
16. Veronese FM, Harris JM (2002) Adv Drug Deliv Rev 54:453
17. Yokoyama M, Miyauchi M, Yamada N, Okano T, Sakurai Y, Kataoka K, Inoue S (1990) Cancer Res 50:1693
18. Kabanov AV, Felgner PL, Seymour LW (1998) Wiley, Chichester, UK
19. Abuchowski A, Fuertges F (1990) J Control Release 11:139
20. Duncan R (1992) Anticancer Drugs 3:175
21. Matsumura Y, Maeda H (1986) Cancer Res 46:6387
22. Jain RK (1997) Adv Drug Deliv Rev 26:71
23. Satchi-Fainaro R, Mamluk R, Wang L, Short SM, Nagy JA, Feng D, Dvorak AM, Dvorak HF, Puder M, Mukhopadhyay D, Folkman J (2005) Cancer Cell 7:251
24. Hanahan D, Folkman J (1996) Cell 86:353
25. Hida K, Hida Y, Amin DN, Flint AF, Panigrahy D, Morton CC, Klagsbrun M (2004) Cancer Res 64:8249
26. Carmeliet P, Jain RK (2000) Nature 407:249
27. Carmeliet P, Collen D (2000) Ann N Y Acad Sci 902:249
28. Folkman J, Kalluri R (2003) In: Kufe DW, Pollock RE, Weichselbaum RR, Bast RCJ, Gansler TS, Holland JF, Frei EI (eds) Cancer medicine. BC Decker Inc, Hamilton, Ontario, Canada, p 161
29. Tozer GM, Kanthou C, Parkins CS, Hill SA (2002) Int J Exp Pathol 83:21
30. Folkman J, Kalluri R (2003) Tumor angiogenesis. In: Kufe DW, Pollock RE, Weichselbaum RR, Bast RCJ, Gansler TS, Holland JF, Frei EI (eds) Cancer medicine. BC Decker Inc, Hamilton, Ontario, Canada pp 161–194
31. Ingber D, Fujita T, Kishimoto S, Sudo K, Kanamaru T, Brem H, Folkman J (1990) Nature 348:555
32. Gabizon A, Shmeeda H, Barenholz Y (2003) Clin Pharmacokinet 42:419
33. Kopecek J, Kopeckova P, Minko T, Lu Z (2000) Eur J Pharm Biopharm 50:61
34. Hood JD, Bednarski M, Frausto R, Guccione S, Reisfeld RA, Xiang R, Cheresh DA (2002) Science 296:2404
35. Nagle T, Berg C, Nassr R, Pang K (2003) Nat Rev Drug Discov 2:75
36. Brekke OH, Sandlie I (2003) Nat Rev Drug Discov 2:52
37. Davis FF (2002) Adv Drug Deliv Rev 54:457
38. Harris JM, Chess RB (2003) Nat Rev Drug Discov 2:214
39. Delgado C, Francis GE, Fisher D (1992) Crit Rev Ther Drug Carrier Syst 9:249
40. Francis GE, Delgado C, Fisher D (1992) In: Ahern TJ, Manning MC (eds) Stability of proteins pharmaceuticals (Part B). Plenum, New York, p 235
41. Goodson RJ, Katre NV (1990) Biotechnology (N Y) 8:343
42. Chapman AP, Antoniw P, Spitali M, West S, Stephens S, King DJ (1999) Nat Biotechnol 17:780
43. Sato H (2002) Adv Drug Deliv Rev 54:487
44. Lee S, Greenwald RB, McGuire J, Yang K, Shi C (2001) Bioconjug Chem 12:163

45. Satchi-Fainaro R, Hailu H, Davies JW, Summerford C, Duncan R (2003) Bioconjug Chem 14:797
46. Satchi R, Connors TA, Duncan R (2001) Br J Cancer 85:1070
47. Kamada H, Tsutsumi Y, Tsunoda S, Kihira T, Kaneda Y, Yamamoto Y, Nakagawa S, Horisawa Y, Mayumi T (1999) Biochem Biophys Res Commun 257:448
48. Kamada H, Tsutsumi Y, Yamamoto Y, Kihira T, Kaneda Y, Mu Y, Kodaira H, Tsunoda SI, Nakagawa S, Mayumi T (2000) Cancer Res 60:6416
49. Tsunoda S, Kamada H, Yamamoto Y, Ishikawa T, Matsui J, Koizumi K, Kaneda Y, Tsutsumi Y, Ohsugi Y, Hirano T, Mayumi T (2000) J Control Release 68:335
50. Levy Y, Hershfield MS, Fernandez-Mejia C, Polmar SH, Scudiery D, Berger M, Sorensen RU (1988) J Pediatr 113:312
51. Davis FF (2003) Adv Exp Med Biol 519:51
52. Abshire TC, Pollock BH, Billett AL, Bradley P, Buchanan GR (2000) Blood 96:1709
53. Agrawal NR, Bukowski RM, Rybicki LA, Kurtzberg J, Cohen LJ, Hussein MA (2003) Cancer 98:94
54. Molineux G (2004) Curr Pharm Des 10:1235
55. Kinstler O, Molineux G, Treuheit M, Ladd D, Gegg C (2002) Adv Drug Deliv Rev 54:477
56. Holmes FA, Jones SE, O'Shaughnessy J, Vukelja S, George T, Savin M, Richards D, Glaspy J, Meza L, Cohen G, Dhami M, Budman DR, Hackett J, Brassard M, Yang BB, Liang BC (2002) Ann Oncol 13:903
57. Heil G, Hoelzer D, Sanz MA, Lechner K, Liu Yin JA, Papa G, Noens L, Szer J, Ganser A, O'Brien C, Matcham J, Barge A (1997) Blood 90:4710
58. Huang SF, Kim SJ, Lee AT, Karashima T, Bucana C, Kedar D, Sweeney P, Mian B, Fan D, Shepherd D, Fidler IJ, Dinney CP, Killion JJ (2002) Cancer Res 62:5720
59. Gutterman JU (1994) Proc Natl Acad Sci USA 91:1198
60. Brouty-Boye D, Zetter BR (1980) Science 208:516
61. Singh RK, Gutman M, Bucana CD, Sanchez R, Llansa N, Fidler IJ (1995) Proc Natl Acad Sci USA 92:4562
62. Singh RK, Gutman M, Llansa N, Fidler IJ (1996) J Interferon Cytokine Res 16:577
63. Gohji K, Fidler IJ, Tsan R, Radinsky R, von Eschenbach AC, Tsuruo T, Nakajima M (1994) Int J Cancer 58:380
64. Thomas H, Balkwill FR (1991) Pharmacol Ther 52:307
65. Slaton JW, Perrotte P, Inoue K, Dinney CP, Fidler IJ (1999) Clin Cancer Res 5:2726
66. Singh Y, Shikata N, Kiyozuka Y, Nambu H, Morimoto J, Kurebayashi J, Hioki K, Tsubura A (1997) Breast Cancer Res Treat 45:15
67. Chang E, Boyd A, Nelson CC, Crowley D, Law T, Keough KM, Folkman J, Ezekowitz RA, Castle VP (1997) J Pediatr Hematol Oncol 19:237
68. Ezekowitz A, Mulliken J, Folkman J (1991) Br J Haematol 79(Suppl 1):67
69. Kaban LB, Mulliken JB, Ezekowitz RA, Ebb D, Smith PS, Folkman J (1999) Pediatrics 103:1145
70. Marler JJ, Rubin JB, Trede NS, Connors S, Grier H, Upton J, Mulliken JB, Folkman J (2002) Pediatrics 109:E37
71. Takahashi K, Mulliken J, Kozakewich H, Rogers R, Folkman J, Ezekowitz R (1994) J Clin Invest 93:2357
72. Bielenberg D, Bucana C, Sanchez R, Mulliken J, Folkman J, Fidler I (1999) Int J Oncol 14:401
73. Salmon P, Le Cotonnec JY, Galazka A, Abdul-Ahad A, Darragh A (1996) J Interferon Cytokine Res 16:759
74. Einhorn S, Grander D (1996) J Interferon Cytokine Res 16:275

75. Glue P, Fang JW, Rouzier-Panis R, Raffanel C, Sabo R, Gupta SK, Salfi M, Jacobs S (2000) Clin Pharmacol Ther 68:556
76. Reddy KR, Wright TL, Pockros PJ, Shiffman M, Everson G, Reindollar R, Fried MW, Purdum PP 3rd, Jensen D, Smith C, Lee WM, Boyer TD, Lin A, Pedder S, DePamphilis J (2001) Hepatology 33:433
77. Heathcote EJ, Shiffman ML, Cooksley WG, Dusheiko GM, Lee SS, Balart L, Reindollar R, Reddy RK, Wright TL, Lin A, Hoffman J, De Pamphilis J (2000) N Engl J Med 343:1673
78. Fried MW, Shiffman ML, Reddy KR, Smith C, Marinos G, Goncales FL, Haussinger JrD, Diago M, Carosi G, Dhumeaux D, Craxi A, Lin A, Hoffman J, Yu J (2002) N Engl J Med 347:975
79. Wang YS, Youngster S, Grace M, Bausch J, Bordens R, Wyss DF (2002) Adv Drug Deliv Rev 54:547
80. Bukowski R, Ernstoff MS, Gore ME, Nemunaitis JJ, Amato R, Gupta SK, Tendler CL (2002) J Clin Oncol 20:3841
81. Maeda H (2001) Adv Drug Deliv Rev 46:169
82. Konno T, Maeda H, Iwai K, Maki S, Tashiro S, Uchida M, Miyauchi Y (1984) Cancer 54:2367
83. Maeda H, Konno T (1997) In: Maeda H, Edo K, Ishida N (eds) Neocarzinostatin: The past, present, and future of an anticancer drug. Springer, Berlin Heidelberg New York, p 227
84. Deguchi Y, Kurihara A, Pardridge WM (1999) Bioconjug Chem 10:32
85. Seymour LW, Ferry DR, Anderson D, Hesslewood S, Julyan PJ, Poyner R, Doran J, Young AM, Burtles S, Kerr DJ (2002) J Clin Oncol 20:1668
86. De Duve C (1974) Biochem Pharmacol 23:2495
87. Ringsdorf H (1975) J Polym Sci Polym Symp 51:135
88. Huang PS, Oliff A (2001) Curr Opin Genet Dev 11:104
89. Duncan R, Spreafico F (1994) Clin Pharmacokinet 27:290
90. Seymour LW, Ulbrich K, Steyger PS, Brereton M, Subr V, Strohalm J, Duncan R (1994) Br J Cancer 70:636
91. Gianasi E, Wasil M, Evagorou EG, Keddle A, Wilson G, Duncan R (1999) Eur J Cancer 35:994
92. Mendichi R, Rizzo V, Gigli M, et al. (1998) J Appl Polym Sci 70:329
93. Duncan R, Cable HC, Lloyd JB, Rejmanova P, Kopecek J (1984) Makromol Chem 184:1997–2008
94. Mendichi R, Rizzo V, Gigli M et al. (1998) J Liq Chromatogr R T 21:1295
95. Duncan R, Seymour LW, O'Hare KB et al. (1992) J Control Release 19:331
96. Vasey PA, Kaye SB, Morrison R, Twelves C, Wilson P, Duncan R, Thomson AH, Murray LS, Hilditch TE, Murray T, Burtles S, Fraier D, Frigerio E, Cassidy J (1999) Clin Cancer Res 5:83
97. Duncan R, Coatsworth JK, Burtles S (1998) Hum Exp Toxicol 17:93
98. Seymour LW, Ulbrich K, Strohalm J, Kopecek J, Duncan R (1990) Biochem Pharmacol 39:1125
99. Thomson AH, Vasey PA, Murray LS, Cassidy J, Fraier D, Frigerio E, Twelves C (1999) Br J Cancer 81:99
100. Duncan R, Kopeckova-Rejmanova P, Strohalm J, Hume I, Cable HC, Pohl J, Lloyd JB, Kopecek J (1987) Br J Cancer 55:165
101. Duncan R, Kopeckova P, Strohalm J, Hume IC, Lloyd JB, Kopecek J (1988) Br J Cancer 57:147
102. Duncan R, Hume IC, Kopeckova P, Strohalm J, Lloyd JB, Kopecek J (1989) J Control Release 10:51

103. Ashwell G, Harford J (1982) Annu Rev Biochem 51:531
104. Duncan R, Seymour LW, Scarlett L et al. (1986) Biochim Biophys Acta 880:62
105. Seymour LW, Ulbrich K, Wedge SR, Hume IC, Strohalm J, Duncan R (1991) Br J Cancer 63:859
106. Julyan PJ, Ferry DR, Seymour LW et al. (1999) J Control Release 57:281
107. Rihova B, Strohalm J, Prausova J, Kubackova K, Jelinkova M, Rozprimova L, Sirova M, Plocova D, Etrych T, Subr V, Mrkvan T, Kovar M, Ulbrich K (2003) J Control Release 91:1
108. Rihova B, Strohalm J, Kubackova K, Jelinkova M, Rozprimova L, Sirova M, Plocova D, Mrkvan T, Kovar M, Pokorna J, Erytch T, Ulbrich K, Maeda H, Kabanov A, Kataoka K, Okano T (2003) Polymer drugs in the clinical stage: Advantages and prospects. Kluwer Academic/Plenum, New York, p 145
109. Meerum Terwogt JM, ten Bokkel Huinink WW, Schellens JH, Schot M, Mandjes IA, Zurlo MG, Rocchetti M, Rosing H, Koopman FJ, Beijnen JH (2001) Anticancer Drugs 12:315
110. Slichenmyer WJ, Rowinsky EK, Donehower RC, Kaufmann SH (1993) J Natl Cancer Inst 85:271
111. Cersosimo RJ (1998) Ann Pharmacother 32:1334
112. Cersosimo RJ (1998) Ann Pharmacother 32:1324
113. Caiolfa VR, Zamal M, Fiorini A et al. (2000) J Control Release 65:105
114. Bissett D, Cassidy J, de Bono JS, Muirhead F, Main M, Robson L, Fraier D, Magne ML, Pellizzoni C, Porro MG, Spinelli R, Speed W, Twelves C (2004) Br J Cancer 91:50
115. Greenwald RB, Gilbert CW, Pendri A, Conover CD, Xia J, Martinez A (1996) J Med Chem 39:424
116. Schoemaker NE, Frigerio E, Fraier D, Schellens JH, Rosing H, Jansen S, Beijnen JH (2001) J Chromatogr B 763:173
117. Rademaker-Lakhai JM, Terret C, Howell SB, Baud CM, De Boer RF, Pluim D, Beijnen JH, Schellens JH, Droz JP (2004) Clin Cancer Res 10:3386
118. Gianasi E, Buckley RG, Latigo J, Wasil M, Duncan R (2002) J Drug Target 10:549
119. Li C, Yu DF, Newman RA, Cabral F, Stephens LC, Hunter N, Milas L, Wallace S (1998) Cancer Res 58:2404
120. Singer JW, Baker B, De Vries P, Kumar A, Shaffer S, Vawter E, Bolton M, Garzone P (2003) Adv Exp Med Biol 519:81
121. Auzenne E, Donato NJ, Li C, Leroux E, Price RE, Farquhar D, Klostergaard J (2002) Clin Cancer Res 8:573
122. Li C, Price JE, Milas L, Hunter NR, Ke S, Yu DF, Charnsangavej C, Wallace S (1999) Clin Cancer Res 5:891
123. Li C, Ke S, Wu QP, Tansey W, Hunter N, Buchmiller LM, Milas L, Charnsangavej C, Wallace S (2000) Clin Cancer Res 6:2829
124. Langer CJ (2004) Expert Opin Investig Drugs 13:1501
125. Sabbatini P, Aghajanian C, Dizon D, Anderson S, Dupont J, Brown JV, Peters WA, Jacobs A, Mehdi A, Rivkin S, Eisenfeld AJ, Spriggs D (2004) J Clin Oncol 22:4523
126. Markman M (2004) J Exp Ther Oncol 4:131
127. Langer CJ (2004) Clin Lung Cancer 6 Suppl 2:S85
128. Bhatt R, de Vries P, Tulinsky J, Bellamy G, Baker B, Singer JW, Klein P (2003) J Med Chem 46:190
129. Singer JW, De Vries P, Bhatt R, Tulinsky J, Klein P, Li C, Milas L, Lewis RA, Wallace S (2000) Ann N Y Acad Sci 922:136
130. Singer JW, Bhatt R, Tulinsky J, Buhler KR, Heasley E, Klein P, de Vries P (2001) J Control Release 74:243

131. Zou Y, Wu QP, Tansey W, Chow D, Hung MC, Charnsangavej C, Wallace S, Li C (2001) Int J Oncol 18:331
132. Rowinsky EK, Rizzo J, Ochoa L, Takimoto CH, Forouzesh B, Schwartz G, Hammond LA, Patnaik A, Kwiatek J, Goetz A, Denis L, McGuire J, Tolcher AW (2003) J Clin Oncol 21:148
133. Greenwald RB, Choe YH, McGuire J, Conover CD (2003) Adv Drug Deliv Rev 55:217
134. Li C, Yu D, Inoue T, Yang DJ, Milas L, Hunter NR, Kim EE, Wallace S (1996) Anticancer Drugs 7:642
135. Schacht E, Ruys L, Vermeersch J, Remon JP, Duncan R (1985) Ann N Y Acad Sci 446:199
136. Brocchini S, Duncan R (1999) In: Mathiovitz E (ed) Encyclopaedia of controlled drug delivery. Wiley, New York, p 786
137. Danhauser-Riedl S, Hausmann E, Schick HD, Bender R, Dietzfelbinger H, Rastetter J, Hanauske AR (1993) Invest New Drugs 11:187
138. Batrakova EV, Dorodnych TY, Klinskii EY, Kliushnenkova EN, Shemchukova OB, Goncharova ON, Arjakov SA, Alakhov VY, Kabanov AV (1996) Br J Cancer 74:1545
139. Kwon GS, Yokoyama M, Okano T, Sakurai Y, Kataoka K (1993) Pharm Res 10:970
140. Nakanishi T, Fukushima S, Okamoto K, Suzuki M, Matsumura Y, Yokoyama M, Okano T, Sakurai Y, Kataoka K (2001) J Control Release 74:295
141. Bae Y, Nishiyama N, Fukushima S, Koyama H, Yasuhiro M, Kataoka K (2005) Bioconjug Chem 16:122
142. Peppas NA, Langer R (1994) Science 263:1715
143. Brem H, Piantadosi S, Burger PC, Walker M, Selker R, Vick NA, Black K, Sisti M, Brem S, Mohr G et al. (1995) Lancet 345:1008
144. Valtonen S, Timonen U, Toivanen P, Kalimo H, Kivipelto L, Heiskanen O, Unsgaard G, Kuurne T (1997) Neurosurgery 41:44
145. Westphal M, Hilt DC, Bortey E, Delavault P, Olivares R, Warnke PC, Whittle IR, Jaaskelainen J, Ram Z (2003) Neuro-oncol 5:79
146. Ewend MG, Sampath P, Williams JA, Tyler BM, Brem H (1998) Neurosurgery 43:1185
147. Laws ER Jr, Morris AM, Maartens N (2003) Neurosurgery 53:255
148. Lesniak MS, Langer R, Brem H (2001) Curr Neurol Neurosci Rep 1:210
149. Kumazawa E, Ochi Y (2004) Cancer Sci 95:168
150. Ochi Y, Shiose Y, Kuga H, Kumazawa E (2004) Cancer Chemother Pharmacol
151. Oguma T, Morikawa H, Iwasaki D, Atsumi R (2004) Biomed Chromatogr 19:19
152. Hreczuk-Hirst D, Chicco D, German L, Duncan R (2001) Int J Pharm 230:57
153. Chabot GG (1997) Clin Pharmacokinet 33:245
154. Takimoto CH, Wright J, Arbuck SG (1998) Biochim Biophys Acta 1400:107
155. Harada M, Imai J, Okuno S, Suzuki T (2000) J Control Release 69:389
156. Harada M, Murata J, Sakamura Y, Sakakibara H, Okuno S, Suzuki T (2001) J Control Release 71:71
157. Harada M, Sakakibara H, Yano T, Suzuki T, Okuno S (2000) J Control Release 69:399
158. Okuno S, Harada M, Yano T, Yano S, Kiuchi S, Tsuda N, Sakamura Y, Imai J, Kawaguchi T, Tsujihara K (2000) Cancer Res 60:2988
159. Vicent MJ, Tomlinson R, Brocchini S, Duncan R (2004) J Drug Target 12:491
160. Vicent MJ, Manzanaro S, de la Fuente JA, Duncan R (2004) J Drug Target 12:503
161. Kerbel R, Folkman J (2002) Nat Rev Cancer 2:727
162. Bauer JA, Morrison BH, Grane RW, Jacobs BS, Borden EC, Lindner DJ (2003) J Interferon Cytokine Res 23:3
163. Sawa T, Wu J, Akaike T, Maeda H (2000) Cancer Res 60:666
164. Alessi P, Ebbinghaus C, Neri D (2004) Biochim Biophys Acta 1654:39

165. Friedlander M, Brooks PC, Shaffer RW, Kincaid CM, Varner JA, Cheresh DA (1995) Science 270:1500
166. Sipkins DA, Cheresh DA, Kazemi MR, Nevin LM, Bednarski MD, Li KC (1998) Nat Med 4:623
167. Brooks PC, Montgomery AM, Rosenfeld M, Reisfeld RA, Hu T, Klier G, Cheresh DA (1994) Cell 79:1157
168. Zardi L, Carnemolla B, Siri A, Petersen TE, Paolella G, Sebastio G, Baralle FE (1987) EMBO J 6:2337
169. Borsi L, Carnemolla B, Nicolo G, Spina B, Tanara G, Zardi L (1992) Int J Cancer 52:688
170. Ran S, Downes A, Thorpe PE (2002) Cancer Res 62:6132
171. Seon BK, Takahashi N, Haba A, Matsuno F, Haruta Y, She XW, Harada N, Tsai H (2001) Rinsho Byori 49:1005
172. Brekken RA, Overholser JP, Stastny VA, Waltenberger J, Minna JD, Thorpe PE (2000) Cancer Res 60:5117
173. Huston JS, Levinson D, Mudgett-Hunter M, Tai MS, Novotny J, Margolies MN, Ridge RJ, Bruccoleri RE, Haber E, Crea R et al. (1988) Proc Natl Acad Sci USA 85:5879
174. Birchler M, Viti F, Zardi L, Spiess B, Neri D (1999) Nat Biotechnol 17:984
175. Demartis S, Tarli L, Borsi L, Zardi L, Neri D (2001) Eur J Nucl Med 28:534
176. Tarli L, Balza E, Viti F, Borsi L, Castellani P, Berndorff D, Dinkelborg L, Neri D, Zardi L (1999) Blood 94:192
177. Viti F, Tarli L, Giovannoni L, Zardi L, Neri D (1999) Cancer Res 59:347
178. Santimaria M, Moscatelli G, Viale GL, Giovannoni L, Neri G, Viti F, Leprini A, Borsi L, Castellani P, ZardiL, Neri D, Riva P (2003) Clin Cancer Res 9:571
179. Bigner DD, Brown MT, Friedman AH, Coleman RE, Akabani G, Friedman HS, Thorstad WL, McLendon RE, Bigner SH, Zhao XG, Pegram CN, Wikstrand CJ, Herndon JE 2nd, Vick NA, Paleologos N, Cokgor I, Provenzale JM, Zalutsky MR (1998) J Clin Oncol 16:2202
180. Brown MT, Coleman RE, Friedman AH, Friedman HS, McLendon RE, Reiman R, Felsberg GJ, Tien RD, Bigner SH, Zalutsky MR, Zhao XG, Wikstrand CJ, Pegram CN, Herndon JE 2nd, Vick NA, Paleologos N, Fredericks RK, Schold SC Jr, Bigner DD (1996) Clin Cancer Res 2:963
181. Cokgor I, Akabani G, Kuan CT, Friedman HS, Friedman AH, Coleman RE, McLendon RE, Bigner SH, Zhao XG, Garcia-Turner AM, Pegram CN, Wikstrand CJ, Shafman TD, Herndon JE 2nd, Provenzale JM, Zalutsky MR, Bigner DD (2000) J Clin Oncol 18:3862
182. Reardon DA, Akabani G, Coleman RE, Friedman AH, Friedman HS, Herndon JE 2nd, Cokgor I, McLendon RE, Pegram CN, Provenzale JM, Quinn JA, Rich JN, Regalado LV, Sampson JH, Shafman TD, Wikstrand CJ, Wong TZ, Zhao XG, Zalutsky MR, Bigner DD (2002) J Clin Oncol 20:1389
183. Carnemolla B, Castellani P, Ponassi M, Borsi L, Urbini S, Nicolo G, Dorcaratto A, Viale G, Winter G, Neri D, Zardi L (1999) Am J Pathol 154:1345
184. Carver LA, Schnitzer JE (2003) Nat Rev Cancer 3:571
185. Schnitzer JE (1998) N Engl J Med 339:472
186. Schnitzer JE (2001) Adv Drug Deliv Rev 49:265
187. Carson-Walter EB, Watkins DN, Nanda A, Vogelstein B, Kinzler KW, St Croix B (2001) Cancer Res 61:6649
188. Wyder L, Vitaliti A, Schneider H, Hebbard LW, Moritz DR, Wittmer M, Ajmo M, Klemenz R (2000) Cancer Res 60:4682

189. Ruoslahti E (2000) Semin Cancer Biol 10:435
190. Ruoslahti E (2004) Biochem Soc Trans 32:397
191. Kolonin M, Pasqualini R, Arap W (2001) Curr Opin Chem Biol 5:308
192. Mutuberria R, Satijn S, Huijbers A, Van Der Linden E, Lichtenbeld H, Chames P, Arends JW, Hoogenboom HR (2004) J Immunol Methods 287:31
193. Pasqualini R, Koivunen E, Kain R, Lahdenranta J, Sakamoto M, Stryhn A, Ashmun RA, Shapiro LH, Arap W, Ruoslahti E (2000) Cancer Res 60:722
194. Orlando RA, Cheresh DA (1991) J Biol Chem 266:19543
195. Nanda A, Carson-Walter EB, Seaman S, Barber TD, Stampfl J, Singh S, Vogelstein B, Kinzler KW, St Croix B (2004) Cancer Res 64:817
196. Rmali KA, Watkins G, Harrison G, Parr C, Puntis MC, Jiang WG (2004) Eur J Surg Oncol 30:948
197. Li J, Ji J, Holmes LM, Burgin KE, Barton LB, Yu X, Wagner TE, Wei Y (2004) Cancer Gene Ther 11:363
198. Haubner R, Wester HJ, Burkhart F, Senekowitsch-Schmidtke R, Weber W, Goodman SL, Kessler H, Schwaiger M (2001) J Nucl Med 42:326
199. Haubner R, Wester HJ, Weber WA, Mang C, Ziegler SI, Goodman SL, Senekowitsch-Schmidtke R, Kessler H, Schwaiger M (2001) Cancer Res 61:1781
200. Pasqualini R, Arap W, McDonald DM (2002) Trends Mol Med 8:563
201. Folkman J (1999) Nat Biotechnol 17:749
202. Koivunen E, Arap W, Valtanen H, Rainisalo A, Medina OP, Heikkila P, Kantor C, Gahmberg CG, Salo T, Konttinen YT, Sorsa T, Ruoslahti E, Pasqualini R (1999) Nat Biotechnol 17:768
203. Oku N, Asai T, Watanabe K, Kuromi K, Nagatsuka M, Kurohane K, Kikkawa H, Ogino K, Tanaka M, Ishikawa D, Tsukada H, Momose M, Nakayama J, Taki T (2002) Oncogene 21:2662
204. Arap W, Haedicke W, Bernasconi M, Kain R, Rajotte D, Krajewski S, Ellerby HM, Bredesen DE, Pasqualini R, Ruoslahti E (2002) Proc Natl Acad Sci USA 99:1527
205. Satchi-Fainaro R (2002) J Drug Target 10:529
206. Wan K-W, Vicent MJ, Duncan R (2003) In: Proc Int Symp Controlled Release of Bioactive Materials, vol 30, 23rd June 2003, Glasgow, UK, p 491
207. Mitra A, Mulholland J, Nan A, McNeill E, Ghandehari H, Line BR (2005) J Control Release 102:191
208. Chen X, Hou Y, Tohme M, Park R, Khankaldyyan V, Gonzales-Gomez I, Bading JR, Laug WE, Conti PS (2004) J Nucl Med 45:1776
209. Chen X, Park R, Shahinian AH, Bading JR, Conti PS (2004) Nucl Med Biol 31:11
210. Janssen AP, Schiffelers RM, ten Hagen TL, Koning GA, Schraa AJ, Kok RJ, Storm G, Molema G (2003) Int J Pharm 254:55
211. Kudelka AP, Levy T, Verschraegen CF, Edwards CL, Piamsomboon S, Termrungruanglert W, Freedman RS, Kaplan AL, Kieback DG, Meyers CA, Jaeckle KA, Loyer E, Steger M, Mante R, Mavligit G, Killian A, Tang RA, Gutterman JU, Kavanagh JJ (1997) Clin Cancer Res 3:1501
212. Kudelka AP, Verschraegen CF, Loyer E (1998) N Engl J Med 338:991
213. Bhargava P, Marshall JL, Rizvi N, Dahut W, Yoe J, Figuera M, Phipps K, Ong VS, Kato A, Hawkins MJ (1999) Clin Cancer Res 5:1989
214. Dezube BJ, Von Roenn JH, Holden-Wiltse J, Cheung TW, Remick SC, Cooley TP, Moore J, Sommadossi JP, Shriver SL, Suckow CW, Gill PS (1998) J Clin Oncol 16:1444
215. Udagawa T, Fernandez A, Achilles EG, Folkman J, D'Amato RJ (2002) FASEB J 16:1361
216. Folkman J, Kalluri R (2004) Nature 427:787

217. Browder T, Butterfield CE, Kraling BM, Shi B, Marshall B, O'Reilly MS, Folkman J (2000) Cancer Res 60:1878
218. Papahadjopoulos D, Allen TM, Gabizon A, Mayhew E, Matthay K, Huang SK, Lee KD, Woodle MC, Lasic DD, Redemann C et al. (1991) Proc Natl Acad Sci USA 88:11460
219. Symon Z, Peyser A, Tzemach D, Lyass O, Sucher E, Shezen E, Gabizon A (1999) Cancer 86:72
220. Maeda N, Takeuchi Y, Takada M, Namba Y, Oku N (2004) Bioorg Med Chem Lett 14:1015
221. Satchi R, Connors TA, Duncan R (2001) Br J Cancer 85:1070
222. Satchi-Fainaro R, Wrasidlo W, Lode HN, Shabat D (2002) Bioorg Med Chem 10:3023
223. Shabat D, Lode HN, Pertl U, Reisfeld RA, Rader C, Lerner RA, Barbas CF 3rd (2001) Proc Natl Acad Sci USA 98:7528
224. Shabat D, Rader C, List B, Lerner RA, Barbas CF 3rd (1999) Proc Natl Acad Sci USA 96:6925
225. Yen HC, Oberley TD, Vichitbandha S, Ho YS, St Clair DK (1996) J Clin Invest 98:1253
226. Fang J, Sawa T, Akaike T, Maeda H (2002) Cancer Res 62:3138
227. Fang J, Sawa T, Maeda H (2003) Adv Exp Med Biol 519:29
228. Fang J, Sawa T, Akaike T, Greish K, Maeda H (2004) Int J Cancer 109:1
229. Fang J, Sawa T, Akaike T, Akuta T, Sahoo SK, Khaled G, Hamada A, Maeda H (2003) Cancer Res 63:3567
230. Sahoo SK, Sawa T, Fang J, Tanaka S, Miyamoto Y, Akaike T, Maeda H (2002) Bioconjug Chem 13:1031
231. Haddow A, De Lamirande G, Bergel F, Bray RC, Gilbert DA (1958) Nature 182:1144
232. Yoshikawa T, Kokura S, Tainaka K, Naito Y, Kondo M (1995) Cancer Res 55:1617
233. Doi K, Akaike T, Fujii S, Tanaka S, Ikebe N, Beppu T, Shibahara S, Ogawa M, Maeda H (1999) Br J Cancer 80:1945
234. Motterlini R, Foresti R, Bassi R, Calabrese V, Clark JE, Green CJ (2000) J Biol Chem 275:13613
235. Keyse SM, Tyrrell RM (1990) Carcinogenesis 11:787
236. Maines MD (1997) Annu Rev Pharmacol Toxicol 37:517
237. Maines MD (1988) FASEB J 2:2557
238. Jeney V, Balla J, Yachie A, Varga Z, Vercellotti GM, Eaton JW, Balla G (2002) Blood 100:879
239. Baranano DE, Rao M, Ferris CD, Snyder SH (2002) Proc Natl Acad Sci USA 99:16093
240. Maeda H, Konno T (1997) In: Maeda H, Edo K, Ishida N (eds) Neocarzinostatin: The past, present, and future of an anticancer drug. Springer, Berlin Heidelberg New York, p 227
241. Graham ML (2003) Adv Drug Deliv Rev 55:1293
242. Pockros PJ, Carithers R, Desmond P, Dhumeaux D, Fried MW, Marcellin P, Shiffman ML, Minuk G, Reddy KR, Reindollar RW, Lin A, Brunda MJ (2004) Am J Gastroenterol 99:1298
243. Zeuzem S, Diago M, Gane E, Reddy KR, Pockros P, Prati D, Shiffman M, Farci P, Gitlin N, O'Brien CB, Lamour F, Lardelli P (2004) Gastroenterology 127:1724
244. Mukherjee A, Monson JP, Jonsson PJ, Trainer PJ, Shalet SM (2003) J Clin Endocrinol Metab 88:5865
245. Kellner K, Tessmar J, Milz S, Angele P, Nerlich M, Schulz MB, Blunk T, Gopferich A (2004) Tissue Eng 10:429
246. Calceti P, Salmaso S, Walker G, Bernkop-Schnurch A (2004) Eur J Pharm Sci 22:315

247. Schoemaker NE, van Kesteren C, Rosing H, Jansen S, Swart M, Lieverst J, Fraier D, Breda M, Pellizzoni C, Spinelli R, Grazia Porro M, Beijnen JH, Schellens JH, ten Bokkel Huinink WW (2002) Br J Cancer 87:608
248. Tsukioka Y, Matsumura Y, Hamaguchi T, Koike H, Moriyasu F, Kakizoe T (2002) Jpn J Cancer Res 93:1145
249. Nilsson F, Kosmehl H, Zardi L, Neri D (2001) Cancer Res 61:711
250. Halin C, Rondini S, Nilsson F, Berndt A, Kosmehl H, Zardi L, Neri D (2002) Nat Biotechnol 20:264
251. Carnemolla B, Borsi L, Balza E, Castellani P, Meazza R, Berndt A, Ferrini S, Kosmehl H, Neri D, Zardi L (2002) Blood 99:1659
252. Matsuno F, Haruta Y, Kondo M, Tsai H, Barcos M, Seon BK (1999) Clin Cancer Res 5:371
253. Beck MT, Chen NY, Franek KJ, Chen WY (2003) Cancer Res 63:3598
254. Dings RP, van der Schaft DW, Hargittai B, Haseman J, Griffioen AW, Mayo KH (2003) Cancer Lett 194:55
255. Arap W, Pasqualini R, Ruoslahti E (1998) Science 279:377
256. Schraa AJ, Kok RJ, Moorlag HE, Bos EJ, Proost JH, Meijer DK, de Leij LF, Molema G (2002) Int J Cancer 102:469
257. Asai T, Fukatsu H, Kuromi K, Ogino K, Tanaka M, Oku N, Taki T (2002) Biol Pharm Bull 25:904
258. Asai T, Nagatsuka M, Kuromi K, Yamakawa S, Kurohane K, Ogino K, Tanaka M, Taki T, Oku N (2002) FEBS Lett 510:206
259. Marty C, Odermatt B, Schott H, Neri D, Ballmer-Hofer K, Klemenz R, Schwendener RA (2002) Br J Cancer 87:106

Nanostructured Devices Based on Block Copolymer Assemblies for Drug Delivery: Designing Structures for Enhanced Drug Function

Nobuhiro Nishiyama[1] · Kazunori Kataoka[1,2] (✉)

[1]Center for Disease Biology and Integrative Medicine, Graduate School of Medicine, The University of Tokyo, 7-3-1 Hongo, Bunkyo-ku, 113-0033 Tokyo, Japan
nishiyama@bmw.t.u-tokyo.ac.jp, kataoka@bmw.t.u-tokyo.ac.jp

[2]Department of Materials Engineering, Graduate School of Engineering, The University of Tokyo, 7-3-1 Hongo, Bunkyo-ku, 113-8656 Tokyo, Japan
kataoka@bmw.t.u-tokyo.ac.jp

1	Introduction	69
2	Drug delivery systems	71
3	Preparation and characterization of block copolymer micelles	72
3.1	Synthesis of block copolymers	72
3.2	Properties of polymeric micelles	76
3.3	Blood circulation and tissue distribution	78
4	Polymeric micelles for drug delivery	80
4.1	Encapsulation of hydrophobic drugs	80
4.2	Encapsulation of other drugs	83
4.3	Stimuli-triggered drug release	84
4.4	Surface-functionalized polymeric micelles	86
4.5	Intracellular localization of micelles	87
5	Polyion complex (PIC) micelles	88
5.1	Properties of the PIC micelles	88
5.2	PIC Micelles for gene delivery	90
6	Polymer vesicles	95
7	Concluding remarks	96
References		97

Abstract Block copolymers spontaneously assemble into nanoscaled polymeric micelles, which have significant potential as drug carriers. Following much work in related fields over the last decade, a drug carrier based on polymeric micelles has been created. Advances in polymer chemistry have significantly contributed to progress in polymeric micelle-based drug carrier research, because the micelle parameters that are most important to successful drug delivery can be modulated by engineering the structures of the micelle-forming block copolymers. More intelligent polymeric micelles enabling environmentally-sensitive drug release and cell type-specific targeting have also recently emerged and have attracted increasing interest in the field of polymer chemistry. This paper therefore reviews recent progress in research on polymeric micelles for drug delivery, focusing upon the relationship between the chemical design of polymeric micelles and their physicochemical and biological properties.

Keywords Block copolymer · Drug delivery system (DDS) · Drug targeting · Gene delivery · Polymeric micelle

Abbreviations

ADR	Adriamycin
AFM	Atomic force microscopy
AmB	Amphotericin B
ASGP	Asialoglycoprotein
BPEI	Branched polyethyleneimine
CDDP	*cis*-Diamminedichloroplatinum(II), cisplatin
CL	ε-Caprolactone
CMC	Critical micelle concentration
CNV	Choroidal neovascularization
DDS	Drug delivery system
DMAc	N,N-Dimethylacetamide
DPT	Dipropylene triamine
EO	Ethylene oxide
EPR effect	Enhanced permeability and retention effect
FDA	Food and Drug Administration
GA	Glycolic acid
LCST	Lower critical solution temperature
LPEI	Linear polyethyleneimine
NCA	N-carboxyanhydride
ODN	Oligonucleotides
PCL	Poly(ε-caprolactone)
PDLLA	Poly(D,L-lactide)
pDNA	Plasmid DNA
PEO	Poly(ethylene oxide)
PEO-*b*-ODN	PEO-*b*-oligonucleotide
PEO-*b*-DPT	PEO-*b*-poly(3-[(3-aminopropyl)amino]propylaspartamide)
PEO-*b*-PAAs	PEO-*b*-poly(amino acids)
PEO-*b*-PAMA	PEO-*b*-poly[(2-dimethylamino)ethyl methacrylate]
PEO-P(Asp-ADR)	PEO-*b*-poly(α, β-aspartic acid)
PEO-*b*-P(Asp)	Adriamycin-conjugated PEO-*b*-P(Asp)
PEO-*b*-P(Asp-Hyd)	PEO-*b*-P(Asp) processing the hydrazide groups in the side chains
PEO-*b*-PBLA	PEO-*b*-poly(β-benzyl L-aspartate)
PEO-*b*-PCL	PEO-*b*-poly(ε-caprolactone)
PEO-*b*-PDLLA	PEO-*b*-poly(D,L-lactide)
PEO-*b*-PE	PEO-*b*-phosphatidylethanolamine
PEO-*b*-PEE	PEO-*b*-polyethylethylene
PEO-*b*-P(Glu)	PEO-*b*-poly(glutamic acid)
PEO-*b*-P(His)	PEO-*b*-poly(L-histidine)
PEO-*b*-PLGA	PEO-*b*-poly(lactide-*co*-glycolic acid)
PEO-*b*-PLA	PEO-*b*-poly(lactide)

PEO-*b*-PLLA	PEO-*b*-poly(L-lactide)
PEO-*b*-P(Lys)	PEO-*b*-poly(L-lysine)
PEO-*b*-P(Lys-IM)	PEO-*b*-P(Lys) that has the side chain partially substituted with a 1-imino-4-mercaptobutyl group
PEO-*b*-P(Lys-MP)	PEO-*b*-P(Lys) that has the side chain partially substituted with a 3-mercaptopropionyl group
PEO-*b*-P(Lys(Z))	PEO-*b*-poly(ε-benzyloxycarbonyl L-lysine)
PGA	Poly(glycolic acid)
P(His)	Poly(L-histidine)
PIC	Polyion complex
PLA	Poly(lactide)
PLLA	Poly(L-lactide)
Pluronic	PEO-*b*-poly(propylene oxide)-*b*-PEO
P(Lys)	Poly(L-lysine)
pNP	*p*-Nitrophenylcarbonyl group
RES	Reticuloendothelial system
siRNA	Small interfering RNA
SPDP	3-(2-Pyridyldithio)propionate
TEM	Transmission electron microscopy
T_g	Glass transition temperature
32(−)DP	Third generation dendrimer with 32 carboxylic acid groups on the periphery
32(+)DP	Third generation dendrimer with 32 quaternary trimethylamino groups on the periphery

1
Introduction

Block copolymers, which are composed of two or more covalently-linked polymers with different physicochemical properties, have attracted growing interest from academia and industry since the 1960s [1]. Amphiphilic block copolymers spontaneously form spherical, rodlike, lamellar or vesicular aggregates in selective solvents depending on parameters such as the chemical structure, the composition, and the concentration [2, 3]. One typical form of block copolymer assembly is the polymeric micelle, which has a hydrophobic core surrounded by a hydrophilic coronal shell. Recently, there has been increasing interest in the application of polymeric micelles as drug delivery systems (DDS) [4–10, 12, 13]. Polymeric micelles have several advantages as drug carriers, such as (1) their applicability to a variety of therapeutic agents (such as hydrophobic substances, metal complexes and charged macromolecules such as polypeptides and polynucleotides), (2) ease of physical loading of drugs without chemical modification, (3) simplicity of micelle preparation, (4) high drug loading capacity, and (5) controlled drug release. These properties can be optimized by modulating the micelle core-forming blocks depending on the chemical properties of the drugs. It should also be noted that the pharmacokinetic properties of polymeric micelles are not affected by the properties of the loaded drugs but instead are mainly governed by their sizes and surface properties. Furthermore, recent advances in synthetic chemistry have allowed us to design smart polymeric micelles with functions such as molecule-specific targeting and stimuli-responsive drug release (Fig. 1). The number of articles related to the application of polymeric micelles to gene and drug delivery has increased remarkably over the past decade, as seen in Fig. 2.

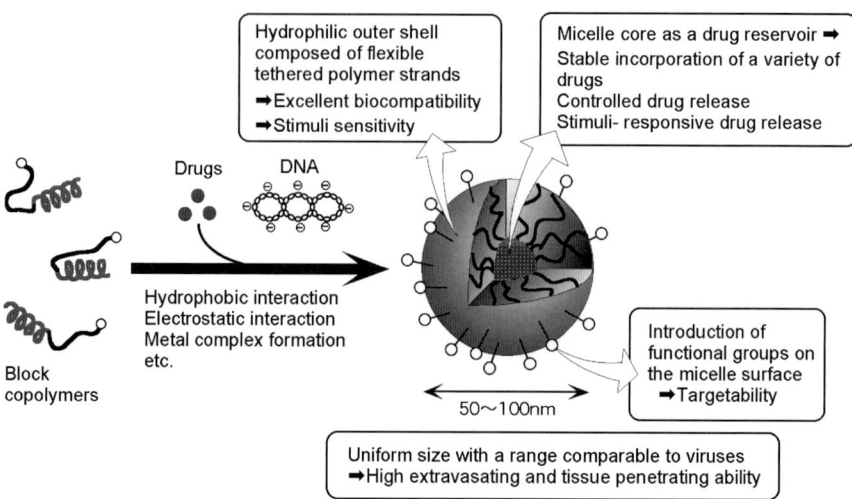

Fig. 1 Smart block copolymer micelles for drug delivery

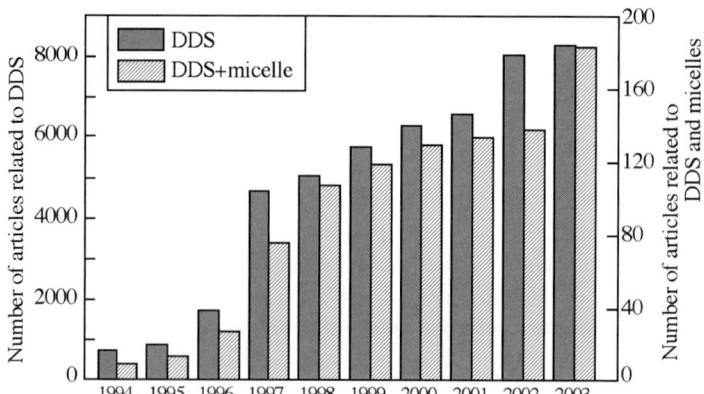

Fig. 2 Trend graph of the number of articles related to drug delivery systems (DDS) and micelles published in the last decade. This literature search was performed using SciFinder Scholar software (American Chemical Society)

2
Drug delivery systems

The major objective of using drug carriers is to modulate drug disposition in the body. The long-circulating nature (stealth property) of drug carriers is a requisite for successful drug targeting. The main obstacles to long circulation of drug carriers are considered to be glomerular excretion in the kidney and recognition by the reticuloendothelial system (RES) located in the liver, spleen and lung [14] (Fig. 3). Since a threshold molecular weight exists for glomerular filtration (42 000–50 000 for water-soluble synthetic polymers [15]), this can be avoided by increasing the molecular weight of the carriers. On the other hand, drug carriers with low biocompatibility can be recognized by the RES and then eliminated from blood circulation. Surface modification of the carriers with biocompatible polymers can impair or even avoid recognition by the RES [16, 17]. Among such biocompatible polymers, poly(ethylene oxide) (PEO) is probably the most commonly used, due to the high flexibility of its structure, its high degree of hydration, its nontoxicity and its weak immunogenecity, and as such it has been approved by the Food and Drug Administration (FDA) [18]. PEO chains where one end is attached to the surface are particularly effective at preventing the adsorption of proteins and the adhesion of cells due to high steric repulsion effects [19].

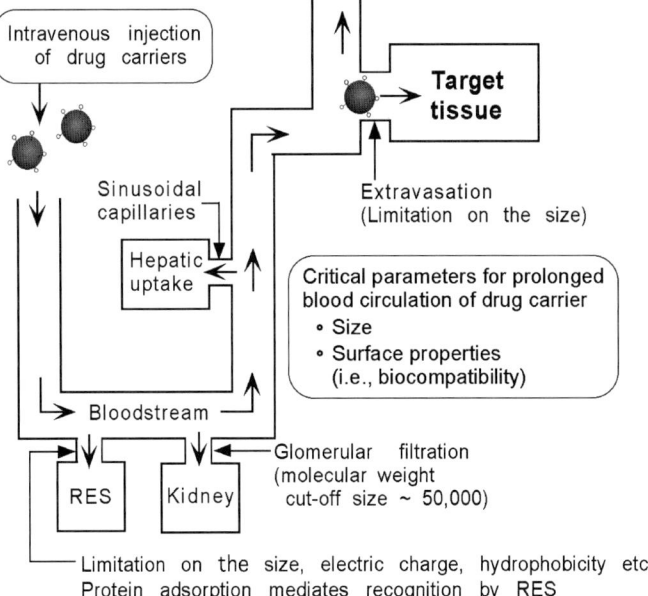

Fig. 3 Itinerary of a drug carrier after intravenous administration

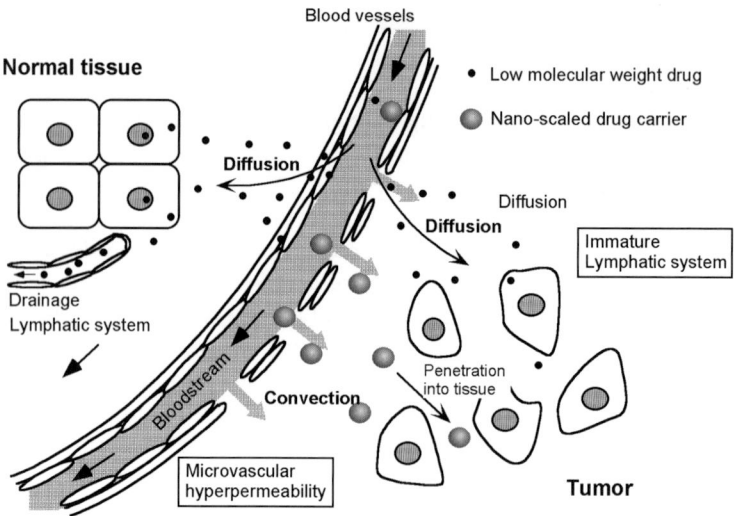

Fig. 4 Anatomical differences between normal tissue and solid tumor

One of the most important reasons for using macromolecular carriers is their preferential accumulation in solid tumors. This elevated macromolecule accumulation in tumors is currently explained by their microvascular hyperpermeability to circulating macromolecules and the impaired lymphatic drainage of macromolecules in tumor tissues. This phenomenon was termed the "enhanced permeability and retention (EPR) effect" by Maeda and Matsumura [20, 21] (Fig. 4). It has been suggested that tumor microvascular hyperpermeability is due to overexpression of the vascular pemeability factor (VPF)/vascular endothelial growth factor (VEGF) [22, 23] as well as the secretion of other factors such as the basic fibroblast growth factor (bFGF) [24], bradykinin, nitric oxide and peroxynitrate in tumor tissues [25, 26]. To date, an increasing number of studies have reported that biocompatible carriers, including synthetic polymers, liposomes and polymeric micelles, accumulate in various types of tumors due to the EPR effect [27, 28].

3
Preparation and characterization of block copolymer micelles

3.1
Synthesis of block copolymers

In this section we describe the syntheses of functional PEO-*b*-poly(lactide) (PEO-*b*-PLA) and PEO-*b*-poly(amino acids) (PEO-*b*-PAAs) block copolymers. The synthesis routes for them have been mainly established by our group. Although a variety of block copolymers forming polymeric micelles have been studied, these two types of block copolymers are particularly interesting from the standpoint of biomedical applications because of their safety for clinical use and the wide variety of chemical designs available for a specific functionality.

Poly(lactide) (PLA) is a synthetic polymer widely used for biodegradable biomaterials such as implants due to the desirable biodegradable and nontoxic properties, which led to it being approved by the FDA [29]. Since there are two stereoisomeric forms for lactic acid, D and L, two types of PLAs, poly(L-lactide) (PLLA) and poly(D,L-lactide) (PDLLA), have been extensively studied. PLLA has a semicrystalline nature and its mechanical toughness makes it suitable for implant devices for fixing fractures [29]. On the other hand, amorphous PDLLA is applicable to drug delivery devices due to its capacity for drug incorporation and subsequent drug release, as mediated by its nonenzymatic hydrolysis, which is autocatalyzed by the carboxylic group at the PLA end [29]. To achieve a desirable degradation rate, LA is often copolymerized with glycolic acid (GA) and ε-caprolactone (CL), since PGA and PCL give faster and slower degradation rates than PLA, respectively [30–35]. The scheme for the facile and one-pot synthesis of the end-functionalized PEO-PDLLA is shown in Fig. 5a. Ethylene oxide (EO) polymerization is initiated by potassium 3,3-diethoxypropanolate, forming a heterofunctional PEO (which means that the PEO possesses different functional groups at the α- and ω-ends), without any side reaction [36]. In this reaction, a series of potassium alkoxides with protected functional groups can be used as initiators, yielding various heterotelechelic PEOs possessing aldehyde [36] and benzaldehyde groups [37], a primary amino group [38], and monosaccharide residues [39] at the α-end. The polymerization is followed by an anionic ring opening polymerization of D,L-lactide from the ω-end of the heterofunctional PEO to obtain α-acetal-PEO-b-PDLLA [40], and occasionally the ω-end of the PLA segment is functionalized using an end-capping reagent such as methacrylic anhydride [41]. This polymerization proceeds almost quantitatively, and the molecular weight of each block can be controlled by the initial monomer/initiator ratio. The synthesized copolymers have a narrow molecular weight distribution ($M_w/M_n < 1.1$). This narrow molecular weight distribution is a critical factor in the constituent PEO-b-PDLLA forming monodispersive polymeric micelles with a distinct core-shell structure.

The acetal moiety at the α-end of PEO can be easily converted into a reactive aldehyde group by gentle treatment with a weak acid solution (pH $\sim$ 2). No scission of the PLA segment occurs during the conversion of the acetal group into the aldehyde group, since the PLA segment is segregated in the hydrophobic core of the micelles. The reactive aldehyde group at the distal end of the PEO chain is available for conjugation with targetable ligand molecules such as monosaccharide derivatives [42] and peptides [43] via a reductive amination reaction using $NaBH_3CN$ (Fig. 5b). This reaction permits

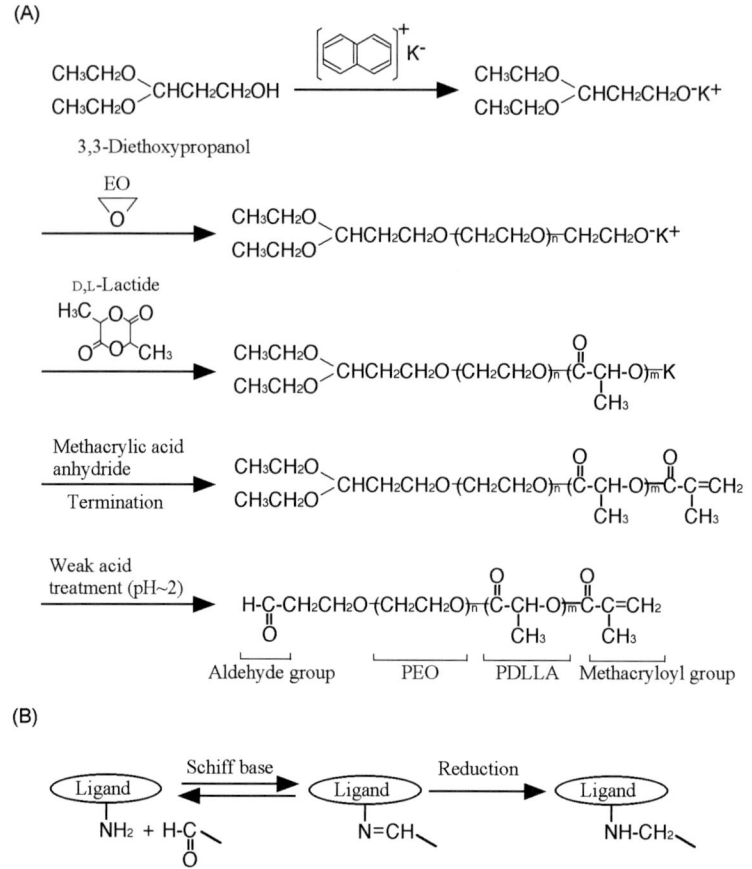

Fig. 5 Synthesis of end-functionalized PEO-PDLLA block copolymer (**A**) and conjugation of ligand molecule to the distal end of PEO via reductive amination reaction (**B**)

high functionalization (> 50%) of the PEO end of the copolymer. On the other hand, the methacryloyl group at the ω-end of PLA allows the polymerization of the micellar core, which is initiated by azobis(2,4-dimethylvaleronitrile) (V-65) at 60 °C [41]. The core-polymerized polymeric micelles were quite stable regardless of the concentrations, solvents (organic solvents) and the presence of surfactant, providing stable nanospheres with a range of applications in biomedical applications (for example, they can be used as a platform for the electrophoretic separation of DNA [44]).

PEO-b-PAAs are synthesized by the ring-opening polymerization of an N-carboxyanhydride (NCA) of an amino acid with a protected side chain, initiated from the primary amino group of α-methoxy-ω-aminoPEO [45]. For example, PEO-b-poly(β-benzyl L-aspartate) (PEO-b-PBLA) and PEO-b-poly(ε-benzyloxycarbonyl L-lysine) (PEO-b-P(Lys(Z))) are synthesized by

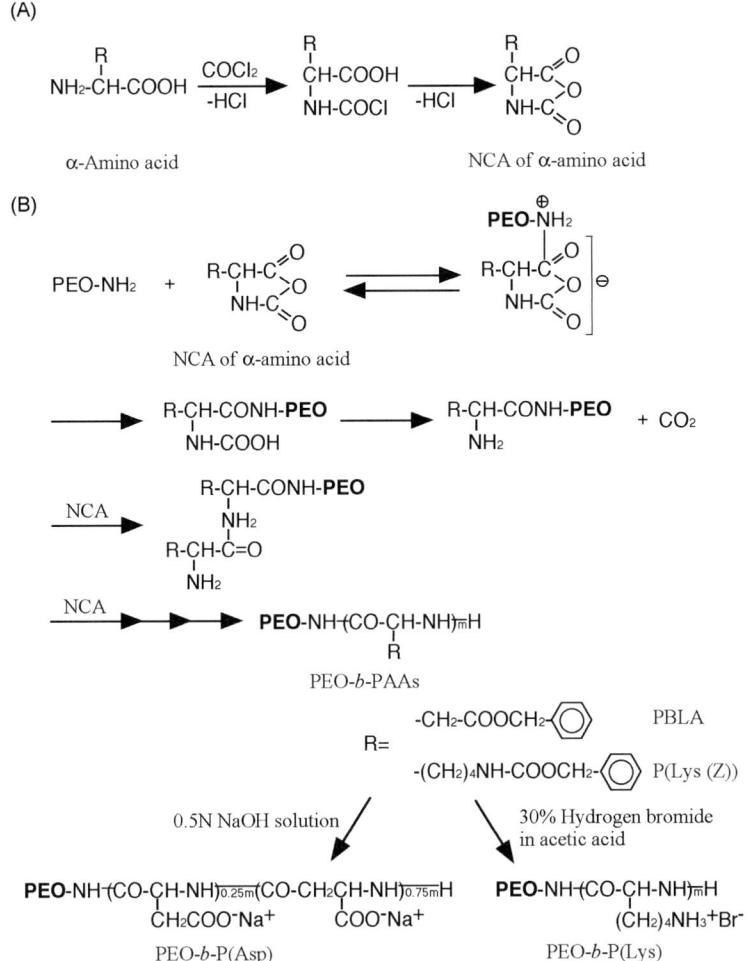

Fig. 6 Synthesis of amino acid NCAs (**A**) and PEG-b-PAA block copolymers (**B**)

polymerizing the NCAs of β-benzyl L-aspartate and ε-benzyloxycarbonyl L-lysine, respectively [46–48] (Fig. 6). Deprotection of the benzyl ester and Z groups of PEO-*b*-PBLA and PEO-*b*-P(Lys(Z)) resulted in the formation of PEO-*b*-poly(α, β-aspartic acid) (PEO-*b*-P(Asp)) and PEO-*b*-poly(L-lysine) (PEO-*b*-P(Lys)), respectively (Fig. 6). The polymerization of the NCAs permits the synthesis of PEO-*b*-PAAs with a narrow molecular weight distribution ($M_w/M_n < 1.2$), and the degree of polymerization of the PAAs is controllable up to around 100. The advantages of using PEO-*b*-PAAs are that a variety of functional groups are available on the side chains of the PAAs for the formation of polymeric micelles through various interactions, including hydrophobic interactions, electrostatic interactions, hydrogen bonding and metal complexation, as well as chemical modification. All of these are feasible ways to conjugate the drugs and to control the physicochemical properties of the formed polymeric micelles [6, 7]. The secondary structure of the PAA (α-helix and β-sheet) can also be a critical parameter for the formation [49] and stabilization of the micelles [50].

3.2
Properties of polymeric micelles

Amphiphilic block copolymers in selective solvents undergo self-assembly into various nanosized morphologies [2, 3]. Typical aggregates formed by the block copolymers are polymeric micelles with spherical shapes consisting of a core and a coronal shell. These spherical polymeric micelles are several tens of nanometers across with a narrow size distribution, and they are characterized by their unique core-shell structure, in which a core composed of insoluble blocks is surrounded by a palisade of soluble (hydrophilic) blocks. So far, polymeric micelles intended for biomedical use have been prepared from a variety of amphiphilic block copolymers including PEO-b-PBLA [51–56], poly(ethylene oxide)-b-poly(propylene oxide)-b-poly(ethylene oxide) (PEO-b-PPO-b-PEO) (Pluronic) [9], PEO-b-phosphatidylethanolamine (PEO-b-PE) [12, 13], PEO-b-PDLLA [40–43], PEO-b-poly(ε-caprolactone) (PEO-b-PCL) [11], and PEO-b-poly(lactide-co-glycolic acid) (PEO-b-PLGA) [57].

Polymeric micelles are typically prepared from amphiphilic block copolymers using the dialysis method [11, 58]. The polymers are dissolved in a good solvent (such as N,N-dimethylacetamide (DMAc)) (polymer concentration: $1 \sim 2$ mg/mL), and then the polymer solutions are dialyzed against distilled water (selective solvent) to remove the good solvent. This method allows for kinetic control of the micellization. Since the good solvent is removed from the core, the equilibrium between the unimers and micelles is frozen due to the glassy nature of the core-forming blocks. Alternatively, polymeric micelles are prepared by the direct dissolution method [11]. This method is employed when block copolymers are soluble in water at a specific temperature and over a certain concentration range (as for Pluronic). Treatment by heating or ultrasonication is often used to obtain thermodynamically stable polymeric micelles.

Compared to surfactant micelles, polymeric micelles have a much higher thermodynamic stability. It is well-known that micelles have a critical micelle concentration (CMC) below which only unimers exist but above which both micelles and unimers are present. It was reported that polymeric micelles have a CMC around 10^{-6}–10^{-7} M [58, 59], which is 1000-fold lower than that of surfactant micelles (10^{-3}–10^{-4}). The CMC of polymeric micelles is affected by many factors, including the properties of the core-forming blocks, such as hydrophobicity, the glass transition temperature (T_g), the degree of crystallinity, and the length and ratio of the hydrophilic and hydrophobic blocks. Also, polymeric micelles are assumed to have a higher kinetic stability (slower dissociation rate into unimers) than surfactant micelles due to the integrated molecular effect and the entangling of the core-forming blocks. In our previous study, the dissociation rate of the micelles of adriamycin (ADR)-conjugated PEO-b-P(Asp) into unimers was estimated to be on the order of days in phosphate-buffered saline as well as in the presence of 50% rabbit serum [60]. These slow dissociating properties of polymeric micelles are certainly an advantage for their use as nanocarrier systems, since they lead to the sustained release of encapsulated drugs.

To stabilize the micelle structure, core and shell crosslinking has been studied [41, 61–63]. Recently, Shen et al. reported that amphiphilic brush copolymers composed of PEO and PCL chains, which were synthesized by macromonomer copolymerization, formed polymeric micelles in which the core-shell interface was crosslinked [64]. The diameters of the crosslinked micelles increased with increasing PCL/PEO chain ratio in the range 27.4–198 nm. The crosslinked micelles were 100 times more stable against dilution compared with micelles from corresponding amphiphilic block copolymers.

Here, the physicochemical properties of PEO-b-PDLLA micelles are described in detail as a typical polymeric micelle useful for drug delivery [59, 65, 66]. PEO-b-PDLLA, where the M_w of PEO and that of PDLLA were 5700 and 5400, respectively, self-assembled into narrowly-distributed polymeric micelles with approximately 35 nm across in aqueous media after a dialysis or ultrasonication-aided dispersion method was applied. The important aspect of the PEO-b-PDLLA micelles is that the thermal properties of the micelles changed at the T_g of the PDLLA segment ($\sim 42\,°C$) [59]. The CMC of the PEO-b-PDLLA micelles was determined at various temperatures (25–55 °C) using physically entrapped pyrene as a fluorescent probe to monitor the change in the polarity of the microenvironment in the micelle core. At temperatures above T_g, the CMC value increased with the temperature according to the following equation, $\Delta G^0 \sim RT\ln(\text{CMC})$, where G^0 and R are the Gibbs standard free energy and the universal gas constant, respectively. In contrast, an almost constant CMC (6.0×10^{-7} M) was observed regardless of the temperature change below the T_g. This result was associated with a gradual increase in the chain mobility of the PDLLA segment in the core of the micelles at temperatures above T_g, which was estimated by ^{1}H-NMR measurements of the PEO-b-PDLLA micelles in D$_2$O. The mobility of the core-forming segment was also related to the exchange behavior of the constituent block copolymers between the micelles. The chain exchange rates between the micelles were appreciably accelerated by increasing the temperature from 25 °C to 40 °C. The frequency of the chain exchange rate may correlate with the possible interaction of the block copolymers with biological components, including proteins and cellular membranes. Thus, the characteristics of the core-forming blocks significantly affect the crucial properties of the micelles as drug carriers.

3.3
Blood circulation and tissue distribution

The biodistributions of polymeric micelles are predominantly affected by their sizes and the properties of the outer shell. To obtain polymeric micelles that exhibit stable circulation in the bloodstream, the hydrophilic shell-forming block of PEO needs to be regulated at an M_w of 5000 to 12 000, and the length of it should preferably be greater than that of the core-forming block [67]. The outer PEO shell of the micelle inhibits the surface adsorption of proteins and other biological components, which is related to the micelle's elimination from the blood circulation through a mechanism involving the RES [18, 19]. The particle size of the micelle should be less than 100–150 nm, since larger particles may be susceptible to clearance by the RES. Also, it appears that smaller micelles might show a high accessibility to tumor tissues through the porous tumor capillaries [68] and deeper tissue penetration into the tumors [69].

Polymeric micelles may not dissociate when they are strongly diluted after intravenous injection into a human or animal body. In our previous study [70], 0.4 mg of PEO-b-PDLLA micelles were intravenously injected into mice (body weight: 20 g). Assuming that the volume of plasma in mice is 45.6 mL/kg, the initial concentration of the polymeric micelle in the plasma was calculated to be 440 mg/L, which is more than 50-fold higher than the CMC value of the PEO-b-PDLLA micelles (8 mg/L at 40 °C). Even if the initial micelle concentration in the plasma is below the CMC, micelle dissociation may be kinetically slow, allowing the micelles to circulate in the bloodstream for an appreciable time period. Indeed, the PEO-b-PDLLA micelles showed prolonged blood circulation ($t_{1/2} \sim 18$ h) after intravenous injection, and 25% of the injected dose remained in circulation at 25 h-post injection [70]. During the circulation, the micelles did not interact with blood cells. The stable circulation of the micelles was confirmed by a gel filtration assay [70]. During tissue distribution, the polymeric micelles were localized in the vascular spaces of the normal organs including the lung, kidney, liver and spleen for up to 24 h [70]. The minimal accumulation of the micelles in the liver and spleen suggests that they are able to avoid recognition by the RES and entrapment by hepatic sinusoidal capillaries, as characterized by large interendothelial cell junctions (~ 100 nm) and the absence of basement membranes despite the relatively small size of the micelles (~ 40 nm). It is worth mentioning that the PEO-b-PDLLA micelles were slowly excreted into the urine (24% of injected dose at 24 h-post injection), since the molecular weight of the block copolymers is much lower than the threshold of glomerular excretion [70]. These results indicate little cumulative accumulation of the drug carriers in the body.

To study the effects of the surface charge of the polymeric micelle on its blood circulation and tissue accumulation, small neutral and negatively-charged peptides (Try and Try-Glu, respectively) were installed on the micelle surface by placing a reactive aldehyde group to the PEO end, resulting in the preparation of neutral and negatively-charged polymeric micelles with zeta potentials of 1.3 mV and – 10.6 mV, respectively [43]. Both of the micelles showed almost the same profiles in the blood circulation, but a slightly lower accumulation in the liver and spleen was observed for the negatively-charged micelles [70]. The lower cellular uptake of the negatively-charged micelles may be due to electrostatic repulsion by the negatively-charged cell membranes. Nevertheless, there was no appreciable difference between the biodistributions of the neutral and negatively-charged micelles [70], suggesting that surface modification of the micelles with small ligand molecules hardly affects their circulation and disposition in the body.

Long-circulating polymeric micelles can preferentially accumulate in solid tumors due to the aforementioned EPR effect. It has been suggested that the EPR effect is universally observed in malignant tumors. Jain et al. reported that vascular cut-off sizes in solid tumors range from 380 to 780 nm, depending on the type and size of the tumor and its microenvironment [22–24]. Therefore, the vascular pore cut-off size of the tumor is unlikely to be an obstacle to transvascular transport of polymeric micelles, which are less than 100 nm across. Such anatomical characteristics of solid tumors appear to allow polymeric micelles to accumulate in a tumor in a passive manner. Indeed, we have demonstrated that antitumor drug-incorporated micelles achieve a 10 ~ 20-fold higher accumulation in tumors than a free drug does, while showing reduced accumulation in normal tissues [67, 71–73]. On the other hand, polymeric micelles with targeting moieties may be of further interest due to their ability to specifically bind to tumor cells. However, the tumor cells are generally located outside of the microvasculature, so that extravasation of the micelles, which is a passive process governed by particle size and longevity in blood circulation, is a prerequisite for specific interactions with tumor cells.

More recently, it has been suggested that polymeric micelles accumulate not only in solid tumors but also in other diseased sites. We have recently found that a balloon injury in the rat carotid artery caused a marked and sustained increase in vascular permeability, allowing significant accumulation of polymeric micelles in the balloon-injured artery [74]. As a result, ADR-incorporated micelles inhibited neointimal formation in the injured artery, offering a new approach to the prevention of restenosis after percutaneous coronary intervention during clinical cardiosurgery [74]. Similarly, Torchilin et al. reported that PEO-b-PE micelles selectively accumulated in the area of an experimental myocardial infarction after intravenous injection [75]. It appears that the infarct area might have anatomical characteristics that are similar to those of solid tumors (the EPR effect). On the other hand, exudative age-related macular degeneration, which is a major cause of visual loss in aged adults in developed countries, is characterized by choroidal neovascularization (CNV). We have recently reported that polymeric micelles also accumulate in CNV sites in rat models, which were experimentally created by laser photocoagulation, after intravenous injection [76]. This accumulation of micelles at the CNV sites was maintained for as long as 168 h. This result may offer a new approach to the treatment of CNV, via drug targeting with polymeric micelles.

4
Polymeric micelles for drug delivery

4.1
Encapsulation of hydrophobic drugs

The encapsulation of a drug by polymeric micelles is achieved by the dialysis method or the oil in water (o/w) emulsion method [58] (Fig. 7). In the former method, the drug and block copolymers are dissolved in a good solvent and then dialyzed against a selective solvent. As polymeric micelles form during the dialysis process, the drug is incorporated into the cores of the micelles. Excess drug is also removed during the dialysis process. In the latter method,

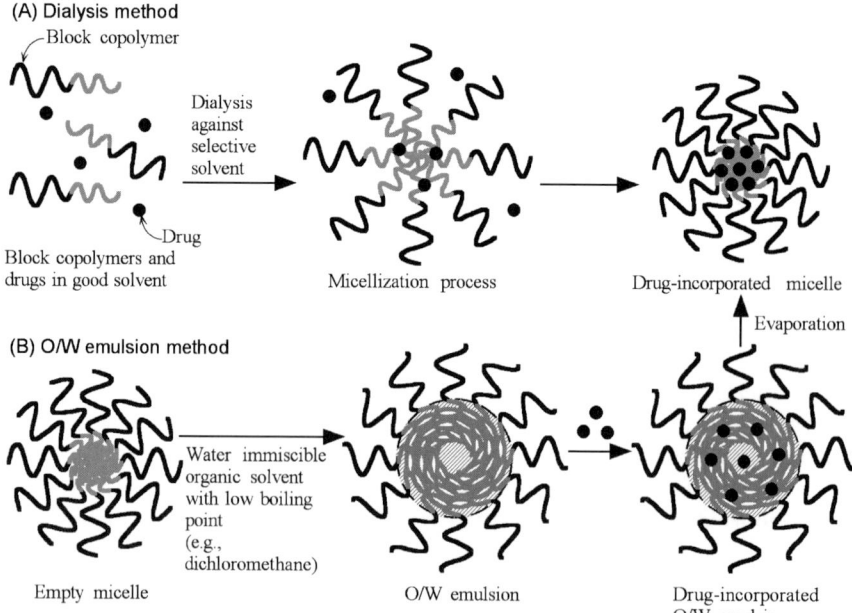

Fig. 7 Methods for physical encapsulation of drugs in polymeric micelles: **A** dialysis method, and **B** O/W emulsion method

a drug dissolved in an organic solvent (such as dichloromethane) is added dropwise to a solution of micelles in water. The cores of the micelles entrap an emulsion containing the drug, and the drug is incorporated as the organic solvent evaporates. The free drug can be removed by purification using ultrafiltration. Both methods ensure a high incorporation ratio of drug to block copolymers, achieving around 20% (w/w) [58]. The effect of a high loading of drugs on the biodistribution of the micelles may be reduced due to the segregated core-shell structure and compacted inner core.

The structures of the block copolymers significantly influence the ability of polymeric micelles to carry hydrophobic drugs, affecting key properties such as the stability, size, size distribution, loading capacity, release kinetics and biodistribution of the micelles. In other words, these critical parameters can be changed or controlled by tailoring the block copolymers for a particular task. The hydrophobic cores of the polymeric micelles made from amphiphilic block copolymers serve as a nano-reservoir for hydrophobic drugs. Therefore, the chemical structures and properties of the core-forming blocks significantly affect the drug loading efficiency and drug release rate. In many cases, polymeric micelles produce a sustained release of the drug over a period of hours or days. The drugs appear to be released by a mechanism of drug diffusion from the micelle core or by micelle disintegration. Micelles with a frozen core (in other words the T_g values of the core-forming blocks are above the physiological temperature) can give a slower core drug diffusion rate. Also, the diffusion rate can be slowed down if a favorable interaction exists between the drugs and the core-forming blocks (as in the case of PEO-P(Asp-ADR) micelles, *described below*). On the other hand, the amount of the drug loaded affects its release kinetics. It was reported that higher loadings of hydrophobic drugs (such as lidocaine and clonazepam) cause drug crystallization inside the micellar core, leading to a reduced drug release rate [57, 77]. In any case, considering the harsh in vivo conditions, a slower drug release seems to be required to achieve efficient drug delivery to the target.

The compatibility between the core-forming blocks and the drugs to be loaded is one of the most important factors to take into account when designing block copolymers. In our previous study, ADR was covalently introduced into the side chain of PEO-*b*-P(Asp) via an amide bond between the carboxyl group in P(Asp) and the primary amino group of the glycosidyl residue in ADR in order to obtain the micelle-forming amphiphilic block copolymers PEO-*b*-P(Asp-ADR) [46, 47, 71, 72, 78, 79] (Fig. 8). An inner core of PEO-P(Asp-ADR) micelles can be charged with free ADR, and the physically entrapped ADR was found to produce more cytotoxic activity than the chemically conjugated one [80]. However, it was found that the chemically conjugated ADR contributes to the stable physical entrapment of unconjugated drugs through a π-π interaction between the anthracycline structures in conjugated and unconjugated drugs [60]. Indeed, the efficiency of entrap-

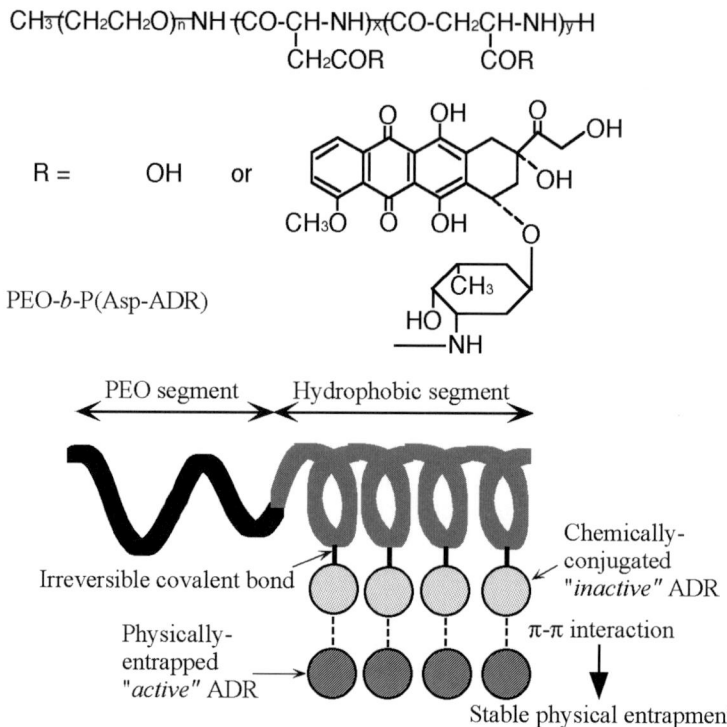

Fig. 8 Chemical structure of PEO-*b*-P(Asp-ADR) block copolymers. The micelle can physically entrap free ADR

ment of the unbound ADR into the micelles increased with the ratio of the chemically conjugated ADR to the P(Asp) segment, and increased amounts of both the physically entrapped and chemically conjugated ADRs were shown to improve the micellar structure based on a column chromatography assay [60]. The PEO-P(Asp-ADR) micelles showed a stable blood circulation and preferential accumulation in solid tumors after intravenous injection into tumor-bearing mice, consequently achieving significantly enhanced antitumor activity [71, 72]. The optimized formulation of the PEO-P(Asp-ADR) micelles is currently being studied in a phase II clinical trial in Japan [81]. Kwon et al. reported that amphotericin B (AmB) was efficiently encapsulated into polymeric micelles possessing AmB compatible moieties (saturated fatty acid esters) in the side chains of the core-forming blocks [82, 83]. Also, Yokoyama et al. reported the efficient encapsulation of KRN5500, a spicamycin derivative composed of a long-chain fatty acid, glycine, aminoheptose and adenine, into the micelles from PEO-*b*-PBLA derivatives with a substituted cetyl ester group in the side chain [84].

Recently, Kissel et al. reported that the core crosslinking of the PEO-*b*-PCL micelles not only improved the thermodynamic stability of the micelles against dilution but also increased the loading efficiency of paclitaxel in the micelles [85]. Despite the core crosslinking, the biodegradability of the PCL segments ensures no long-term accumulation of the drug carriers in the body.

4.2
Encapsulation of other drugs

In addition to hydrophobic molecules, metal complexes can be incorporated into polymeric micelles. In our recent studies [73, 86, 87], cis-diamminedichloroplatinum(II) (cisplatin, CDDP), a metal complex antitu-

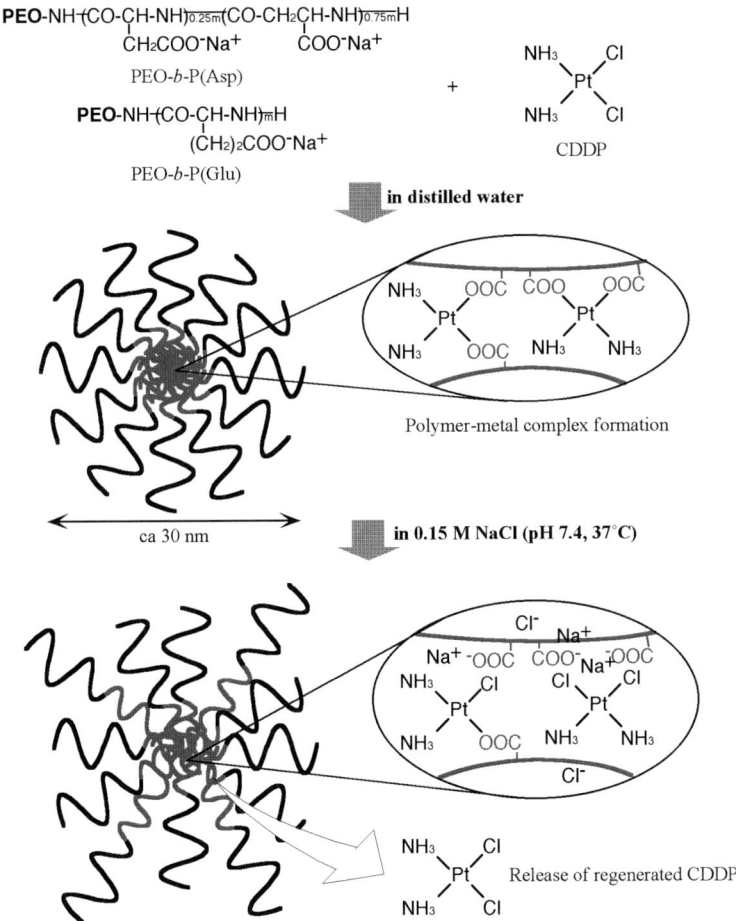

Fig. 9 Chemical structure and formation of CDDP-incorporated micelles, and their dissolution behavior in physiological saline at 37 °C

mor drug, was complexed with a PEO-*b*-P(Asp) or PEO-*b*-poly(glutamic acid) (PEO-*b*-P(Glu)) copolymer through a ligand substitution reaction of the platinum(II) in distilled water, forming CDDP-incorporated polymeric micelles of size ~ 30 nm, with a remarkably narrow size distribution (Fig. 9). The CDDP-incorporated micelles were stable in distilled water, but showed a sustained release of CDDP in physiological saline, which might be attributed to an inverse ligand substitution of the platinum (II) from the carboxylates in the micelle core to the chloride ions in the medium. Of interest is the fact that the CDDP-incorporated micelles gradually dissociated into small aggregates or unimers synchronized with the release of CDDP, because the driving force of the micelles is polymer–metal complex formation. When intravenously injected into tumor-bearing mice, the CDDP-incorporated micelles showed a remarkably prolonged blood circulation (13% dose at 24 h post-injection) and they accumulated effectively in solid tumors (in a 20-fold higher concentration than free CDDP). An in vivo antitumor activity assay revealed that the micelles had significant antitumor activity, achieving a 40% complete cure without significant body weight loss in colon 26-bearing mice, whereas the free CDDP showed no cure of solid tumors at the dose in which 20% body weight loss was observed [73]. Thus, the CDDP-incorporated micelles could be a promising formulation of CDDP for the targeted therapy of solid tumors. Hydrogen bonding interactions also appear to be available for encapsulation of the drugs into polymeric micelles. Recently, an antidiuretic hormone oligopeptide, (Arg^8)-vasopressin, was successfully incorporated into polymeric micelles by hydrogen bonding with PEO-*b*-P(Asp) in a free acid form [88].

4.3
Stimuli-triggered drug release

The drug release from polymeric micelles can be triggered by several physical stimuli. Rapoport et al. reported that ultrasound irradiation enhanced ADR release from Pluronic micelles and cellular uptake of both the free and encapsulated drugs [89]. This technique has been proven to be applicable for in vivo use, and it was observed that the application of ultrasound significantly reduced the tumor size in rats compared with nonsonicated controls [89], suggesting the possibility of using this method to enhance the drug activity at the tumor site. On the other hand, polymeric micelles with thermoresponsive coronal shells were prepared from poly(*N*-isopropylacrylamide-*co*-*N*,*N*-dimethylacrylamide)-*b*-PDLLA copolymers [90]. The thermally responsive polymeric micelles had a low critical solution temperature (LCST) at 40 °C, and the ADR-loaded micelles gave faster ADR release at 42.5 °C (above the LCST) than that at 37 °C (below the LCST), resulting in enhanced cytotoxicity against bovine aorta endothelial cells at 42.5 °C [90]. A combination

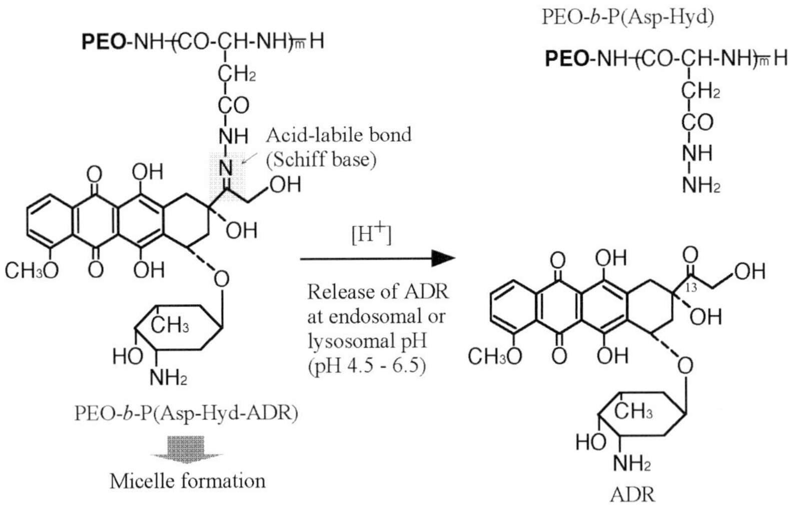

Fig. 10 pH-sensitive release of ADR from PEO-*b*-(Asp-Hyd-ADR) micelles

of selective delivery to solid tumors by polymeric micelles and local heating at the tumor sites might offer a new strategy for an effective tumor-targeted therapy.

We recently reported on pH-sensitive polymeric micelles where drug release was triggered by decreasing the pH. PEO-*b*-P(Asp) possessing hydrazide groups in the P(Asp) side chains (PEO-*b*-P(Asp-Hyd)), which was synthesized by a modified acid anhydride reaction, was coupled with ADR through an acid-labile hydrazone bond between the C13 of ADR and the hydrazide groups (PEO-*b*-P(Asp-Hyd-ADR)) [91] (Fig. 10). The pH-sensitive polymeric micelles showed an appreciable ADR release at low pH (4.5–5.5), whereas they were stable at physiological pH (~ 7.0). Confocal microscopy observations revealed the localization of the micelles in acidic lysosomal compartments in living cells and the subsequent time-dependent intracellular release of ADR. On the other hand, Bae et al. recently reported pH-sensitive polymeric micelles from PEO-*b*-poly(L-histidine) (PEO-*b*-P(His)) block copolymers [92]. Poly(L-histidine), which has pK_a values similar to physiological pH, behaves as a hydrophobic segment at comparatively high pH (~ 8.0) while showing hydrophilicity, biodegradability and fusogenic activity at low pH (< 7.0). The PEO-*b*-P(His) micelles gave an increased ADR release rate as the pH decreased from 8.0. To shift the pH range so that the drug release occurs at more acidic conditions, PEO-*b*-P(His) was blended with PEO-*b*-PLLA to form mixed micelles. These mixed micelles gave accelerated ADR release in the pH range 6.6–7.2, the exact pH depending on the mixing ratio of the two kinds of block copolymers. These pH-sensitive micelles could achieve stable blood circulation with minimal drug leakage but accelerated drug release inside the tumor cells. This strategy ensures safe and effective drug delivery.

4.4
Surface-functionalized polymeric micelles

Acetal-PEO-b-PDLLA, which is synthesized according to the scheme shown in Fig. 5, formed polymeric micelles with reactive aldehyde groups on their surfaces. Sugar moieties (lactose, galactose and mannose) [42, 93] or peptidyl ligands [43] were installed on the surfaces of the polymeric micelles through Schiff base formation and further reductive amination reactions. The functionality of the ligand molecules, defined as the number of ligand molecules per 100 copolymer chains, was maintained as high as 80%, while the particle size and size distribution remained unchanged [93, 94]. Such targetable polymeric micelles might be useful for active targeting by specifically binding to the targeted tissues or cells as well as facilitating cellular uptake by receptor-mediated endocytosis. Indeed, the lactose-PEO-b-PDLLA micelles specifically interacted with RCA-I lectin, whereas the mannose-installed micelles were specifically bound to the Con A lectin [93]. The interaction of the lactose-installed micelles with the RCA-I lectin protein bed simulating the glycoreceptors on hepatocytes was further studied by surface plasmon resonance [94]. In this study, enhanced binding was observed for lactose functionalities higher than 40%, suggesting the importance of ligand molecule multivalency on the surfaces of the polymeric micelles.

Bae et al. have conjugated a folic acid to the mixed micelles of PEO-b-P(His) and PEO-b-PLLA (see Sect. 4.3), and demonstrated that the folate-conjugated micelles show a higher in vitro cytotoxicity of loaded ADR against human breast adenocarcinoma MCF-7 cells than nontargeted micelles, probably due to a combination of folate receptor-mediated endocytosis and subsequent accelerated drug release in the endosomal compartment [92]. Torchilin et al. recently reported antibody-conjugated polymeric micelles (immunomicelles) [95]. Amphiphilic block copolymers(PEO-b-PE) with a reactive p-nitrophenylcarbonyl (pNP) group on the distal end of PEO were synthesized by coupling PEO-(pNP)$_2$ with PE [96], and the pNP moiety was used to conjugate the tumor-specific antibody onto the micellar surface. They found that the immunomicelles recognized and bound various tumor cells in vitro, and showed a higher accumulation in experimental tumors than the nontargeted micelles. The immunomicelles improved the in vivo antitumor activity of paclitaxel [95]. It should be noted that the immunomicelles were prepared from a mixture of pNP-PEO-b-PE and PEO-b-PE, of which the M_w values of PEO were 3500 and 2000, respectively. Conjugation of the antibody to the longer PEO end on the micellar surface might ensure its specific binding to the target.

Wooley et al. have extensively studied shell crosslinked nanoparticles with functional ligands on their surface [63, 97–99]. The shell crosslinked nanoparticles were prepared using amidation-based crosslinking of the acrylic acid residues of the micelles from poly(acrylic acid)-b-PCL [98] or poly(acrylic acid)-b-poly(methyl acrylate) [99]. In the former case, hydrolysis of the PCL segments resulted in the formation of shell crosslinked nanocapsules. The surfaces of the shell crosslinked nanoparticles were then modified with a cell membrane penetrating the Tat peptide through the coupling of the amino chain end of the Tat peptide with the acrylic acid residues present on the surfaces of the nanoparticles [98, 99]. Fluorescence microscopy observations revealed that transduction of nanoparticles labeled with a fluorescent probe occurred into cells, suggesting that the membrane-penetrating peptides on the nanoparticle surfaces were functioning properly.

4.5
Intracellular localization of micelles

In general, polymeric micelles are assumed to enter the cell by endocytosis, resulting in localization in the endosome and lysosome. However, Maysinger et al. has recently reported that fluorescence-labeled PEO-*b*-PCL micelles localized not only in the lysosome but also in the mitochondrion, Golgi apparatus and endoplasmic reticulum [100]. It is hypothesized that the micelles may disassemble into single chains in the lysosome and permeabilize the lysosomal membranes to relocalize the micelles. If this is the case, intracellular localization of polymeric micelles may depend on their thermodynamic and kinetic stabilities, which are associated with the properties of the block copolymers. A shell of PEO chains may also contribute to the localization of the micelles in cytoplasmic organelles, since amphiphilic PEO possibly interacts with the cellular membranes. Indeed, Hasan et al. reported that the introduction of PEO graft chains into the poly(L-lysine)-drug conjugates resulted in their localization in mitochondrion [101].

From the viewpoint of pharmaceutical applications, polymeric micelles may alter the internalization route, intracellular localization and concentration gradient of a drug, leading to an improvement in its pharmacological activity. Kavanov et al. have studied Pluronic micelles as a carrier for hydrophobic drugs, and found that Pluronic micelles inhibit the P-glycoprotein-mediated drug efflux system in several cell lines [102]. Here, the Pluronic unimers are likely to play an essential role in the inactivation of P-glycoprotein. It was also found that Pluronic micelles enhanced drug accumulation in the brains of mice through the blood-brain barrier [103]. Recently, we have studied the gene expression profiles of cells (human non-small-cell lung cancer PC-14 cells) treated with CDDP-incorporated micelles using a gene expression array [104]. The expression patterns of the genes related to cell cycle regulation, apoptosis-related proteins, detoxification and DNA repair enzymes, which are considered to be associated with free CDDP-induced cytotoxicity, were similar in both the cells treated with the free CDDP and the CDDP-incorporated micelles. However, the CDDP-incorporated micelles down-regulated the genes encoding integrins and matrix metalloproteinases (MMPs), which play an integral role in tumor invasion, metastasis and angiogenesis, whereas the free CDDP up-regulated them. These results suggest the possibility that the micellar carriers may improve the pharmacological activity of the free CDDP. A better understanding of the universal effects of micellar carriers may be helpful for their clinical use, and may lead to the creation of more effective micellar formulations.

5
Polyion complex (PIC) micelles

5.1
Properties of the PIC micelles

Recently, new types of block copolymer micelles formed through via electrostatic interactions between a pair of oppositely charged block copolymers or a pair of charged block copolymers and oppositely charged macromolecules, which are termed "polyion complex (PIC) micelles", have attracted much attention in various fields, including drug delivery [48, 105–110]. For example, mixing negatively charged PEO-b-P(Asp) and positively charged PEO-b-P(Lys) [48] or P(Lys) homopolymers [105] in stoichiometric charge ratios in aqueous media resulted in the spontaneous formation of PIC micelles with sizes of 30–50 nm and significantly narrow size distributions, comparable to natural viruses. The structures of the PIC micelles, where the PIC core is surrounded by a hydrophilic PEO segment shell, prevents precipitation or aggregate formation over a wide range of concentrations. The physicochemical properties of the PIC micelles are significantly affected by many factors such as the molecular weights, charge densities and flexibility of charged segments, the pK_a values, the bulkinesses and the mobilities of the charged groups. In particular, it is known that the PIC micelles are destabilized with increasing ionic strength due to electrostatic shielding [61]. Recently, PIC micelles were prepared from a combination of charged block copolymers (PEO-b-P(Lys) or PEO-b-P(Asp)) and oppositely charged third-generation dendrimers with 32 charged groups on the periphery (32(–)DP or 32(+)DP, respectively) [111]. The 32(+)DP/PEO-b-P(Asp) micelles showed higher stability against NaCl concentration than the P(Lys)$_{27}$/PEO-b-P(Asp) micelles, which may be explained by assuming that the PIC from a rigid dendrimer may produce a smaller entropy gain upon dissociation than that from a flexible P(Lys) homopolymer. It was also suggested that hydrogen bonds may form between the carboxylic acid groups of the dendrimers and the primary amine groups of the P(Lys) blocks in the 32(–)DP/PEO-b-P(Lys) micelles, providing further stabilization of the micelle structure against an increase in NaCl concentration up to 1500 mM, ten times higher than physiological concentration.

Interestingly, we recently found that the PIC micelles formed from a pair of PEO-b-P(Asp) and PEO-b-P(Lys) exhibit a unique phenomenon – chain length recognition – during their formation [112, 113]. A pair of oppositely charged block copolymers form PIC micelles that possess the same chain length of charged segments in the PIC core even if two kinds of block copolymers with matched and unmatched chain lengths exist in the system. For example, adding PEO-b-P(Asp) (the degree of polymerization of P(Asp) ($DP_{P(Asp)}$) = 18) to the mixture of PEO-b-P(Lys) ($DP_{P(Lys)}$ = 18) and PEO-b-P(Lys) ($DP_{P(Lys)}$ = 78) resulted in the selective formation of the PIC micelles from PEO-b-P(Asp) ($DP_{P(Asp)}$ = 18) and PEO-b-P(Lys) ($DP_{P(Lys)}$ = 18) with a stoichiometric mixing ratio, leaving the PEO-b-P(Lys) ($DP_{P(Lys)}$ = 78) free. This chain length recognition during the formation of the PIC micelles appears to be due to the strict phase separation between the PEO shell and PIC core domains, which requires regular alignment of junctions between PEO and the charged segments at the interface.

PIC micelles also form from a combination of charged block copolymers and oppositely charged biologically active molecules, including proteins [62, 114–116] and nucleic acids [106–110]. For example, the mixing of PEO-b-P(Asp) and a lysozyme, a cationic enzyme, at the charge stoichiometric ratio (the ratio of aspartic acid residue in PEO-b-P(Asp) to the total number of lysine and arginine residues in lysosome ($R = 1.0$)) led to the formation of PIC micelles of size ~ 50 nm. In terms of the stoichiometry of PIC micelle formation, stoichiometric micelles and free lysozymes exist in the system at $R < 1.0$, indicating micelle formation in a cooperative manner [114, 115]. On the other hand, nonstoichiometric micelles possessing excess PEO-b-P(Asp) were formed at $R > 1.0$, and the micelle size increased linearly with increasing R [114, 115]. It appears that the existence of free block copolymers in the system is unfavorable at these conditions due to the thermodynamic penalty, which gives them their strong tendency to assemble into the PIC micelles. Lysozymes in PIC micelles showed 100-fold increased enzymatic activity against a small substrate, 4-nitrophenyl-di-N-acetyl-β-chitodioside (($NAG)_2$), due to a remarkable decrease in the Michaelis constant (K_m), which indicates an increase in the affinity of a substrate for the lysozymes in the micelle core [116]. This result led to a new idea of on-off regulation of the enzymatic activity of lysozymes in the PIC micelles using an external electric field, which should affect the structure of the micelles [117]. Indeed, quick and distinct on-off switching of the enzymatic reaction was achieved by applying a pulsed electric field with a voltage of 65–70 V/cm to the PIC micelles. On the other hand, the micellization of the lysozymes resulted in reduced lytic activity against a large substrate, *Micrococcus luteus* cells, because cell access to the PIC core was prevented by the outer PEO layer. This phenomenon facilitated the on-off switching of lysozyme cell lysis activity by changing the salt concentration in the media, which significantly affected the stability of the PIC micelles [118].

5.2
PIC Micelles for gene delivery

Recently, enormous effort have been devoted to the development of nonviral gene carriers based on cationic polymers (polyplexes) in order to achieve gene transfer into the target cells through topical or systemic administration [6, 7, 119–121]. Nonviral gene carriers have many advantages compared with recombinant viral vectors, such as safety (due to the lack of specific immunogenecity), simplicity of use, and ease of large-scale production. So far, a variety of polycations including P(Lys) [122, 123], linear or branched polyethyleneimine (LPEI or BPEI, respectively) [124–127], cationic dendrimers [128], and chitosan [129, 130] have been studied as nonviral gene carriers due to their ability to mask the negative charge of the plasmid DNA (pDNA) and package it into a small particle (< 200 nm) for effective cellular uptake via endocytosis and the protection of DNA from enzymatic or hydrolytic degradation, and their ability to transfer genes to cultured cells have been demonstrated. However, these polyplexes may not be useful for in vivo gene delivery due to their cationic nature, which might lead to uncontrollable biodistribution in the body [131] and may even cause fatal toxicity [132]. It has been suggested that this in vivo polyplex toxicity might be associated with erythrocyte aggregation, causing occlusion of the lung capillaries [132]. In this regard, surface modification o cationic polyplexes with hydrophilic polymers such as PEO is a promising way to improve their biocompatibility [133]. A typical core–shell type polyplex (polyplex micelle) can be formed from PEO-b-P(Lys) block copolymers [110, 134]. PEO-b-P(Lys) and pDNA formed polyplex micelles with a low absolute zeta-potential value, where a PIC core consisting of P(Lys) and a single pDNA molecule was covered with dense PEO palisades, at a charge ratio (Lys/nucleotide unit ratio) of 2 [134]. Atomic force microscopy (AFM) revealed that the polyplex micelles consisted of a toroidal structure of size 60–100 nm and a rod-like structure with a long axis of length 150–300 nm [135]. The PEO-b-P(Lys)/pDNA polyplex micelles showed high serum tolerance, maintaining the condensed state of pDNA in the PIC core due to decreased nonspecific interactions with biological components. Eventually, the cellular uptake and the transfection efficiency of the polyplex micelles were not hampered by incubation with serum, whereas the structures and functions of the P(Lys)/pDNA polyplexes and the cationic liposome/pDNA complexes (lipoplexes) were substantially spoiled by the serum incubation [134, 136]. When the polyplex micelles were injected intravenously, pDNA was observed in the blood circulation after as long as 3 h, whereas the naked pDNA was degraded in the blood within 5 min [137], indicating that the polyplex micelles appear to have excellent blood circulating properties.

Despite modifying the surfaces with hydrophilic polymers, the polyplexes can become destabilized under harsh in vivo conditions, especially by polyelectrolyte exchange reactions with negatively charged biomolecules. However, the introduction of disulfide crosslinks into P(Lys) further stabilized the polyplexes. The disulfide bonds are quite stable in the extracellular milieu, whereas they are cleavable in intracellular reductive environments, allowing the polyplexes to be stabilized with minimal loss of gene transfer activity. It was reported that the introduction of the disulfide crosslinks into the P(Lys)/pDNA polyplexes coated with PEO chains increased the blood concentration of pDNA from 6% to 40% of the dose at 30 min post-injection [138]. In our recent study [135], PEO-b-P(Lys) was thiolated using either of two thiolation reagents, N-succinimidyl 3-(2-pyridyldithio)propionate (SPDP) or 2-iminothiolane (Traut's reagent), to give the thiolated PEO-b-P(Lys) with decreased charge density (PEO-b-P(Lys-MP), MP: 3-mercaptopropionyl group) or that with the charge density maintained (PEO-b-P(Lys-IM)), respectively (Fig. 11). Although both of the crosslinked polyplex micelles made from PEO-b-P(Lys-MP) and PEO-b-P(Lys-IM) showed no dissociation upon a counter polyanion exchange under nonreductive conditions, only the crosslinked micelles from PEO-b-P(Lys-MP) gave efficient release of pDNA under reductive conditions mimicking the intracellular environment. A decrease in the cationic charge density of the P(Lys) segment is likely to promote pDNA release from the micelles upon cleavage of the disulfide crosslinks. It was also demonstrated that the optimal substitution degree of P(Lys) with thiol groups exists in the PEO-b-P(Lys-MP) system, which allows it to achieve such an environmentally sensitive pDNA release. Consequently, the crosslinked polyplex micelles with optimized cationic charge densities and disulfide crosslinking produced better gene transfer than noncrosslinked polyplex micelles. These results strongly suggest that it may be important to optimize a balance between the cationic charge density and the disulfide crosslinking in the crosslinked micelles in order to achieve pDNA release under a specific intracellular condition for enhanced gene expression.

To achieve cell type-specific in vivo gene delivery, the polyplexes coated with hydrophilic polymers were modified with targetable ligands such as

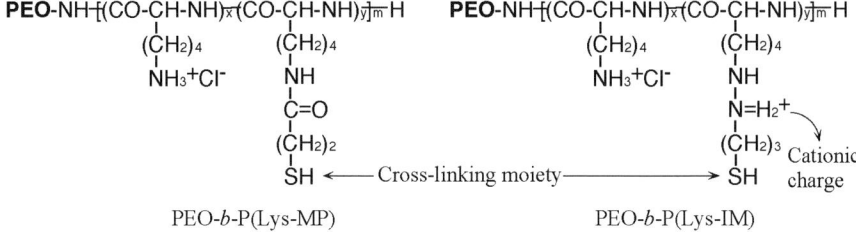

Fig. 11 Chemical structures of thiolated PEO-b-P(Lys) with decreased charge density [PEO-b-P(Lys-MP)] and maintained charge density [PEO-b-P(Lys-IM)], used in disulfide crosslinked polyplex micelles

peptides [139] and antibodies [140, 141]. We recently synthesized α-lactosyl-PEO-*b*-poly(2-dimethylamino)ethyl methacrylate block copolymers (lactose-PEO-*b*-PAMA) as a targetable in vivo gene carrier [142]. The synthetic scheme for lactose-PEO-*b*-PAMA is shown in Fig. 12 [142–144]. The lactosylated polyplex micelles from lactose-PEO-*b*-PAMA resulted in more efficient gene transfer to asialoglycoprotein (ASGP) receptor-expressing human hepatoma HepG2 cells than nontargetable micelles [142]. The transfection efficiency of the lactosylated polyplex micelles was significantly attenuated by the addition of excess asialofetuin, a natural ligand against the ASGP receptor [142], suggesting the internalization of the lactosylated micelles via receptor-mediated endocytosis. The targetable polyplex micelles possess great potential for in vivo site-specific gene transfer via systemic administration.

In addition to pDNA delivery, the development of in vivo carrier systems for a new class of nucleic acid medicines such as antisense oligonucleotides (ODN) and small interfering RNA (siRNA), which can repress or silence the gene expression in a sequence-specific manner, is strongly desired. Recently,

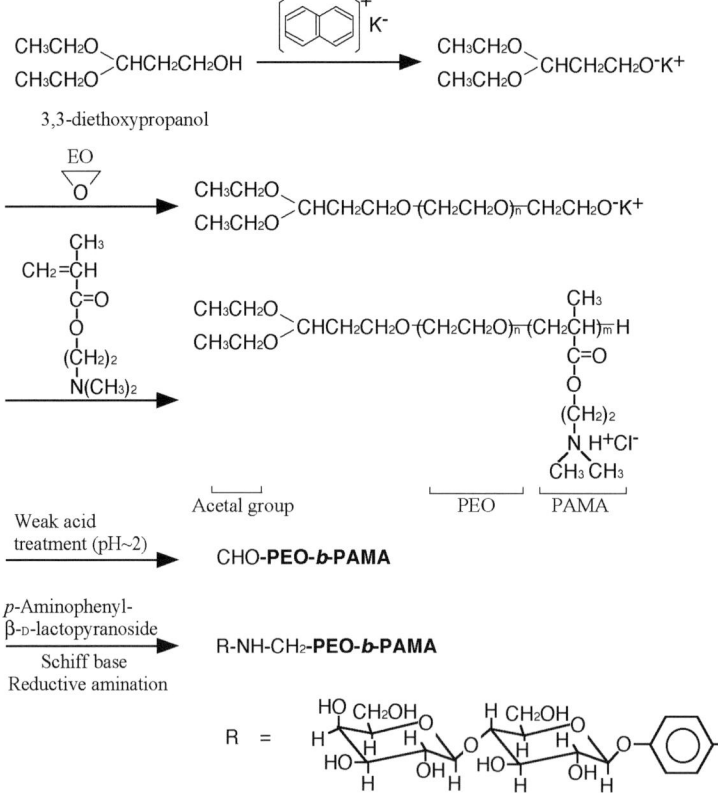

Fig. 12 Synthesis of lactose-PEO-*b*-PAMA block copolymers

Park et al. and our group have separately reported PEO-*b*-oligonucleotide (PEO-*b*-ODN) conjugates with an acid-cleavable linkage, a new class of PIC micelle-forming block copolymers [145, 146]. In our synthetic scheme, the PEO-*b*-ODN was synthesized by a Michael reaction of the 3′-thiol-modified ODN with an ω-acrylate group of a heterobifunctional PEO that has an acetal group at the α-end for the subsequent conjugation of the ligand molecules (Fig. 13) [146]. The PEO-*b*-ODN and LPEI spontaneously formed PIC micelles of size 103 nm at the stoichiometric charge ratio. The PIC micelles showed remarkable stability against degradation by nuclease and minimal interaction with serum proteins. The cleavage of the β-thiopropionate linkage (ester linkage), eliminating PEO chains from the PIC micelles, was selectively observed at endosomal pH (= 5.5), suggesting intracellular release of free ODN. Park's group has already observed that PIC micelles composed of the PEO-*b*-ODN conjugate and BPEI possess in vitro and in vivo antiproliferative activity against ovarian cancer cells [147].

The drawback of the nonviral DNA or RNA carriers is their inefficient expression of drug activity inside cells. The major obstacle to successful DNA or RNA delivery to cells is assumed to be their inefficient transport from the endosome to the cytoplasm [119–121]. Many previous studies have described that polyplexes made from polycations with a comparatively low pK_a undergo efficient gene transfer to cultured cells [124–128, 148, 149]. Such efficient transfection might be explained by the "proton sponge effect" [150], where the protonation of polycations with low pK_a in the endosomal compartment (pH = 5.5 ∼ 6.0) causes osmotic swelling of the endosome, leading to the disruption

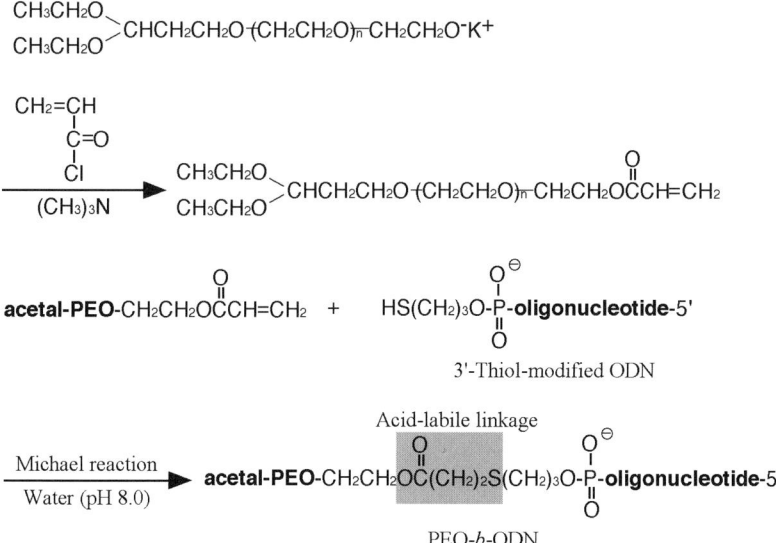

Fig. 13 Synthesis of PEO-*b*-ODN block copolymer with an acid-labile linkage

of the endosomal membrane and the subsequent release of DNA or RNA into the cytoplasm. However, there are several problems to be solved before using such polycations for the in vivo delivery of DNA or RNA. First, polycations with a low pK_a have weak affinities to nucleic acids, resulting in the formation of PIC that is easily dissociated under physiological conditions. Second, the proton sponge effect may be hampered by the protonation of polycations due to the zipper effect or the neighboring group effect during the formation of PIC with DNA or RNA [151]. These problems could be solved by adding excess polycations to form polyplexes with cationically-deviated compositions; however, such polyplexes may not be practical for in vivo use because excess polycations are likely to be liberated from the polyplexes under the highly diluted and harsh conditions in the body, resulting in the decreased efficiency of the endosomal escape of DNA or RNA, as well as the emergence of toxicity derived from free polycations. Ideal in vivo DNA or RNA delivery systems require the following properties: (i) high biocompatibility, to achieve longevity during blood circulation as well as no systemic toxicity; (ii) high stability against enzymatic and hydrolytic degradation as well as exchange by ionic molecules, and; (iii) high buffering capacity for the proton sponge effect. To fulfill these requirements, we have recently reported a novel type of block copolymer, PEO-*b*-poly(3-[(3-aminopropyl)amino]propylaspartamide) (PEO-*b*-DPT), for siRNA delivery [152]. PEO-*b*-DPT, which was synthesized by an aminolysis (ester–amide exchange) reaction of the side chain of PEO-*b*-PBLA with dipropylene triamine (DPT), has primary amine (pK_a 9.9) and secondary amine groups (pK_a 6.4) at the distal end of the polymer side chain and at the position close to the polymer backbone, respectively, in one monomer unit (Fig. 14). This unique structure of PEO-*b*-DPT may allow only the primary amino group to be involved in the PIC formation, thereby maintaining the buffering capacities of the secondary amino groups for the proton sponge effect. Indeed, the PEO-*b*-DPT/siRNA complexes showed remarkably enhanced

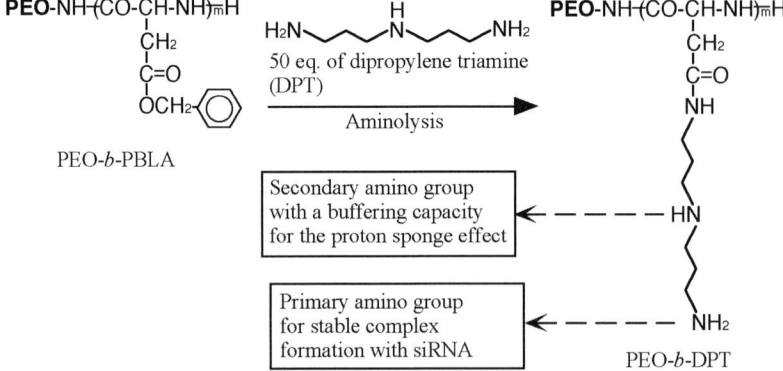

Fig. 14 Block copolymers with spatially regulated cationic moieties for siRNA delivery

siRNA activity against not only the transfected luciferase but also the endogenous gene (Lamin A/C) compared with complexes made from the cationic block copolymers with either primary amine or tertiary amino groups [152]. The gene silencing activity of the PEO-b-DPT/siRNA complexes was not affected by serum incubation due to the stable PIC structure formed between the primary amines in the polymer and siRNA, suggesting their feasibility for application to in vivo siRNA delivery.

6
Polymer vesicles

Associates of block copolymers other than typical core–shell type polymeric micelles have also been studied recently [2, 3]. In particular, polymer vesicles have received considerable attention due to their possible applications as drug-delivering vehicles and stable microstructured biomaterials [153, 154]. It is well known that phospholipids form vesicles consisting of lipid bilayers. The vesicles can effectively entrap various soluble compounds and they can also accumulate hydrophobic substances within the lipid bilayers. Thus, the vesicles from phospholipids (liposomes) are currently used as vehicles for various drugs. Similarly, some block copolymers can also self-assemble into bilayered or more interdigitated membranes in dilute solution, forming polymer vesicles (polymersomes) [153, 154]. The formation of polymer vesicles and their biomedical applications have been described in other review articles [153, 154]. Polymer vesicles can efficiently incorporate hydrophobic substances into the vesicle bilayers and encapsulate hydrophilic low molecular weight and macromolecular compounds inside the vesicle, in a comparable way to liposomes [155]. From the standpoint of drug release, polymer vesicles have decreased fluidity and permeability and greater thickness of the vesicle membranes in comparison with liposomes [156], leading to a slower release rate of encapsulated molecules. In principle, the release of the encapsulated substances, which is usually triggered by external stimuli (such as changes in pH or temperature) or hydrolytic and enzymatic degradation of the polymers, can be achieved by careful chemical design and modification of the vesicle-forming block copolymers, in similar design approach to that used for polymeric micelles, which have been extensively reviewed in this paper.

Stability of polymer vesicles in biological media is another essential property of drug carriers. One important feature of polymer vesicles that is distinct from liposomes is the remarkable stability of the vesicles against dissociation by surface-active molecules such as surfactants and amphiphilic proteins. It was reported that polymer vesicles formed by PEO-polyethylethylene (PEO-b-PEE) were stable for at least five days in 50% blood plasma [157]. Interestingly, Meier et al. have recently reported that naturally occurring membrane proteins can be reconstituted into the membranes of polymer vesicles formed from poly(2-methyloxazoline)-*b*-poly(dimethylsiloxane)-*b*-poly(2-methyloxazoline) copolymers [158]. Reconstituted channel proteins allowed the virus-assisted loading of virus DNA into polymer vesicles. Thus, polymer vesicles can be endowed with biomimetic functions.

Recently, Discher et al. reported that giant and stable worm micelles formed from PEO-*b*-PEE block copolymers, which resemble filamentous phages, showed a remarkably prolonged circulation of several days after intravenous injection into rats [159, 160]. To our knowledge, such a remarkably long blood circulation of the polymer assembly, which was termed the "super-stealth property", has not been reported. Since the worm micelles can be charged with a large amount of hydrophobic drugs (such as paclitaxel), they also have great potential as drug delivery systems.

Thus, polymer vesicles have very interesting properties, which may not be achieved by polymeric micelles and liposomes, which makes them potentially very useful as drug carriers and tools for studying the mechanisms of biosystems. This new technology will continue to attract great interest from researchers in the fields of chemistry, biology and medicine.

7
Concluding remarks

Nanoscaled block copolymer micelles have significant potential for the systemic delivery of various therapeutic agents, including hydrophobic drugs and genetic medicines. The micelle core, which serves as a drug reservoir, is surrounded by the coronal shell, which is composed of biocompatible and hydrophilic polymers, providing excellent structural stability in the body, irrespective of the properties or loading amount of the drug. The key parameters related to drug delivery (such as the CMC, the loading capacity and the release kinetics of the drug, stability in the bloodstream and tissue accumulation) are significantly affected by the chemical structures and compositions of the micelle-forming block copolymers, as well as the compatibility between the core-forming blocks and the drugs to be loaded. Thus, tailor-made block copolymers for a particular drug may be the key to successful drug delivery based on polymeric micelles. Besides, recent advances in synthetic polymer chemistry have allowed new functional block copolymers to be synthesized. Environmentally-sensitive or stimuli-responsive polymeric micelles can activate the loaded drugs selectively at the target site without any drug leakage or loss of activity during blood circulation. Also, targetable polymeric micelles can ensure efficient uptake or internalization of the loaded drugs by the targeted cells, which may be useful when delivering macromolecular drugs such as plasmid DNA or drugs impermeable to the cell membrane. Thus, polymeric micelles will continue to attract much interest from both a theoretical and a practical standpoint.

References

1. Webber SE, Munk P, Tuzar Z (1996) Solvents and self-organization of polymers. NATO ASI Series. Kluwer Academic, London
2. Zhang L, Eisenberg A (1995) Science 268:1728
3. Moffitt M, Khougaz K, Eisenberg A (1996) Acc Chem Res 29:95
4. Schmolka IR (1977) J Am Oil Chem Soc 54:110
5. Bader H, Ringsdorf H, Schmidt B (1984) Angew Makromol Chem 123–124:457
6. Kataoka K, Kwon GS, Yokoyama M, Okano T, Sakurai Y (1993) J Control Release 24:119
7. Kataoka K, Harada A, Nagasaki Y (2001) Adv Drug Deliv Rev 47:113
8. Kakizawa Y, Kataoka K (2002) Adv Drug Deliv Rev 54:203
9. Oh KT, Bronich TK, Kabanov AV (2004) J Control Release 94:411
10. Lavasanifar A, Samuel J, Kown G (2002) Adv Drug Deliv Rev 54:169
11. Allen C, Maysinger D, Eisenberg A (1999) Colloid Surf B Biointerfaces 16:3
12. Lukyanov AN, Torchilin VP (2004) Adv Drug Deliv Rev 56:1273
13. Torchilinn VP (1995) Adv Drug Deliv Rev 16:295
14. Kataoka K (1996) Targetable polymeric drugs (Ch 4). In: Park K (ed), Controlled drug delivery: the next generation. ACS, Washington, DC
15. Seymour LW, Duncan R, Strohalm J, Kopecek J (1987) J Biomed Mater Res 21:1341
16. Stolnik S, Illum L, Davis SS (1995) Adv Drug Deliv Rev 16:195
17. Mosqueria VCF, Legrand P, Gulik A, Bourdon O, Gref R, Labarre D, Barratt G (2001) Biomaterials 22:2967
18. Harris JM (1992) Introduction to biotechnical and biomedical applications of poly(ethylene glycol) (Ch 1). In: Harris JM (ed) Poly(ethylene glycol) chemistry: biotechnical and biomedical applications. Plenum, New York
19. Jeon SI, Lee JH, Andrade JD, De Gennes PG (1991) J Colloid Interf Sci 142:149
20. Matsumura Y, Maeda H (1986) Cancer Res 46:6387
21. Maeda H (2001) Adv Drug Deliv Rev 46:169
22. Dvorak HF, Brown LF, Detmar M, Dvorak AM (1995) Am J Pathol 146:1029
23. Hobbs SK, Monsky WL, Yuan F, Roberts WG, Griffith L, Torchilin VP, Jain RK (1998) Proc Natl Acad Sci USA 95:4607
24. Jain RK (2001) Adv Drug Deliv Rev 46:149
25. Maeda H (2001) Adv Enzyme Regul 41:189
26. Wu J, Akaike T, Maeda H (1998) Cancer Res 58:159
27. Duncan R (2003) Nat Rev Drug Discov 2:347
28. Allen TM (2002) Nat Rev Cancer 2:750
29. Zhang X, Goosen MFA, Wys UP, Pichora D (1993) J Macromol Sci Rev Macromol Chem Phys C33:81
30. Morita T, Horiki Y, Suzuki T, Yoshino H (2001) Eur J Pharm Biopharm 51:45
31. Hiljanen-Vainio MP, Orava PA, Sepala JV (1997) J Biomed Mater Res 34:39
32. Ural E, Kesenci K, Fambri L, Migliaresi C, Piskin E (2000) Biomaterials 21:2147
33. Piskin E (1994) J Biomater Sci Polym Ed 6:775
34. Nakamura T, Hitomi S, Watanabe S, Shimizu Y, Jamshidi K, Hyon SH, Ikada Y (1989) J Biomed Mater Res 23:1115
35. Jamshidi K, Hyon SH, Ikada Y (1988) Polymer 29:2229
36. Nagasaki Y, Kutsuna T, Iijima M, Kato M, Kataoka K (1995) Bioconjug Chem 6:231
37. Akiyama Y, Nagasaki Y, Kataoka K (2004) Bioconjug Chem 15:424
38. Nagasaki Y, Iijima M, Kato M, Kataoka K (1995) Bioconjug Chem 6:702
39. Nakamura T, Nagasaki Y, Kataoka K (1998) Bioconjug Chem 9:300

40. Nagasaki Y, Okada T, Scholz C, Iijima M, Kato M, Kataoka K (1998) Macromolecules 31:1473
41. Iijima M, Nagasaki Y, Okada T, Kato M, Kataoka K (1999) Macromolecules 32:1140
42. Yasugi K, Nakamura T, Nagasaki Y, Kato M, Kataoka K (1999) Macromolecules 32:8024
43. Yamamoto Y, Nagasaki Y, Kato M, Kataoka K (1999) Colloids Surf B Biointerfaces 16:135
44. Tabuchi M, Ueda M, Kaji N, Yamasaki Y, Nagasaki Y, Yoshikawa K, Kataoka K, Baba Y (2004) Nat Biotechnol 22:337
45. Hernandez JR, Klok HA (2003) J Polym Sci A Polym Chem 41:1167
46. Yokoyama M, Inoue S, Kataoka K, Yui N, Sakurai Y (1987) Makromol Chem Rapid Commun 8:431
47. Yokoyama M, Inoue S, Kataoka K, Yui N, Sakurai Y (1989) Makromol Chem 190:2041
48. Harada A, Kataoka K (1995) Macromolecules 28:5294
49. Harada A, Cammas S, Kataoka K (1996) Macromolecules 29:6183
50. Kataoka K, Ishihara A, Harada A, Miyazaki H (1998) Macromolecules 31:6071
51. Kwon GS, Naito M, Yokoyama M, Okano T, Sakurai Y, Kataoka K (1993) Langmuir 9:945
52. Kwon GS, Naito M, Kataoka K, Yokoyama M, Sakurai Y, Okano T (1994) Colloids Surf B Biointerfaces 2:429
53. Kwon GS, Naito M, Yokoyama M, Okano T, Sakurai Y, Kataoka K (1995) Pharm Res 12:192
54. Kwon GS, Naito M, Yokoyama M, Okano T, Sakurai Y, Kataoka K (1997) J Control Release 48:195
55. Liaw J, Aoyagi T, Kataoka K, Sakurai Y, Okano T (1998) Pharm Res 15:1721
56. Kataoka K, Matsumoto T, Yokoyama M, Okano T, Sakurai Y, Fukushima S, Okamoto K, Kwon GS (2000) J Control Release 64:143
57. Gref R, Minamitake Y, Peracchia MT, Trubetskoy V, Torchilin VP, Langer R (1994) Science 263:1600
58. La SB, Okano T, Kataoka K (1996) J Pharm Sci 85:85
59. Yamamoto Y, Yasugi K, Harada A, Nagasaki Y, Kataoka K (2002) J Control Release 82:359
60. Yokoyama M, Sugiyama T, Okano T, Sakurai Y, Naito M, Kataoka K (1993) Pharm Res 10:895
61. Kakizawa Y, Harada A, Kataoka K (1999) J Am Chem Soc 121:11247
62. Jaturanpinyo M, Harada A, Yuan X, Kataoka K (2004) Bioconjug Chem 15:344
63. Thurmond KBII, Kowalewski T, Wooley KL (1996) J Am Chem Soc 118:7239
64. Xu P, Tang H, Li S, Ren J, Kirk EV, Murdoch WJ, Radosz M, Shen Y (2004) Biomacromolecules 5:1736
65. Yasugi K, Nagasaki K, Kato M, Kataoka K (1999) J Control Release 62:89
66. Jule E, Yamamoto Y, Thouvenin M, Nagasaki Y, Kataoka K (2004) J Control Release 97:407
67. Kwon GS, Kataoka K (1995) Adv Drug Deliv Rev 16:295
68. Weissig V, Whiteman KR, Torchilin VP (1998) Pharm Res 15:1552
69. Lukyanov AN, Gao Z, Mazzola L, Torchilin VP (2002) Pharm Res 19:1424
70. Yamamoto Y, Nagasaki Y, Kato Y, Sugiyama Y, Kataoka K (2001) J Control Release 77:27
71. Kwon GS, Suwa S, Yokoyama M, Okano T, Sakurai Y, Kataoka K (1994) J Contr Release 29:17
72. Yokoyama M, Okano T, Sakurai Y, Fukushima S, Okamoto K, Kataoka K (1999) J Drug Target 7:171

73. Nishiyama N, Okazaki S, Cabral H, Miyamoto M, Kato Y, Sugiyama Y, Nisho K, Matsumura Y, Kataoka K (2003) Cancer Res 63:8977
74. Uwatoku T, Shimokawa H, Abe K, Matsumoto Y, Hattori T, Oi K, Matsuda T, Kataoka K, Takeshita A (2003) Circ Res 92:e62
75. Lukyanov AN, Hartner WC, Torchilin VP (2004) J Control Release 94:187
76. Ideta R, Yanagi Y, Tamaki Y, Tasaka F, Harada A, Kataoka K (2004) FEBS Lett 557:21
77. Jeong YI, Cheon JB, Kim SH, Nah JW, Lee YM, Sung YK, Akaike T, Cho CS (1998) J Control Release 51:169
78. Yokoyama M, Miyauchi M, Yamada N, Okano T, Sakurai Y, Kataoka K, Inoue S (1990) Cancer Res 50:1693
79. Yokoyama M, Okano T, Sakurai Y, Ekimoto H, Shibazaki C, Kataoka K (1991) Cancer Res 51:3229
80. Yokoyama M, Okano T, Sakurai Y, Kataoka K (1994) J Control Release 32:269
81. Hamaguchi T, Matsumura Y, Shirao K, Shimada Y, Yamada Y, Muro K, Okusaka T, Ueno H, Ikeda M, Watanabe N (2003) 39th ASCO Annual Meeting, 31 May – 3 June 2003, Chicago, IL, Abs No. 571
82. Lavasanifar A, Samuel J, Kwon GS (2000) J Biomed Mater Res 52:831
83. Lavasanifar A, Samuel J, Kwon GS (2001) Colloids Surf B Biointerfaces 22:115
84. Yokoyama M, Satoh A, Sakurai Y, Okano T, Matsumura Y, Kakizoe T, Kataoka K (1998) J Control Release 55:219
85. Shuai X, Merdan T, Schaper AK, Xi F, Kissel T (2004) Bioconjug Chem 15:441
86. Nishiyama N, Yokoyama M, Aoyagi T, Okano T, Sakurai Y, Kataoka K (1999) Langmuir 15:377
87. Nishiyama N, Kato Y, Sugiyama Y, Kataoka K (2001) Pharm Res 18:1035
88. Aoyagi T, Sugi K, Sakurai Y, Okano T, Kataoka K (1999) Colloids Surf B Biointerfaces 16:237
89. Rapoport N, Pitt WG, Sun H, Nelson JL (2003) J Control Release 91:85
90. Kohori F, Sakai K, Aoyagi T, Yokoyama M, Yamato M, Sakurai Y, Okano T (1999) Colloids Surf B Biointerfaces 16:195
91. Bae YS, Fukushima S, Harada A, Kataoka K (2003) Angew Chem Int Edit 42:4640
92. Lee ES, Na Kun, Bae YH (2003) J Control Release 91:103
93. Nagasaki Y, Yasugi K, Yamamoto Y, Harada A, Kataoka K (2001) Biomacromolecules 2:1067
94. Jule E, Nagasaki Y, Kataoka K (2002) Langmuir 18:10334
95. Torchilin VP, Lukyanov AN, Gao Z, Papahadjopoulos-Sternberg B (2003) Proc Natl Acad Sci USA 100:6039
96. Torchilin VP, Levchenko TS, Lukyanov AN, Khaw BA, Klibanov AL, Rammohan R, Samokhin GP, Whiteman KR (2001) Biochim Biophys Acta 1511:397
97. Pan D, Turner JL, Wooley KL (2003) Chem Comm 19:2400
98. Liu J, Zhang Q, Remsen EE, Wooley KL (2001) Biomacromolecules 2:362
99. Becker ML, Remsen EE, Pan D, Wooley KL (2004) Bioconjug Chem 15:699
100. Savic R, Luo L, Eisenberg A, Maysinger D (2003) Science 300:615
101. Hamblin MR, Miller JL, Hasan T (2001) Cancer Res 61:7155
102. Batrakova EV, Han HY, Miller DW, Kabanov AV (1999) Pharm Res 15:1525
103. Batrakova EV, Miller DW, Li S, Alakhov VY, Kabanov AV, Elmquist WF (2001) J Pharmcol Exp Ther 296:551
104. Nishiyama N, Koizumi F, Okazaki S, Matsumura Y, Nishio K, Kataoka K (2003) Bioconjug Chem 14:449
105. Harada A, Kataoka K (1997) J Macromol Sci Pure Appl Chem A34:2119
106. Kabanov AV, Vinogradov SV, Suzdaltseva YG, Alakhov VY (1995) Bioconjug Chem 6:639

107. Kabanov AV, Kabanov VA (1998) Adv Drug Deliv Rev 30:49
108. Wolfert MA, Schacht EH, Toncheva V, Ulbrich K, Nazarova O, Seymour LW (1996) Hum Gene Ther 7:2123
109. Kataoka K, Togawa H, Harda A, Yasugi K, Matsumoto T, Katayose S (1996) Macromolecules 29:8556
110. Katayose K, Kataoka K (1997) Bioconjug Chem 8:702
111. Stapert HR, Nishiyama N, Jiang DL, Aida T, Kataoka K (2000) Langmuir 16:8182
112. Harada A, Kataoka K (1999) Science 283:65
113. Harada A, Kataoka K (2003) Macromolecules 36:4995
114. Harada A, Kataoka K (1998) Macromolecules 31:288
115. Harada A, Kataoka K (1999) Langmuir 15:4208
116. Harada A, Kataoka K (2001) J Control Release 72:85
117. Harada A, Kataoka K (2003) J Am Chem Soc 125:15306
118. Harada A, Kataoka K (1999) J Am Chem Soc 121:9241
119. Zauner W, Ogris M, Wagner E (1998) Adv Drug Deliv Rev 30:97
120. Merdan T, Kopecek J, Kissel (2002) Adv Drug Deliv Rev 54:715
121. Kabanov AV, Felgner PL, Seymour LW (1998) Self-assembling complexes for gene delivery from laboratory to clinical trial. Wiley, Chichester
122. Wu GY, Wu CH (1988) J Biol Chem 263:14621
123. Erbacher P, Roche AC, Monsigny M, Midoux P (1996) Exp Cell Res 225:186
124. Boussif O, Lezoualc'h F, Zanta MA, Mergny MD, Scherman D, Demeneix B, Behr JP (1995) Proc Natl Acad Sci USA 92:7297
125. Ferrari S, Moro E, Pettenazzo A, Behr JP, Zacchello F, Scarpa M (1997) Gene Ther 4:1100
126. Fischer D, Bieber T, Li Y, Elsasser HP, Kissel (1999) Pharm Res 16:1273
127. Itaka K, Harada H, Yamasaki Y, Nakamura K, Kawaguchi H, Kataoka K (2003) J Gene Med 6:76
128. Tang MX, Redemann CT, Szoka FC (1996) Bioconjug Chem 7:703
129. Sato T, Ishii T, Okahata Y (2001) Biomaterials 22:2075
130. Erbacher P, Zou S, Bettinger T, Steffan AM, Remy JS (1998) Pharm Res 15:1332
131. Ward CM, Read ML, Seymour LW (2001) Blood 97:2221
132. Boeckle S, von Gersdorff K, van der Piepen S, Culmsee C, Wagner E, Ogris M (2004) J Gene Med 6:1102
133. Toncheva V, Wolfert MA, Dash PR, Oupicky D, Ulblich K, Seymour LW, Schacht E (1998) Biochim Biophys Acta 1380:354
134. Itaka K, Yamauchi K, Harada A, Nakamura K, Kawaguchi H, Kataoka K (2003) Biomaterials 24:4495
135. Miyata K, Kakizawa Y, Nishiyama N, Harada A, Yamasaki Y, Koyama H, Kataoka K (2004) J Am Chem Soc 126:2355
136. Itaka K, Harada A, Nakamura K, Kawaguchi H, Kataoka K (2002) Biomacromolecules 3:841
137. Harada-Shiba M, Yamauchi K, Harada A, Takamisawa I, Shimokado K, Kataoka K (2002) Gene Ther 9:407
138. Oupicky D, Carlisle RC, Seymour LW (2001) Gene Ther 8:713
139. Nah JW, Yu L, Han SO, Ann CH, Kim SW (2002) J Control Release 78:273
140. Vinogradov S, Batrakova E, Li S, Kabanov A (1999) Bioconjug Chem 10:851
141. Merdan T, Callahan J, Peterson H, Kunath K, Brakowsky U, Kopeckova P, Kissel T, Kopecek J (2003) Bioconjug Chem 14:989
142. Wakebayashi D, Nishiyama N, Yamasaki Y, Itaka K, Kanayama N, Harada A, Nagasaki Y, Kataoka K (2004) J Control Release 95:653

143. Kataoka K, Harada A, Wakebayashi D, Nagasaki Y (1999) Macromolecules 32:6892
144. Wakebayashi D, Nishiyama N, Itaka K, Miyata K, Yamasaki Y, Harada A, Koyama H, Nagasaki Y, Kataoka K (2004) Biomacromolecules 5:2128
145. Jeong JH, Kim SW, Park TG (2003) Bioconjug Chem 14:473
146. Oishi M, Sasaki S, Nagasaki Y, Kataoka K (2003) Biomacromolecules 4:1426
147. Jeong JH, Kim SW, Park TG (2003) J Control Release 93:183
148. Cheng JY, Wetering P, Talsma H, Crommelin DJA, Hennink WE (1996) Pharm Res 13:1038
149. Midoux P, Monsigny M (1999) Bioconjug Chem 10:406
150. Behr JP (1997) Chimia 51:34
151. Kabanov AV, Bronich TK, Kabanov VA, Yu K, Eisenberg A (1996) Macromolecules 29:6797
152. Itaka K, Kanayama N, Nishiyama N, Jang WD, Yamasaki Y, Nakamura K, Kawaguchi H, Kataoka K (2004) J Am Chem Soc 126:13612
153. Antonietti M, Forster S (2003) Adv Mater 15:1323
154. Discher DE, Eisenberg A (2002) Science 297:967
155. Ahmed F, Discher DE (2004) J Control Release 96:37
156. Lee JCM, Santore M, Bates FS, Discher DE (2002) Macromolecules 35:323
157. Lee JCM, Bermudez H, Discher BM, Sheehan MA, Won YY, Bates FS, Discher DE (2001) Biotechnol Bioeng 73:135
158. Graff A, Sauer M, Gelder PV, Meier W (2002) Proc Natl Acad Sci USA 99:5064
159. Dalhaimer P, Engler AJ, Parthasarathy R, Discher DE (2004) Biomacromolecules 5:1714
160. Dalhaimer P, Discher DE (2004) Abstracts, 228th ACS National Meeting, 22–26 August 2004, Philadelphia, PA, PMSE-431

ns Polym Sci (2006) 193: 103–121
DOI 10.1007/12_026
© Springer-Verlag Berlin Heidelberg 2005
Published online: 30 November 2005

The EPR Effect and Polymeric Drugs: A Paradigm Shift for Cancer Chemotherapy in the 21st Century

H. Maeda[1,2] (✉) · K. Greish[1,3] · J. Fang[2,4]

[1] BioDynamics Research Laboratory,
Kumamoto University Cooperative Research Center 2081-7 Tabaru,
Mashiki-machi, 861-2202 Kumamoto, Japan
msmaedah@gpo.kumamoto-u.ac.jp

[2] Faculty of Pharmaceutical Sciences, Sojo University, 860-0082 Kumamoto, Japan
msmaedah@gpo.kumamoto-u.ac.jp

[3] Department of Cardiovascular Surgery, Graduate School of Medical Sciences,
Kumamoto University, 860-8556 Kumamoto, Japan

[4] Department of Pathology, Duke University Medical Center, Durham, NC, USA

1	Introduction	105
2	EPR Effect: Theory and Principles	106
3	Further Augmentation of Drug Delivery by Modulating the EPR Effect	108
3.1	Angiotensin-Induced Hypertension	108
3.2	Bradykinin	109
3.3	Prostacyclin Agonists	110
3.4	Other Inflammatory Mediators	110
4	Enhancing Intracellular Uptake of Macromolecular Drugs	111
5	Examples of Macromolecular Anticancer Therapeutics	111
5.1	SMANCS	111
5.2	Clinical Status of SMANCS	112
5.3	Quality of Life	114
6	Poly-L-Glutamic Acid Conjugates	115
6.1	Paclitaxel Conjugates	115
6.2	Camptothecin Conjugates	116
6.3	Other Unique Polymer Conjugates for Tumor-Targeted Drug Delivery that Induce Oxystress and Utilize the EPR Effect	116
7	Conclusions	119
	References	119

Abstract Blood vessels in tumors are different to normal blood vessels because they have abnormal architectures and impaired functional regulation. We have studied these abnormalities, in particular vascular permeability in tumors, and found greatly enhanced permeability for macromolecules, which are retained in tumors for extended periods. We named this phenomenon the "*enhanced permeability and retention* (EPR) effect". This effect, related to the transport of macromolecular drugs composed of liposomes, micelles, proteinaceous or polymer-conjugated macromolecules, lipid particles, and nanoparticles into the tumor, is the hallmark of solid tumor vasculature. These macromolecular species are therefore ideal for selective delivery to tumor. The EPR effect has facilitated the development of macromolecular drugs consisting of various polymer-drug conjugates (pendant type), polymeric micelles, and liposomes that exhibit far better therapeutic efficacy and far fewer side effects than the parent low-molecular-weight compounds.

Here, we discuss various aspects of the EPR effect via examples, including the use of polymeric drugs such as SMANCS [poly(styrene-co-maleic acid-half-*n*-butylate) (SMA)-conjugated neocarzinostatin (NCS)]. In addition, we review our new macromolecular drug candidates that generate reactive oxygen species via a novel mode of action. Because solid tumors frequently lack antioxystress enzymes, generating oxystress in tumor tissue may be another unique anticancer strategy. Most tumor cells have a weak or limited defense system against reactive oxygen species, and the oxygen radical-generating techniques that we have developed are primarily endogenous. Consequently, an approach to cancer therapy based on the EPR effect and oxyradical induction in order to produce apoptosis appears promising.

Keywords EPR effect · Drug targeting · Macromolecular drugs · SMANCS · Cancer drug delivery · Plasma half-life · Reactive oxygen radicals

Abbreviations

ACE	Angiotensin-converting enzyme
AT II	Angiotensin II
AUC	Area under the concentration–time curve
BK	Bradykinin
BSA	Bovine serum albumin
COX	Cyclooxygenase
CT	Computed tomography
EPR effect	Enhanced permeability and retention effect in solid tumor
HO-1	Hemoxygenase-1
HPMA	*N*-(2-Hydroxypropyl) methacrylamide copolymer
MDR	Multidrug resistance
MMPs	Matrix metalloproteinases
Mw	Weight-average molecular weight
NCS	Neocarzinostatin
NO	Nitric oxide
NOS	Nitric oxide synthase
PEG-DAO	PEG-Conjugated D-amino acid oxidase
PEG	Poly(ethylene glycol), also called polyoxyethylene
PEG-ZnPP	PEG-Conjugated zinc protoporphyrin IX
PG	Prostaglandin
ROS	Reactive oxygen species
SMA	Poly(styrene-co-maleic acid/anhydride)
SMANCS	Poly(styrene-co-maleic acid-half-*n*-butylate) conjugated with neocarzinostatin
Tax	Paclitaxel, also known as Taxol

1
Introduction

Cancer remains the first or second biggest cause of death in humans, and great efforts have been undertaken to treat it [1]. However, a major problem that limits the success of many anticancer agents is an inability to selectively target tumor cells and tissues; instead, such agents also affect normal tissues and organs. Since the 1970s, many strategies have tried to make anticancer drugs more selective to tumors, particularly those utilizing antibodies directed against specific tumor epitopes or angiogenic effectors. Recently, some success has been achieved in this area, such as the use of trastuzumab to treat HER2-positive breast cancer; rituximab and the radiolabeled antibodies ibritumomab tiuxetan (Zevalin) and ^{131}I tositumomab (Bexxar) to treat non-Hodgkin's lymphoma; imatinib to treat chronic myeloid leukemia; and the application of bevacizumab (Avastin) as an inhibitor of vascular endothelial growth factor [2]. However, if an antibody is directed to the tumor-associated antigens in circulation before it reaches the tumor, its targeting efficiency is reduced. In addition, tumor antigen heterogeneity and the emergence of resistant subclones may present future challenges to this avenue of targeted anticancer therapy.

Another approach to the selective targeting of drugs to tumors is based on the tumor vasculature, and is related to the vascular leakage or permeability in tumor tissues. Indeed, we could selectively deliver large-molecular-size anticancer agents to tumor tissues because of a universal phenomenon that characterizes solid tumors; we coined the term "the enhanced permeability and retention (EPR) effect" of macromolecules and lipids in tumor tissues to describe this phenomenon [3–24]. SMANCS, or poly(styrene-co-maleic acid-half-*n*-butylate) (SMA)-conjugated neocarzinostatin (NCS), is the first clinically approved polymeric drug that takes advantage of the EPR effect: it has a long plasma half-life and pronounced tumor targeting efficiency. The highly selective delivery of SMANCS, given in Lipiodol formulations via the tumor-feeding artery, resulted in a clearly smaller tumor in > 90% of cases of primary hepatoma, with much better survival scores and (especially notable) an improved quality of life, as well as very few side effects compared to conventional avoidable anticancer drugs [25–29].

Many polymeric drugs with excellent EPR effects are now in clinical use, in clinical trials, or in development; these drugs include poly-L-glutamic acid conjugates of paclitaxel (Tax), Camptothecin, cisplatinum, or Supratek's SP series. Doxil (Stealth liposomes containing doxorubicin) is in the market. The purpose of this article is to describe the mechanism of the EPR effect in tumor tissues in the context of macromolecular drug delivery. We will also provide examples of the effectiveness of the polymeric anticancer agent SMANCS and poly-L-glutamic acid conjugates [30, 31], as well as a new class of polymeric anticancer agents that we have developed that exploit the EPR effect and that involve a new mode of action – the generation of oxygen radical species in tumor cells [32–36].

2
EPR Effect: Theory and Principles

When aggregates of tumor cells achieve a diameter of 1–2 mm, they need new blood vessels, or neovasculature, to supply nutrients and oxygen [37]. These newly formed vessels usually have irregular and incomplete architectures and impaired physiological responses [37–41]. Table 1 summarizes the abnormal characteristics of tumor vessels.

The characteristics of vascular pathophysiology listed in Table 1 contribute to the selective, enhanced accumulation and prolonged retention of macromolecular drugs or lipid particles in tumor tissue (EPR effect). Evans blue dye, which forms complexes with albumin (67 kDa) or Lipiodol (a lipid contrast agent), clearly revealed the presence of the EPR effect in rodent tumors after intravenous injection of Evans blue bound albumin, or after arterial

Table 1 Abnormal characteristics and factors of solid tumors that influence the EPR effect of macromolecular drugs in vivo

Unique characteristics and factors	Refs.
(i) Active angiogenesis and high vascular density	[37]
(ii) Extensive production of vascular mediators that facilitate extravasation	
a) Bradykinin	[10–12, 15, 16, 18]
b) Nitric oxide (NO)	[14–18, 52–56]
c) VPF/VEGF*	[41]
d) Prostaglandins	[16, 23, 50, 51]
e) Collagenase (matrix metalloproteinases, or MMPs)	[18]
f) Peroxynitrite	[18, 57]
(iii) Defective vascular architecture, for example, lack of smooth muscle layer cells, lack of or fewer receptors for angiotensin II, large gap in endothelial cell-cell junctions, anomalous conformation of tumor vasculature (such as branching or stretching)	[21, 22, 37–41, 46]
(iv) Impaired lymphatic clearance of macromolecules and lipids from interstitial tissue ($\rightarrow$ prolonged retention of these substances)	[3–5]

* *VPF*, vascular permeability factor, later identified as the same as *VEGF*, vascular endothelial growth factor injection of Lipiodol, by producing an intense blue color only in tumor tissues [3–8, 16–20]. This distribution is quite different from that seen in normal tissues, except for the normal tissues surrounding a tumor or a site of inflammation.

In contrast to the description above, one hypothesis suggested that high interstitial pressure in tumor tissue would impede any macromolecular drugs from entering the tumor, and thus delivery of small as well as macromolecular drugs into tumor tissue would be exceedingly difficult [42]. However, many experimental findings (ours and others) clearly showed that the putative interstitial pressure in tumors did not prevent the transport of various macromolecules, including liposomes, and that these substances do indeed accumulate in tumor tissues. For instance, Duncan and Sat used an N-(2-hydroxypropyl) methacrylamide (HPMA) copolymer-doxorubicin conjugate (PK1) as a probe to study the extent of the EPR effect in different tumor models, and found many mouse and human xenograft tumors with clear tumor size-dependent EPR-mediated targeting of drug to tumor (from $\sim$ 20% dose/g of tumor tissue in small tumors to 1–5% dose/g in large tumors) [43]. This result was consistent with previous reports describing the accumulation and retention of ^{125}I-labeled HPMA copolymer in B16F10 melanoma and sarcoma 180 (S-180) tumors [3–6, 44, 45]. These findings agree with all our data describing the accumulation of SMANCS, ^{51}Cr-labeled bovine serum albumin (BSA), ^{131}I-labeled HPMA copolymer, and Evans blue-albumin complex in various tumor models: Meth A, S-180, and colon 38 in mice; Walker 256 and Yoshida sarcoma in rats; and VX-II carcinoma in rabbits ([4, 14–18, 20, 28], and unpublished data). SMA-doxorubicin micelles, polyglutamate conjugates, and other polymeric drug conjugates also exhibited an EPR effect.

In other studies of the transport of macromolecules into tumor tissues, Yuan et al. measured the size of tumor vessel pores in LS174T human colon adenocarcinoma implanted in dorsal skin chambers in severe combined immunodeficient mice [38]. They showed that tumor vascular pores could be as large as 0.4 μm in diameter [38]. Skinner et al [39] and Suzuki et al [40], as well as Hashizume et al. (who used electron microscopy) [46], in elegant work, identified structural abnormalities in the endothelium of tumor blood vessels. These abnormalities included intercellular openings with a mean diameter of 1.7 μm (range, 0.3–4.7 μm) and transcellular holes with a mean diameter of 0.6 μm in mouse mammary carcinomas [45]. It should be noted that the effective diameter of 67-kDa serum albumin is 7.2 nm. Ohkouchi et al. used the Walker 256 solid tumor system to study the EPR effect of mitomycin C-dextran conjugates, and they confirmed a similarly increased uptake in solid tumors [47].

For the EPR effect to function, the molecular sizes of the macromolecules must be large enough to escape renal clearance (> 40 kDa) [3, 5–8, 20]. Another prerequisite for the EPR effect is that the plasma concentration of the drug, as measured by the area under the concentration–time curve (AUC), must remain

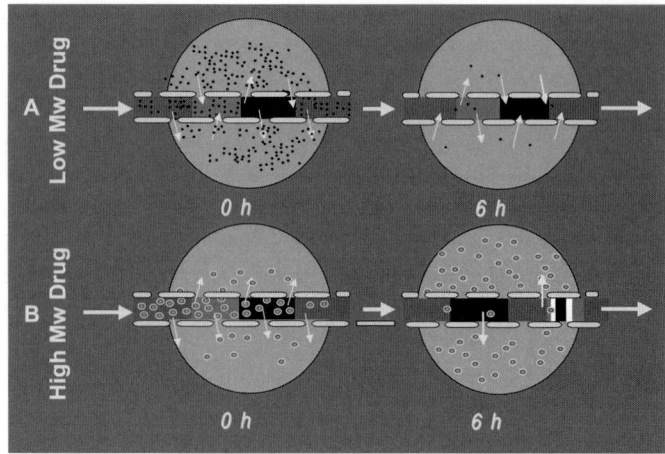

Fig. 1 Comparison of the EPR effect seen with low-molecular-weight drugs (**A**) and with macromolecular drugs (**B**). *Orange circles* represent tumor tissues. Blood vessels are uneven and large gaps exist between endothelial cells, so that almost all molecules can pass through. However, only macromolecules can only enter and remain in the extravascular space. For low-molecular-weight drugs, it is usually difficult to keep the drug concentration in the tumor tissue higher than the blood drug level for long periods (**A**). Macromolecular drugs (**B**) show enhanced uptake by tumor and are retained for a prolonged time at higher concentrations in the tumor tissue

high; preferably, in the case of rodents, for more than 6 hours [3, 5, 6, 20, 45, 48]. As a result of these requirements, extravasation of the polymeric drugs into tumor tissue can increase progressively with time (over several hours or days), but clearance of these macromolecular drugs from the tumor tissues does not occur rapidly if at all (Fig. 1). We obtained the same results with poly(ethylene glycol) (PEG)-conjugated D-amino acid oxidase (PEG-DAO), SMA-doxorubicin micelles, and other agents [32, 33, 35], as described later in this article. The release of active ingredients from polymeric conjugates (or liposomal or micellar drugs) must occur continuously so as to maintain drug levels in tumor tissue above the therapeutic concentration.

3
Further Augmentation of Drug Delivery by Modulating the EPR Effect

3.1
Angiotensin-Induced Hypertension

In 1981, Suzuki et al. found that raising the systolic blood pressure by infusing angiotensin II (AT II) into tumor-bearing rats caused a two- to six-fold selective increase in tumor blood flow, the increase depending on the induced blood pressure [21]. This finding was in great contrast to the blood flow in normal organs and tissues, which remains constant regardless of the induced blood pressure [20–23]. In other words, blood vessels in normal organs, but not those in cancerous tissues, show excellent homeostatic autoregulation of blood flow.

We have examined whether this increased tumor blood flow would influence the EPR effect in an AT II-induced hypertensive state and thereby improve macromolecular drug delivery. We raised systolic blood pressure from 100 to 150 mm Hg for 15 min by infusion of AT II after i.v. injection of ^{51}Cr-labeled SMANCS or the putative macromolecular drug (radiolabeled BSA) [20]. We found that both radiolabeled BSA and SMANCS, which binds to albumin in vivo and thereby forms an apparent molecular mass of about 80 kDa, exhibited a 1.3 to three-fold increased accumulation in tumor tissues. However, the amount of drug delivered to normal organs such as kidney and bone marrow was reduced because of the vasoconstriction occurring in normal organs under AT II-induced hypertension; tighter endothelial intercellular gaps suppressed the transvascular transfer of macromolecules larger than 40 kDa. It is known that the dose of common anticancer agents cannot be increased to more than 2–3 times higher than the recommended dose because of the narrow safety margin. However, the use of macromolecular drugs in the AT II-induced hypertensive state resulted in effects similar to those achieved with higher dosages of the same drug. We demonstrated this result with macromolecular drugs in the clinical setting by arteriography; under the same conditions, however, mitomycin C (334.3 Da) did not produce the same effect [9, 20]. We are currently achieving a better therapeutic effect, with far fewer side effects, by using this method with SMANCS/Lipiodol for tumors such as cholangiocarcinoma, metastatic liver cancer, and renal cell carcinoma [10].

3.2
Bradykinin

Bradykinin (BK), which we have studied extensively in inflammation, infection, and cancer [11–13, 17], is an important mediator of the EPR effect. We have reported that the BK-generating cascade is normally activated in tumor tissues and that BK is involved in the enhancement of vascular permeability and accumulation of malignant ascitic and pleural fluids accumulation [11–13, 16, 17, 19]. Also, both Bhoola's group and our group reported the presence of excessive levels of BK receptor (B2) in various human and rodent solid tumors [19, 49]. BK is degraded by many peptidases, especially angiotensin-converting enzyme (ACE). Therefore, inhibition of ACE by ACE inhibitors will caused locally increased BK levels (at the tumor site, where BK generation is greatest).

On the basis of these data, we used the ACE inhibitors enalapril and temocapril to inhibit BK degradation, and we showed that the elevated BK level resulted in further enhancement of vascular permeability, or the EPR effect [16, 17, 22, 23]. ACE inhibitors thus increase delivery of macromolecular drugs to tumors, even under normotensive conditions.

3.3
Prostacyclin Agonists

Prostaglandin (PG) biosynthesis (via cyclooxygenases COX-1 and 2), particularly PGE_2 production, is markedly elevated in human and experimental tumors [50, 51]. The COX inhibitor indomethacin significantly suppressed vascular permeability in S-180 and other solid tumor models [16]. In addition, we found that injection of a stable prostacyclin (PGI_2) agonist, beraprost sodium which has a much longer $t_{1/2}$ than PGI_2 in vivo (60 min versus a few seconds), resulted in roughly double the EPR effect, similar to the situation for ACE inhibitors and BK [23]. As an incidental finding, these vasoactive mediators reduced downstream blood flow to an almost negligible amount (about 10%) [23]. Thus, it may be possible to enhance the selective accumulation of macromolecular anticancer agents in tumors by using PGI_2 agonists. Likewise, macromolecular tumor-imaging agents may be concentrated in tumor tissues, although this possibility must be demonstrated in clinical settings.

3.4
Other Inflammatory Mediators

Many factors other than the above-described mediators (such as BK and PGs) are implicated in the EPR effect in solid tumors. For instance, nitric oxide (NO), which is synthesized by NO synthase (NOS) from L-arginine, influences tumor vascular permeability [14–18]. Consequently, the NOS inhibitor N^{ω}-monomethyl-L-arginine (L-NMMA) suppressed vascular permeability in solid tumors [15, 16, 18]. Meyer et al. also found that this NOS inhibitor irreversibly attenuated blood flow in R3230Ac rat mammary adenocarcinoma [52]. Similarly, Tozer et al. demonstrated a selective reduction in tumor blood flow with another NOS inhibitor, N^{ω}-nitro-L-arginine, in P22 tumor-bearing rats [53]. Also, NO is now known to play a key role in angiogenesis, cell proliferation, and extravasation or the EPR effect, which facilitates the supply of nutrients and oxygen. Consequently, inhibition of NO generation was found to suppress tumor growth [15, 16, 54–56]. In addition, we found that pro-matrix metalloproteinases (proMMPs) are activated by peroxynitrite ($ONOO^-$), which is produced extensively in tumor and inflammatory tissues [8, 18, 57]. MMPs facilitate cancer metastasis on the one hand, but also enhance the EPR effect, which helps support nutritional supply, angiogenesis and growth of solid tumors [15–18]. The MMP-induced EPR effect was inhibited by many MMP inhibitors [18].

4
Enhancing Intracellular Uptake of Macromolecular Drugs

Macromolecular drugs (MW > 40 kD) usually benefit from the EPR effect during tumor targeting, as discussed earlier. These drugs are less likely to be transported into cells by simple diffusion because of their size; endocytosis appears to be the usual means of internalization, or, alternatively, internalization occurs by specific binding to receptors [58, 59] or by more specific receptor-mediated uptake. After internalization of the drugs into the cytoplasm via phagosomes, the phagosomes fuse with lysosomes, which are rich in proteolytic and hydrolytic enzymes (such as cathepsins) with a low pH. Active principles are then released. For many pendant-type linked drugs, the types of chemical linkages to polymers become important: and peptidyl linkers must be selected to facilitate cleavage of the linkage and thus meet the requirements of cathepsins or other specific enzymes [43, 44, 60]. In the case of SMANCS, an acidic pH environment liberates the active component (NCS) by spontaneous hydrolysis of the maleyl amide bond. NCS then diffuses into the cytoplasm or the nucleus, where the drug action occurs. It was recently reported that more active drug uptake and endocytosis occur during mitosis [61], which should mean greater selectivity for tumor and a higher uptake efficiency of dividing cancer cells for macromolecular drugs, so that tumor tissues are targeted while the EPR effect is achieved.

In a different context, intracellular uptake of macromolecular drugs serves to protect against P-glycoprotein-dependent efflux in multidrug-resistant (MDR) cells [62–64]. This result suggests that the EPR-mediated targeting of polymeric anticancer drugs not only improves the accumulation of drugs in tumor tissues, but also ensures accumulation and higher intracellular activity of these drugs against MDR-positive cancer cells despite efflux of P-glycoprotein. A strategy for managing MDR may thus be possible.

5
Examples of Macromolecular Anticancer Therapeutics

5.1
SMANCS

NCS is a proteinaceous antitumor antibiotic produced by *Streptomyces carzinostaticus* var. F-41 [65]. NCS, at a dose of 0.01 µg/ml or less, shows cytotoxicity against mammalian cells as well as gram positive bacteria by inhibiting DNA synthesis through direct DNA strand scission as well as superoxide generation by cytochrome P450 reductase [65–68]. We demonstrated that chemical conjugation of NCS with the copolymer of styrene-maleic acid-half-

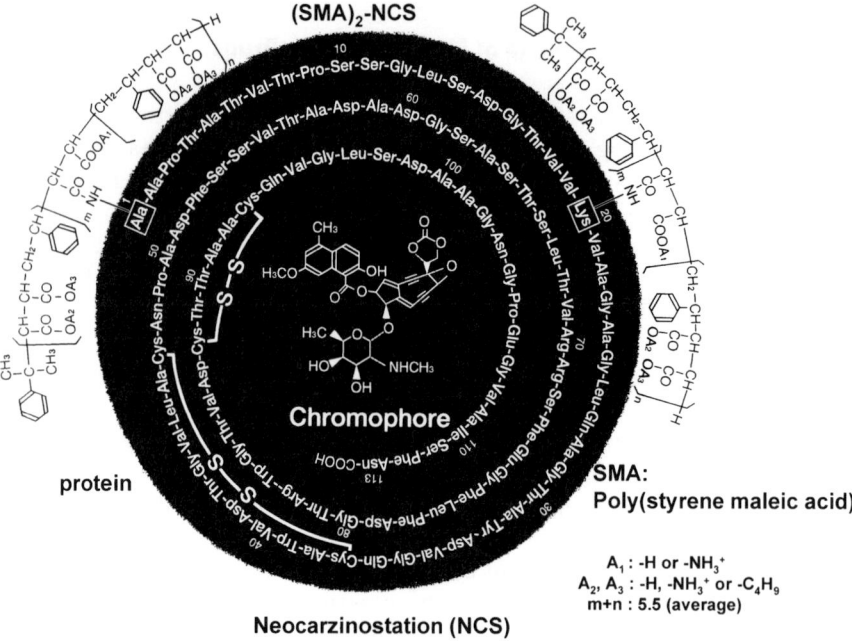

Fig. 2 Chemical structure of SMANCS. Two chains of SMA [copoly(styrene-maleyl-butylate-half-n-ester)] are attached to the N-terminal amino group of alanine 1 and the ε-amino group of lysine 20 of NCS

n-butylate (SMA) greatly improved the physicochemical, biochemical, and pharmacological properties of NCS (Fig. 2).

These advantages are attributed to the macromolecular size of SMANCS, which is further enhanced by binding to plasma albumin, so that the apparent molecular size becomes about 90 kDa in plasma. The hydrophobicity of the styrene ring in SMA confers lipophilicity to SMANCS, which permits its use with the lipid contrast medium Lipiodol. Lipiodol was originally used as a contrast agent in lymphographic imaging. An oily formulation of SMANCS in Lipiodol (SMANCS/Lipiodol), with a tumor-imaging capability, thus becomes possible [7, 9, 25–29, 69]. Both Lipiodol itself and SMANCS/Lipiodol accumulate in an extremely tumor-selective manner because of the EPR effect. Injected SMANCS eventually breaks down via proteolysis, and the SMA polymer is released and eliminated mainly in bile and to a lesser extent by the kidney into urine.

5.2
Clinical Status of SMANCS

SMANCS/Lipiodol, approved in Japan 1993 for treatment of liver cancer, is administered via a catheter according to Seldinger's method into a tumor-feeding artery, such as the hepatic artery for hepatoma. This preparation allows the drug to accumulate in cancer tissue much more effectively or selectively than aqueous formulations [25–29]. In a rabbit model, for example, the concentration of SMANCS in tumor was more than 2000 times higher than that in blood plasma [4]. In addition to an improved therapeutic efficacy or response rate of hepatoma (more than 90%) to SMANCS/Lipiodol, tumor imaging becomes much sharper and more sensitive [4, 26–28]. Tumors, even those smaller than 5 mm in diameter, can be readily identified by conventional computed tomography (CT) [26, 27]. Furthermore, the timing of drug doses and the proper dosing regimen can be determined more accurately, by quantifying the high-density (stained) area of the tumor, which indicates the presence of SMANCS/Lipiodol. In other words, calculation of the required prescription drug dosage is based on tumor size, not on body weight or body surface area. This dose calculation principle is significantly different to that used for common anticancer drugs, which is based on the maximum tolerable dose, which is proportional to the body surface area [70].

We developed a grading system for the filling rate of SMANCS/Lipiodol in the tumor area that includes four grades: grade I indicates less than 10% retention; grade II, 10–50% filling; grade III, 50–75% filling; and grade IV, 75–100% filling with SMANCS/Lipiodol, as judged by CT [27]. When grade III or grade IV is obtained, a positive therapeutic effect can be expected (a 90% response). This dose regimen for SMANCS/Lipiodol that depends on tumor size was substantiated not only by our group in Kumamoto, Japan, but also by a group in Birmingham in the UK [70].

For all cases of primary hepatomas combined (that is, Child's criteria A to C with liver cirrhosis), the five- to seven-year survival rate for i.a. SMANCS/Lipiodol is about 30%. More recent data for Child A and B for five-year survival is about 70% (unpublished). In contrast, during this time frame, no survival is achieved with other treatments. Hepatoma patients with milder liver cirrhosis (such as Child's grade A) and with tumor spread confined to one or two segments of the liver have an approximately 90% survival rate at seven years with SMANCS/Lipiodol [69]. The superiority of the SMANCS compound over conventional low-molecular-weight drugs was confirmed in a randomized clinical trial in which SMANCS/Lipiodol and epirubicin in Lipiodol were both administered by transcatheter arterial infusion [71]. Greater tumor regression was associated with SMANCS/Lipiodol therapy than with epirubicin treatment (47.1% versus 7.7%, respectively). Clinical trials aimed at standardizing the use of SMANCS/Lipiodol in hepatocellular carcinoma are underway at the national level in Japan [72, 73].

5.3
Quality of Life

Use of EPR effect-based targeted delivery of drugs such as SMANCS improves not only the clinical outcomes of patients, but also their quality of life. For example, patients receiving SMANCS usually require only a day or so of bed rest, and can then resume his or her work on the second or third day after

Table 2 Side effects of intra-arterial SMANCS/Lipiodol therapy in patients with hepatoma[a]

Symptoms or sign	Parameter change	% change
Dermatological (exanthema)		0.36
Nausea		5.35 [b]
Vomiting		4.06 [b]
Anorexia		3.63 [b]
Abdominal pain (transitory)		5.53
Liver function		
GOT [c]	Increased	2.16 [b]
GPT [c]	Increased	2.12 [b]
Bilirubin (>1.5 mg/dl)		3.45 [b]
Hypotension		2.22
Blood counts		
WBCs [c]	Decreased	0.38
	Increased	0.83
PMNs [c]	Decreased	0.04
	Increased	0.28
Platelets	Decreased	0.83 [b]
Renal function	Impaired	0.71
BUN [c]	Increased	0.41
Anaphylaxis/shock		0.14
Rigor (transitory)		4.88
Chest pain (transitory)		0.20
Fever (low grade, 2–7 days)		27.80
CRP [c]	Increased	0.67
Ascites formation		1.35 [b]

[a] Based on 3956 patients (post marketing survey data supplied by Yamanuchi Pharmaceutical Co.)
[b] These results are frequently associated with impaired hepatic function (such as caused by liver cirrhosis), and most patients tend to show these effects as liver function deteriorates and disease progresses without the use of SMANCS
[c] *GOT*, glutamic-oxaloacetic transaminase; *GPT*, glutamic-pyruvic transaminase; *WBCs*, white blood cells; *PMNs*, polymorphonuclear neutrophils; *BUN*, blood urea nitrogen; *CRP*, C-reactive protein

drug administration. Most of these patients will require the next treatment after an interval of one month or longer, the time depending on the clearance of the drug from the tumor tissue in the liver as revealed by CT, as well as on how well SMANCS/Lipiodol filled the tumor (grade II or IV for example). This type of therapy does not cause bone marrow suppression or toxicity in the liver or the kidney, which is in great contrast to conventional low-molecular-weight chemotherapeutic agents (Table 2). Also, it usually does not cause loss of appetite. By any parameter used to evaluate the quality of life for SMANCS-treated patients, such as healthy days of life lost, one can clearly show the advantages of this treatment option.

6
Poly-L-Glutamic Acid Conjugates

6.1
Paclitaxel Conjugates

Paclitaxel (Tax) is a natural compound, a member of the taxane diterpenoid family. The drug was first isolated in 1967, and was chemically defined in 1971. Tax was approved by the FDA for treatment of ovarian cancer in December 1992 and for treatment of breast cancer in April 1994. The drug had global sales of more than US$1.5 billion in 2000. The major problem with the use of Tax is its poor water solubility. For this reason, Tax was dissolved in Cremophor and ethanol for clinical use, which resulted in allergic reactions and neurotoxicity in a subset of patients. To improve its solubility in water and intratumor concentration, Tax was conjugated with poly-L-glutamic acid via an ester bond between the γ-carboxylate of poly-L-glutamic acid and the 2'-OH of Tax. This conjugation resulted in an 80-kDa water-soluble drug, as determined by gel permeation chromatography with multi-angle light-scattering. Although the ester bond is usually labile in plasma because of the presence of esterase, rapid release of free drug into plasma did not occur, probably as a result of steric hindrance in this conjugate.

When tumor distribution of Tax after administration of the free ^{3}H-Tax or its conjugates to C57BL/6 mice harboring B-16 melanoma tumors was analyzed, the conjugate had an AUC that was 11 times higher and a Cmax that was 1.7 times higher than free ^{3}H-Tax. This improvement in drug pharmacokinetics and concentration in tumor reflected a three-fold improvement in activity in vivo compared with free drug (the delay in tumor growth achieved by free Tax at the maximal tolerated dose). These encouraging experimental in vivo results led to rapid initiation of clinical trials that are currently underway and are very promising [30]. The clinical dosage regimen via the i.v. route is every four weeks, which offers patients greater freedom than the conventional regimen (which requires confinement to bed).

6.2
Camptothecin Conjugates

Camptothecins are potent cytotoxic agents first isolated from *Camptotheca accuminata*. They exert their effects in dividing cells by inhibition of topoisomerase I. Besides the common problem of little specificity for tumor cells that occurs with low-molecular-weight anticancer drugs, camptothecins possess two major limitations: weak solubility in water, and a high affinity for human serum albumin via a reaction that renders the drug inactive. 20(*S*)-Camptothecin contains a lactone ring that is essential for the cytotoxic drug effect that targets topoisomerase I. The camptothecin with this lactone ring usually exists in equilibrium with an open-ended carboxylate form, which reacts with human albumin at physiological pH, the drug being converted into the inactive form. Similar to the modification of free Tax, a conjugate of camptothecin with poly-L-glutamatic acid was used to enhance the water solubility of this agent as well as to protect the labile lactone ring, so that the conjugate could benefit from the EPR effect. Therefore, this conjugate has improved pharmacokinetics and intratumor accumulation as a result of the EPR effect [31]. A Phase I/II study is now underway in the United States and Europe. This conjugate offers a marked advantage, very similar to that of the Tax conjugate, compared with free camptothecin.

6.3
Other Unique Polymer Conjugates for Tumor-Targeted Drug Delivery that Induce Oxystress and Utilize the EPR Effect

Antioxidant enzymes such as catalase and Cu, Zn-SOD (superoxide dismutase) are highly down-regulated in tumor cells (Fig. 3). To compensate for this, almost all cancer cells possess highly up-regulated hemoxygenase-1 (HO-1) [74–80]. HO-1 induction results in generation of bilirubin, via bilverdin, which functions as a potent antioxidant in the tumor cells [35, 36, 77–84]. Thus, tumor cells could be killed if one could block HO-1, their major anti-reactive oxygen species (ROS) system, with the effect being enhanced when ROS such as H_2O_2 are generated in the tumor tissues [32, 33].

Examples of agents used include PEG-DAO [33] and PEG-conjugated zinc protoporphyrin IX (PEG-ZnPP), an inhibitor of HO-1 [34–36]. These polymeric dugs achieved tumor-targeted delivery on the basis of the EPR-effect. Thus, both generation of ROS selectively in tumor tissue and inhibition of the enzyme HO-1 via PEG-ZnPP would result in a marked increase in both ROS and oxystress at the tumor site, and so these agents have marked antitumor effects.

In one study [36], these conjugates were used to treat SW480 cells, a breast cancer cell line. PEG-ZnPP-treated SW480 cells became more vulnerable

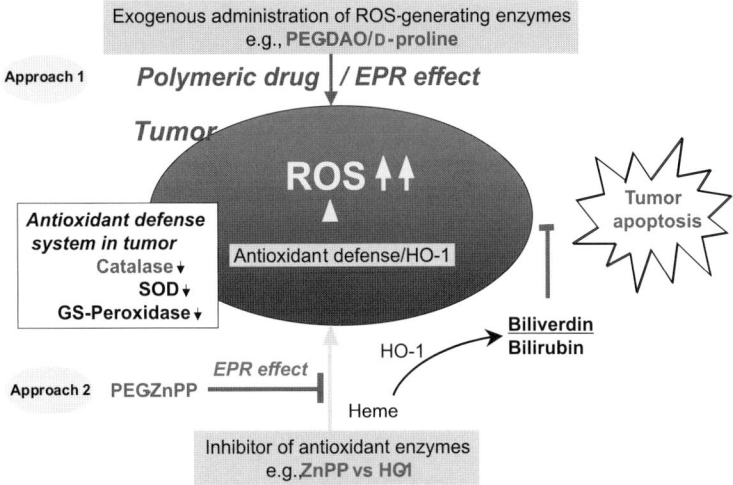

Fig. 3 Schematic showing a new anticancer strategy: oxidation therapy. The generation of ROS in solid tumor tissue is accomplished via two approaches. One approach is through exogenous delivery of ROS-generating enzymes (such as PEG-XO [xanthine oxidase] and PEG-DAO) and by capitalizing on the EPR effect. The second approach is indirect, by inhibiting the antioxidant defense system of the tumor; for example, by using an HO-1 inhibitor. In tumor cells, conventional anti-ROS enzymes such as catalase and superoxide dismutase (SOD) are highly down-regulated, whereas HO-1 is up-regulated and replaces these enzymes in their antioxidative role. Thus the EPR effect-based targeted inhibition of HO-1 in tumor cells by PEG-ZnPP will abrogate the tumor's primary antioxidant defense (biliverdin and bilirubin). Normal cells, however, have the advantage of an adequate amount of catalase, and they are not targeted for destruction. Consequently, the increased ROS in tumor cells will cause tumor apoptotic cell death, which will produce a marked and favorable antitumor effect, but without apparent side effects [32–35]. *GS-peroxidase*, glutathione peroxidase

to various conventional chemotherapeutic agents. In another study, PEG-ZnPP pretreatment significantly reduced tumor growth in mice receiving PEG-DAO plus D-proline compared with either treatment alone (Fig. 4d). More important, each dose in this combination therapy was greatly reduced and the number of treatments were fewer compared with either PEG-DAO or PEG-ZnPP alone; no or very little antitumor effect was found with low doses of the agents given alone (Fig. 4d). In addition, PEG-ZnPP in combination with conventional chemotherapeutic agents such as camptothecin or doxorubicin showed enhanced therapeutic efficacies [36]. These findings suggest that HO-1 is an attractive target for chemotherapeutic intervention and that an inhibitor of HO-1, such as PEG-ZnPP, may be a useful candidate for combination therapy with a broad range of anticancer agents such as cisplatin, camptothecin, doxorubicin, mitomycin C, and etoposide [36].

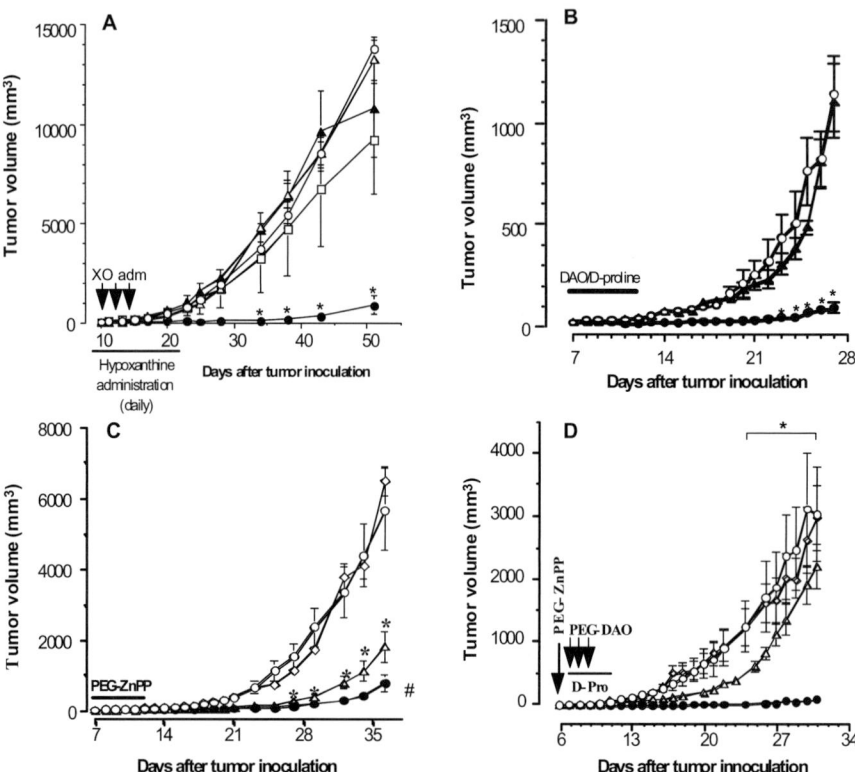

Fig. 4 Antitumor effect of PEG-XO (**A**), PEG-DAO (**B**), PEG-ZnPP (**C**), and the combination of PEG-ZnPP plus PEG-DAO in mice with S-180 solid tumor. S-180 cells were implanted subcutaneously in ddY mice. **A** Arrowheads point to the time of administration of native XO or PEG-XO. ○ control without treatment; ● PEG-XO + hypoxanthine; △ PEG-XO alone; ▲ native XO + hypoxanthine; □ hypoxanthine alone. Data are means ± SE ($n = 6-8$). *$P < 0.001$ in ● versus ▲ or □. **B** Mice were treated with 1.5 U of native DAO (▲) or PEG-DAO (●) followed by D-proline (0.5 mmol, i.p., 2 and 4 h after DAO administration). *Open circles* show control data (mice received no treatment). Data are means ± SE ($n = 4-8$). *$P < 0.001$ in ● versus ▲. **C** Mice received different doses of PEG-ZnPP (●, 5 mg/kg, 1.5 mg/kg) or inactive control PEG-PP (◇, 5 mg/kg) daily for 6 days. Control mice (○) received physiological saline. Data are means ± SE ($n = 6-8$). *$P < 0.001$, PEG-ZnPP groups versus PEG-PP and control groups. **D** PEG-ZnPP plus PEG-DAO group. PEG-ZnPP, 5 mg/kg, was injected on day 6. ○ Control (no treatment); ● PEG-DAO plus PEG-ZnPP; ◇ PEG-DAO but no PEG-ZnPP; △ PEG-ZnPP alone. Data are means ± SE ($n = 8-12$). *$P < 0.01$, control versus each treatment. PEG-DAO, 0.75 U/mouse, was injected i.v. daily from days 7 to 9, with D-proline injected i.p. 4 h after each PEG-DAO administration; in addition, D-proline was injected at 2 days (x2/day) after cessation of PEG-DAO, at a dose of 0.5 mmol/mouse, in the PEG-DAO + PEG-ZnPP group. All figures are from [32–36] with permission (see details in these references)

7
Conclusions

The therapeutic concept based on the EPR effect will have a great impact on cancer chemotherapy. Polymer-conjugates or micellar forms with apparently macromolecular size can be engineered from various low-molecular-weight anticancer agents, and they will have greatly improved characteristics, particularly in terms of their pharmacokinetic properties (prolonged $t_{1/2}$, less toxicity, and much improved distribution to tumor). In addition, they will have enhanced tumor-targeting efficiencies, which will ensure better therapeutic outcomes with far fewer side effects. However, one unsettled question concerns the optimal rate of drug release from the complex or conjugate in the tumor tissue, which differs for each case and depends on the chemical nature or binding of the complex. HO-1 inhibitors and oxidative stress-inducing drugs, which include most of the currently available anticancer agents (such as cisplatin, doxorubicin, and camptothecin), show great potential for future clinical application when used in combination. Also, the EPR effect-based targeting strategy ensures a better quality of life and improved compliance of patients.

We expect that some of the polymeric drugs that are currently in clinical trials will prove to have these great advantages. Data for these trials will be available in 2005.

References

1. WHO (2002) Press release WHO/52 – 28 June 2002. World Health Organization, Geneva
2. Allen TM (2002) Nat Rev Cancer 2:750
3. Matsumura Y, Maeda H (1986) Cancer Res 46:6387
4. Iwai K, Maeda H, Konno T (1984) Cancer Res 44:2115
5. Maeda H, Matsumura Y (1989) Crit Rev Ther Drug Carrier Syst 6:193
6. Noguchi Y, Wu J, Duncan R, Strohalm J, Ulbrich K, Akaike T, Maeda H (1998) Jpn J Cancer Res 89:307
7. Maeda H, Sawa T, Konno T (2001) J Control Release 74:47
8. Maeda H (2001) Adv Enzyme Regul 41:189
9. Maeda H, Fang J, Inutsuka T, Kitamoto Y (2003) Int Immunopharmacol 3:319
10. Greish K, Fang J, Inutsuka T, Nagamitsu A, Maeda H (2003) Clin Pharmacokinet 42:1089
11. Maeda H, Matsumura Y, Kato H (1988) J Biol Chem 263:16051
12. Matsumura Y, Kimura M, Yamamoto T, Maeda H (1988) Jpn J Cancer Res 79:1327
13. Matsumura Y, Maruo K, Kimura M, Yamamoto T, Konno T, Maeda H (1991) Jpn J Cancer Res 82:732
14. Doi K, Akaike T, Horie H, Noguchi Y, Fujii S, Beppu T, Ogawa M, Maeda H (1996) Cancer 77:1598
15. Maeda H, Noguchi Y, Sato K, Akaike T (1994) Jpn J Cancer Res 85:331
16. Wu J, Akaike T, Maeda H (1998) Cancer Res 58:159
17. Maeda H, Wu J, Okamoto T, Maruo K, Akaike T (1999) Immunopharmacology 43:115

18. Wu J, Akaike T, Hayashida K, Okamoto T, Okuyama A, Maeda H (2001) Jpn J Cancer Res 92:439
19. Wu J, Akaike T, Hayashida K, Nakagawa T, Miyakawa K, Muller-Esterl W, Maeda H (2002) Int J Cancer 98:29
20. Li CJ, Miyamoto Y, Kojima Y, Maeda H (1993) Br J Cancer 67:975
21. Suzuki M, Hori K, Abe Z, Saito S, Sato H (1981) J Natl Cancer Inst 67:663
22. Hori K, Saito S, Takahashi H, Sato H, Maeda H, Sato Y (2000) Jpn J Cancer Res 91:261
23. Tanaka S, Akaike T, Wu J, Fang J, Sawa T, Ogawa M, Beppu T, Maeda H (2003) J Drug Target 1:45
24. Greish K, Sawa T, Fang J, Akaike T, Maeda H (2004) J Control Release 97:219
25. Konno T, Maeda H, Iwai K, Tashiro S, Maki S, Morinaga T, Mochinaga M, Hiraoka T, Yokoyama I (1983) Eur J Cancer Clin Oncol 19:1053
26. Konno T, Maeda H, Iwai K, Maki S, Tashiro S, Uchida M, Miyauchi Y (1984) Cancer 54:2367
27. Maki S, Konno T, Maeda H (1985) Cancer 56:751
28. Iwai K, Maeda H, Konno T, Matsumura Y, Yamashita R, Yamasaki K, Hirayama S, Miyauchi Y (1987) Anticancer Res 7:321
29. Konno T, Kai Y, Yamashita R, Nagamitsu A, Kimura M (1994) Acta Oncol 33:133
30. Singer JW, Baker B, De Vries P, Kumar A, Shaffer S, Vawter E, Bolton M, Garzone P (2003) In: Maeda H, Kabanov A, Kataoka K, Okano T (eds) Polymer drugs in the clinical stage. Kluwer Academic/Plenum, New York, p 81
31. Singer JW, Bhatt R, Tulinsky J, Buhler KR, Heasley E, Klein P, de Vries P (2001) J Control Release 74:243
32. Sawa T, Wu J, Akaike T, Maeda H (2000) Cancer Res 60:666
33. Fang J, Sawa T, Akaike T, Maeda H (2002) Cancer Res 62:3138
34. Sahoo SK, Sawa T, Fang J, Tanaka S, Miyamoto Y, Akaike T, Maeda H (2002) Bioconjug Chem 13:1031
35. Fang J, Sawa T, Akaike T, Akuta T, Sahoo SK, Greish K, Hamada A, Maeda H (2003) Cancer Res 63:3567
36. Fang J, Sawa T, Akaike T, Greish K, Maeda H (2004) Int J Cancer 109:1
37. Folkman J (1990) J Natl Cancer Inst 82:4
38. Yuan F, Dellian M, Fukumura D, Leunig M, Berk DA, Torchilin VP, Jain RK (1995) Cancer Res 55:3752
39. Skinner SA, Tutton PJ, O'Brien PE (1990) Cancer Res 50:2411
40. Suzuki M, Takahashi T, Sato T (1987) Cancer 59:444
41. Senger DR, Galli SJ, Dvorak AM, Perruzzi CA, Harvey VS, Dvorak HF (1983) Science 219:983
42. Jain R (1994) Sci Am 271:58
43. Duncan R, Sat YN (1998) Ann Oncol 9:149
44. Duncan R (1999) Pharm Sci Technol Today 2:441
45. Seymour LW, Miyamoto Y, Maeda H, Brereton M, Strohalm J, Ulbrich K, Duncan R (1995) Eur J Cancer 31:766
46. Hashizume H, Baluk P, Morikawa S, McLean JW, Thurston G, Roberge S, Jain RK, McDonald DM (2000) Am J Pathol 156:1363
47. Ohkouchi K, Imoto H, Takakura Y, Hashida M, Sezaki H (1990) Cancer Res 50:1640
48. Maeda H, Matsumoto T, Konno T (1984) J Protein Chem 3:181
49. Plendl J, Snyman C, Naidoo S, Sawant S, Mahabeer R, Bhoola KD (2000) Biol Chem 11:1103
50. Strausser HR, Humes JL (1975) Int J Cancer 15:724
51. Trevisani A, Ferretti E, Capuzzo A, Tomasi V (1980) Br J Cancer 41:341

52. Meyer RE, Shan S, Dan J, Dodge RK, Bonaventura J, Ong ET, Dewhirst MW (1995) Br J Cancer 71:1169
53. Tozer GM, Prise VE, Chaplin DJ (1997) Cancer Res 57:948
54. Gallo O, Masini E, Morbidelli L, Franchi A, Fini-Storchi I, Vergari WA, Ziche M (1998) J Natl Cancer Inst 90:587
55. Garcia-Cardena G, Folkman J (1998) J Natl Cancer Inst 90:560
56. Jackson JR, Seed MP, Kirchen CH, Willoughby DA, Wannan JD (1997) FASEB J 11:457
57. Okamoto T, Akaike T, Nagano T, Miyajima S, Suga M, Ando M, Ichimori K, Maeda H (1997) Arch Biochem Biophys 342:26
58. Maeda H, Aikawa S, Yamashita A (1975) Cancer Res 35:554
59. Oda T, Maeda H (1987) Cancer Res 47:3206
60. Duncan R, Gac-Breton S, Keane R, Musila R, Sat YN, Satchi R, Searle F (2001) J Control Release 74:135
61. Pellegrin P, Fernandez A, Lamb NJ, Bennes R (2002) Mol Biol Cell 13:570
62. Miyamoto Y, Oda T, Maeda H (1990) Cancer Res 50:1571
63. St'astny M, Strohalm J, Plocova D, Ulbrich K, Rihova B (1990) Eur J Cancer 35:459
64. Minko T, Kopeckova P, Pozharov V, Kopecek J (1998) J Control Release 54:223
65. Maeda H (1981) Anticancer Res 1:175
66. Maeda H (1997) In: Maeda H, Edo K, Ishida N (eds) Neocarzinostatin: The past, present, and future of an anticancer drug. Springer, Berlin Heidelberg New York, p 23
67. Ohtsuki K, Ono Y (1997) In: Maeda H, Edo K, Ishida N (eds) Neocarzinostatin: The past, present, and future of an anticancer drug. Springer, Berlin Heidelberg New York, p 129
68. Sato K, Akaike T, Suga M, Ando M, Maeda H (1994) Biochem Biophys Res Commun 205:1716
69. Maeda H (1991) Adv Drug Deliv Rev 6:181
70. Seymour LW, Olliff SP, Poole CJ, De Takats PG, Orme R, Ferry DR, Maeda H, Konno T, Kerr DJ (1998) Int J Oncol 12:1217
71. Abe S, Okubo Y, Ejiri Y, Kume K, Otsuki M (2000) J Gastroenterol 35:28
72. Okusaka T, Okada S, Ishii H, Ikeda M, Nakasuka H, Nagahama H, Iwata R, Furukawa H, Takayasu K, Nakanishi Y, Sakamoto M, Hirohashi S, Yoshimori M (1998) Oncology 55:276
73. Okusaka T, Okada S, Ueno H, Ikeda M, Iwata R, Furukawa H, Takayasu K, Moriyama N, Sato T, Sato K (2002) Oncology 62:228
74. Greenstein JP (1954) Biochemistry of cancer, 2nd edn. Academic, New York
75. Amanaka NY, Deamer D (1974) Physiol Chem Phys 6:95
76. Sato K, Ito K, Kohara H, Yamaguchi Y, Adachi K, Endo H (1992) Mol Cell Biol 12:2525
77. Maines MD, Abrahamsson PA (1996) Urology 47:727
78. Goodman AI, Choudhury M, da Silva JL, Schwartzman ML, Abraham NG (1997) Proc Soc Exp Biol Med 214:54
79. Doi K, Akaike T, Fujii S, Tanaka S, Ikebe N, Beppu T, Shibahara S, Ogawa M, Maeda H (1999) Br J Cancer 80:1954
80. Hasegawa Y, Takano T, Miyauchi A, Matsuzuka F, Yoshida H, Kuma K, Amino N (2002) Cancer Lett 182:69
81. Tanaka S, Akaike T, Fang J, Beppu T, Ogawa M, Tamura F, Miyamoto Y, Maeda H (2003) Br J Cancer 82:902
82. Schacter BA (1988) Semin Hematol 25:349
83. Maines MD (1988) FASEB J 2:2557
84. Minetti M, Mallozzi C, DiStasi AM, Pietraforte D (1998) Arch Biochem Biophys 352:165

Molecular-Scale Studies on Biopolymers Using Atomic Force Microscopy

James S. Ellis · Stephanie Allen · Ya Tsz A. Chim · Clive J. Roberts · Saul J. B. Tendler · Martyn C. Davies (✉)

Laboratory of Biophysics and Surface Analysis, School of Pharmacy, University of Nottingham, University Park, Nottingham NG7 2RD, UK
Martyn.Davies@Nottingham.ac.uk

1	Introduction	125
1.1	AFM Instrumentation	126
1.2	Modes of Operation of AFM	128
1.2.1	Contact Mode	128
1.2.2	Tapping Mode Atomic Force Microscopy	129
1.2.3	Phase Imaging	129
1.2.4	Cryogenic Atomic Force Microscopy	130
1.2.5	Liquid Imaging	130
1.2.6	Tip Geometry and Carbon Nanotubes	131
1.2.7	Nontopographical Applications of AFM	131
1.2.8	Force–Displacement Measurements	132
2	Advances in Single Molecule Biopolymer Investigations Using AFM	134
2.1	DNA	134
2.1.1	DNA Immobilisation	134
2.2	AFM Investigations of DNA Condensation	137
2.2.1	Visualisation of DNA Condensation by Polycations	137
2.3	Direct Visualisation of DNA–Protein Interactions	138
2.4	DNA–Drug Interactions	142
2.4.1	DNA Crosslinkers	142
2.4.2	Intercalators	143
2.4.3	Minor Groove Binders	144
2.5	Direct Imaging of DNA–Drug Interactions	145
2.5.1	Intercalators	145
3	AFM Studies of RNA	146
3.1	Observation of In Situ RNA Synthesis	147
3.2	RNA Tectonics	147
3.3	RNA Crystallization	149
3.4	Force Investigations of RNA	149
4	Polysaccharides by AFM	150
4.1	Cellulose	150
4.2	AFM of Starch Grains	150
4.3	Dextran	152
4.4	Polysaccharide–Enzyme Interactions	153

5	AFM of Proteins and Viral Capsids	153
5.1	Membrane Proteins	154
5.2	Viral Proteins	156
5.3	Imaging Proteins on Substrates	157
5.4	Observation of Amyloid Formation	160
5.5	Single-Molecular Force Interactions of Protein Molecules	162
5.6	SMFS Investigations of the Mechanical Properties of Proteins	162
5.7	The Application of SMFS to Ligand–Receptor Interactions	164
6	Conclusions	166
	References	166

Abstract The atomic force microscope (AFM) is capable of acquiring a range of structural and physicochemical information on a wide range of biopolymers. The ability to achieve nanoscale resolution, coupled with the potential to image and manipulate real-time molecular-scale events, suggests that AFM is one of the most promising techniques available for the study of biopolymers. AFM offers the potential to obtain a wide range of both quantitative and qualitative information on biopolymers, ranging from their conformations in physiological buffers to the forces involved in bond cleavage. This review explores the most common modes of AFM operation including imaging (contact and tapping mode) and force spectroscopy. The application of these modes to biopolymer characterisation will be discussed, with an emphasis on key studies.

Keywords AFM · Biopolymer · DNA · RNA · Proteins

Abbreviations

3-MPA	3-Mercaptopropanoic acid
11-MUA	11-Mercaptoundecanoic acid
AFM	Atomic force microscope or atomic force microscopy
APTES	3-Aminopropyltriethoxysilane (used to create AP-mica)
AP-mica	APTES-mica
Bp	Base pair(s)
BR	Bacteriarhodopsin
BSA	Bovine Serum Albumin
BSA-MPAD	BSA conjugated with a mercaptopropanoic acid derivative of atarazine
CBH I	An exoglucanase enzyme (a cellulase)
CMA	Carboxymethylamylose
CM-AFM	Contact mode AFM
CMC	Carboxymethylcellulose
CNT	Carbon nanotubes
Cryo-AFM	Cryogenic atomic force microscopy
csA	Glycoprotein contact site AS
DMSO	Dimethylsulfoxide
dsDNA	Double-stranded DNA
dsRNA	Double-stranded RNA
EDC	1-Ethyl-3-(3-dimethylaminopropyl)-carbodiimide
EDTA	Ethylenediaminetetraacetic acid
EGII	Endogluconase

EtBr	Ethidium bromide
FDC	Force–distance curve
HPI	Hexagonally packed intermediate layer
Ig	Immunoglobin
LT	Large tectosquares
MoMLV	Moloney murine leukaemia virus
MPADT	mercaptopropanoic acid derivative of atarazine
NCM-AFM	Noncontact mode AFM
NHS	N-hydroxysulfosuccinamide
OT	Optical tweezers
PAMAM	Poly(amidoamine) (a type of dendrimer)
pBR322	An *E. coli* DNA plasmid
PBS	Phosphate-buffered saline
PEGylated	Polyethylene-glycolated
PEI	Polyethylenimine
PID	Proportional-integral-differential
pLL	poly-(L-lysine)
pLL-AsOR	pLL-Asialoorosomucoid
pLL-g-pHPMA	Polyhydroypropylmethacrylamide-graft-pLL
POPC	Palmitoyl-oleoyl phosphatidylcholine
pRSVluc	A luciferase plasmid
PVAL	Poly(vinyl) alcohol
RH	Relative humidity
RNAP	RNA polymerase enzyme
SAM(s)	Self-assembled monolayers
SMFS	Single molecule force spectroscopy
SPM	Scanning probe microscope
ssRNA	Single-stranded RNA
STM	Scanning tunnelling microscope
Ti-O	Titanium oxide
TM-AFM	Tapping mode AFM
TMEV	Theiler's murine encephalomyelitis virus
tRNA	Transfer RNA
WLC	Worm-like coil
YO	Oxide yellow (a dye)
YO-PRO-1	An intercalating dye based on YO
YOYO	Dimerised YO

1
Introduction

Atomic force microscopy (AFM) enables the visualisation of biopolymers in their native hydrated state [1, 2] in liquid, in a partially dehydrated state in air, in a dehydrated state in an ultra-high vacuum, or in their frozen state with cryogenic AFM. Liquid imaging is often preferred where there is a desire to observe dynamic events, or acquire more biologically relevant data [1, 3–6]. The data obtained from AFM may also be more biologically relevant, as no staining [7] or use of heavy metal ions are required [8] (which are typically required for non-cryogenic electron microscopy). In addition, it is possible to use a wide range of buffers and ionic strengths with liquid AFM, enabling us to observe the effects of environmental factors such as pH changes and ionic strength on polymer conformation [3, 8–11]. In this respect, AFM offers a clear advantage over other high-resolution imaging techniques.

The original scanning probe technique, the scanning tunnelling microscope (STM), paved the way for the visualisation of polymers and surfaces to atomic resolution [12]. Whilst STM is still used today in biopolymer characterisation [13], its inability to characterise nonconductive surfaces in ambient conditions [14] is a hindrance. The subsequent development of the AFM [15] enabled the nanoscale visualisation of biopolymers [16] in both ambient [17] and liquid environments [3, 9, 11]. AFM thus facilitates the imaging of dynamic molecular events at interfaces [9, 18–20].

Since the invention of AFM, thousands of papers have been published demonstrating its application across a wide range of disciplines, from studying the surfaces of Martian meteorites [21] to biochemistry, where many natural processes have been studied, including enzyme action [9], disease processes [22], and DNA condensation [10, 23, 24]. In this latter area AFM has been used to investigate how natural phenomena such as DNA packing into viruses [25] can aid DNA condensation to improve DNA delivery in gene therapy [9, 10, 19, 26]. Besides imaging, AFM single molecule force measurements have been obtained to understand fundamental aspects of the structural, mechanical and binding properties of biopolymers. For example, the effects of DNA binders and intercalators on DNA [27] have been studied using atomic force microscopy.

1.1
AFM Instrumentation

The AFM (Fig. 1) has been described in many reviews [14, 28–30] as a relatively straightforward instrument comprising a probing tip and a detection system (laser and photo diode) used to monitor the position of the tip [31]. The tip is located on the apex of a flexible cantilever, and is commonly composed of silicon (Si). This tip–cantilever set-up is often described in mechanical terms as a ball (tip) on the end of a flexible spring (cantilever), which, in theory, is capable of measuring forces with a resolution down to the femtonewton range [16]. However, in reality, forces lower than around ten piconewtons are rarely detected. The AFM tip may hence be described as a nanoscopic force sensor [32], which has a terminal radius that ranges from 2–15 nm [1, 33] for sharp probes to 10–50 nm [34] for unsharpened probes (often used in force–distance work).

The cantilever's flexibility (spring constant) varies between tips, and is particularity important when considering force measurements. Force constant

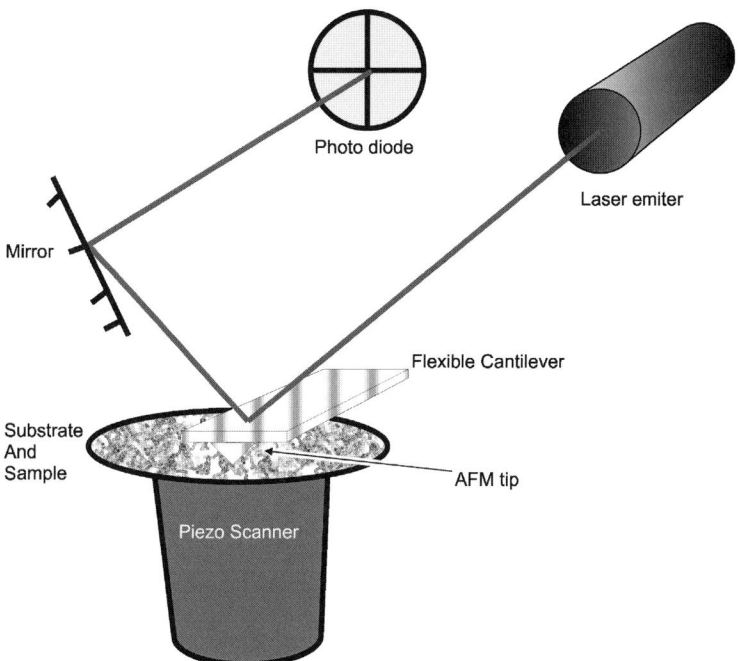

Fig. 1 The mechanism of detection in atomic force microscopy. The surface topography is tracked by the movement of a tip connected to a cantilever. The movement of the cantilever is then monitored by a laser beam, which is bounced off the cantilever surface and then reflected by a mirror into a slit photodiode detector

calibration can be achieved by various means [35], including methods that use the cantilever's free thermal fluctuation [36]. During operation, the beam from a small low-powered diode laser is directed onto the apex of the cantilever (above the tip), which in turn is focused towards a photodiode using a mirror. The laser position enables the position of the cantilever deflection relative to the surface to be determined. In addition, the bending and twisting of the cantilever can be measured using this approach.

The substrate of choice for AFM imaging is dependent upon the biopolymer to be studied. Ultra-flat muscovite or ruby mica are commonly employed for DNA, proteins and RNA [37, 38]. Other common substrates include gold, silicon and glass [39]. To achieve the scale of tip movement required to achieve nanometre resolution, a piezo-ceramic scanner is used. Being voltage-responsive, this piezo scanner is capable of altering dimensions by approximately 1 Å per volt (0.1 nm per volt), enabling accurate control of the tip position relative to the sample surface [32]. This actuator operates with a feedback loop that effectively controls the relative position of the tip. Plotting cantilever deflection as a function of the tip position on the sample surface produces an "image" of the sample (actually an "isoforce" map, which for homogenous samples can be interpreted as an "image" of the sample).

1.2
Modes of Operation of AFM

Different modes of operation of AFM may be applied to the characterisation of biopolymers. As biopolymers differ in their rigidity (Young's modulus), their responses to the normal and sheer forces applied by the AFM tip will also differ. As different modes of AFM are available, the user can forego resolution in order to minimise sample distortion (as in the case of tapping mode atomic force microscopy), or improve the resolution using noncontact or cryogenic AFM.

Soft biopolymers such as DNA and RNA are prone to tip-induced damage, often resulting from the relatively large capillary forces (when imaging in air) or friction forces experienced in contact mode. In this latter case, tapping mode atomic force microscopy (TM-AFM) may be used [40–42]. We will briefly discuss the most commonly employed modes of operation of AFM that are routinely used to obtain sample topography. The most common modes are contact mode AFM (CM-AFM) [2, 43] and TM-AFM [44, 45].

1.2.1
Contact Mode

The simplest of AFM modes, CM-AFM maintains the tip in direct and continuous contact with the sample surface; changes in topography are therefore evident if the cantilever deflection is plotted as a function of the tip movement. We can further subdivide CM-AFM into either constant height or constant force imaging, where the former maintains the cantilever at a constant height (z) relative to the sample, allowing for variation in the force between sample and tip. The latter (constant force microscopy) maintains consistency in the force between the tip and surface, and therefore varies the height (z) of the cantilever relative to the surface (whilst the tip remains in contact with the surface) to achieve this [46]. Such force control is possible by monitoring the deviation of the laser on the photodiode as the cantilever scans and bends; changes in force are translated into deflection of the cantilever. These changes are translated into a signal (error signal) signifying any change from constant force [46]. The information is sent to a computer, which uses a feedback loop termed the "proportional-integral-differential circuit" (PID). This PID causes a voltage to be applied to the piezo scanner to adjust the separation between tip and surface, ensuring that the contact force stays constant.

CM-AFM can produce resolution (~ 0.5 nm) sufficient to resolve submolecular structure [17] in both ambient and liquid conditions [39]. However, the formation of a thin water layer (capillary layer) on the sample surface can cause capillary forces (in ambient conditions) which may lead to both imaging artefacts and degradation of tip and sample [47]. These forces can be alleviated by imaging in liquid [48], or by selecting a cantilever whose effective spring constant is far less than that of the sample [49]. In general, the forces are minimised and kept between 50–100 pN [50], as biopolymer damage may occur if this value is exceeded.

Imaging biopolymers with CM-AFM is often problematic, as there is a tendency for the poorly immobilised molecules to be either swept away or damaged. In this instance, tapping mode AFM may be advantageous, as there is a reduced risk of damaging biopolymers, and sweeping poorly adsorbed polymers off the substrate. This comes, however with a modest reduction in ultimate resolution [50].

1.2.2
Tapping Mode Atomic Force Microscopy

TM-AFM (or "intermittent mode" atomic force microscopy) utilises a combination of both long-range attractive and short-range repulsive forces to monitor the sample topography [39, 51]. It is the most versatile method for analysing biopolymers [48], allowing "faithful" high-resolution imaging of biopolymers [52]. TM-AFM utilises a piezoelectric actuator (often found at the base of the cantilever, or on the tip holder) to oscillate the cantilever close to or at its resonant frequency, such that it overcomes the capillary forces when imaging in air. Here the cantilever is oscillated at typical amplitudes from 20–100 nm [11, 53], where a feedback loop maintains constant amplitude of oscillation. During this oscillation, the cantilever "taps" the sample, touching the surface at the end of its oscillation cycle [50, 54].

Contact with the surface lowers the amplitude of the cantilever compared to the free amplitude when the tip is not in contact [1]. Changes in amplitude are then translated into topography, and such a system is capable of measuring height changes down to the nearest 0.01 nm [1]. Whilst not true noncontact AFM, the reduced lateral force exerted on the sample enables minimal disruption of biopolymers [11]. However, compression forces are still significant, introducing the possibility of reduced height measurements [55].

1.2.3
Phase Imaging

When an AFM tip strikes the surface of the studied biopolymer, energy is transferred to the surface. As a consequence, the cantilever's phase of oscillation will lag behind the driving signal by an amount that depends on the level of interaction with the sample. Energy can be lost due to inelastic processes, capillary forces, and hydrophobic forces that arise during the tip–sample interaction [56, 57]; such energy loss is depicted by the phase lag, and is an extension of TM-AFM, which can be used to provide insights into the surface properties of biopolymers. The phase lag is able to resolve the size, shape and spacing of surface features, and as such is a useful tool in identifying variations in the surface composition [56]. As a result it is possible to obtain useful information when it is difficult to interpret the topography data. Fig. 3 depicts typical information obtained when imaging condensed DNA using AFM; here the left hand image depicts the topography of the sample, whereas the image on the right depicts the phase image.

1.2.4
Cryogenic Atomic Force Microscopy

In the search for routine submolecular resolution of biopolymers, cryo-AFM appears a real possibility. First described in 1991, cryo-AFM studies samples at very low temperatures, using liquid nitrogen vapour [58] to cool the sample. This increases the rigidity and mechanical strength of the biopolymers (increases the Young's modulus) [59] to as great as 1000 to 10 000 times the original hydrated strength [39]. This is believed to greatly reduce the sample deformation and damage induced by the tip [60], facilitating the application of sharper tips to the sample surface and enabling greater sample resolution. Whilst good resolution of certain biopolymers has been achieved (in the 3–5 nm range) [59] the lack of commercial availability of the machines and the high maintenance costs make other methods of microscopy more suitable. With further research, cryo-AFM may be as useful as other forms of AFM.

1.2.5
Liquid Imaging

There are many advantages to using liquid AFM to characterise biopolymers; these advantages include the elimination of capillary forces (responsible for tip-adhesion) and reductions in van der Waals forces [11, 51, 61, 62]. Liquid imaging also allows the degree of biopolymer immobilisation to be adjusted by controlling substrate-polymer interactions [3, 5, 63, 64].

A switch from air imaging to liquid imaging cannot, however, be made without taking the cantilever into account. Changes in both Stokes law (viscosity) and the effective mass of the cantilever will affect the resonant frequency of the cantilever [65]; as a result oscillation is dampened in liquid [51]. Taking tip resolution into account, the cantilever quality factor (Q) is related (Eq. 1) to the effective mass of the cantilever (m), the cantilever's resonant frequency (W_o) and the dampening factor of the cantilever (y) [66]. The value of Q can vary from as high as 10 000 in a vacuum to around 1 in liquid conditions, where a quality factor of 100 is expected in air [66]. Low-quality factors reduce the force sensitivity of the cantilever, and thus affect the resolution, whereas a quality factor that is too high leads to problems with long settle times and reduced imaging speeds.

$$Q = mW_o/y \quad [66]. \tag{1}$$

A method for improving the force sensitivity of cantilevers in liquid environments was proposed [66], where a positive feedback loop enhanced the quality factor of the cantilever by as much as three orders of magnitude. This feedback mechanism was applied to the study of single DNA molecules imaged on mica, where the sample was first dried under nitrogen, then imaged in liquid (butanol). In this instance, the quality factor using the active Q feedback loop was increased from 1 (as expected in liquid) to 1000. Parts of the DNA molecule were more easily defined with this enhanced quality factor. It was also suggested [66] that the height of the DNA molecule was more realistic using the enhanced quality factor than with other AFM methods, suggesting the images were a better representation of the molecule's state in vivo.

1.2.6
Tip Geometry and Carbon Nanotubes

Tip geometry is an important influence on sample resolution [67], which can often be improved by using extra-sharp tips, such as oxide-sharpened tips [68]. As the terminal radius of an AFM tip is usually larger, or of the same order, as the width of the biopolymer under investigation, the observed image will often be larger than the actual area due to tip–sample convolution. Artefacts may also occur due to the presence of an extraneous object protruding from the tip, which creates an image referred to as a "double tip" [69]. Silicon tips generally give superior resolution to silicon nitride tips, since they can be etched so as to develop an extremely sharp apex with a radius of less than 1 nm [70]. Silicon tips are often manufactured so that the crystal is orientated in the [1] direction [14], as this will leave only one single silicon atom at the apex, with three dangling bonds, as opposed to the [111] orientation, which has a wider apex. However, these tips are very fragile, and rapidly become blunt. Carbon nanotubes (CNT) may provide better resolution due to their very narrow dimensions [60, 67], with radii of 0.5 to 7 nm, and they are also more resistant to blunting. In this respect, CNTs are believed to be superior to standard AFM tips. However, no reliable commercial source currently exists for CNT AFM tips, so these must be made by the user, limiting widespread use.

1.2.7
Nontopographical Applications of AFM

In addition to classical topographical images, the AFM can provide a wealth of additional information on the biopolymer under investigation. For example, an individual biopolymer can be tethered between a tip and surface and stretched to obtain a force–distance curve (FDC) during force spectroscopy [71, 72] where a plethora of information on the biopolymer can be derived, including such properties as the Young's modulus [56].

1.2.8
Force–Displacement Measurements

AFM can provide a vast degree of localized mechanical information [73] by utilising probe–sample interactions to obtain a force curve, measured by noting the deflections of the cantilevers, that theoretically has femtonewton sensitivity [16]. However, ambient noise limits the level of detection. The deflection in the cantilever δ_c (measured by the laser on the photodiode) is determined and is used to generate a force–distance curve based on the principles of Hookes Law (shown in Eq. 2).

$$F = -k_c \delta_c . \tag{2}$$

Here F is the force, δ_c is the cantilever deflection in nanometres and k_c is the cantilever spring constant in Nm^{-1}. An FDC is produced by measuring the deflection of the cantilever as it moves toward and away from the sample. The force curve generated can be used to derive properties of the sample, such as elasticity, adhesion and electrostatic forces [72, 74].

Measurements can be conducted both in ambient conditions and liquid, although liquid is more commonly used to avoid capillary forces and allow for environmental control by changing buffers [32], enabling the effect of ionic strength to be determined [75–77]. Where naked tips are preferred for imaging, chemical modification may be required to increase the likelihood of interaction [78, 79] and polymer pick-up for force–distance work. "Chemical force microscopy" may be used to describe the application of a functionalised tip in AFM to map surface properties [80–82].

A typical FDC between an AFM tip and a surface is depicted in Fig. 2, where there are two distinct traces, the approach and retract traces [34, 72]. At the start of the FDC, the distance between tip and surface is so great that there is no interaction between them. As the approach begins, the distance between tip and surface reduces such that a point is reached in which attractive forces (such as van der Waals) cause the tip to instantaneously snap towards the surface. It is possible for long-range repulsive forces to dominate prior to the tip–surface contact, in which case the cantilever bends away from the surface before the snap-in [72].

The occurrence of the snap-in may be minimised by using buffer solutions that prevent excessive interactions [34, 83]. When the tip is in contact with the surface repulsive forces predominate [39, 72], causing the cantilever to bend away from the surface; this is described as the "contact region" of the force curve. Here many properties of the sample and the tip–sample interaction may be assessed, including the Young's modulus and the contour

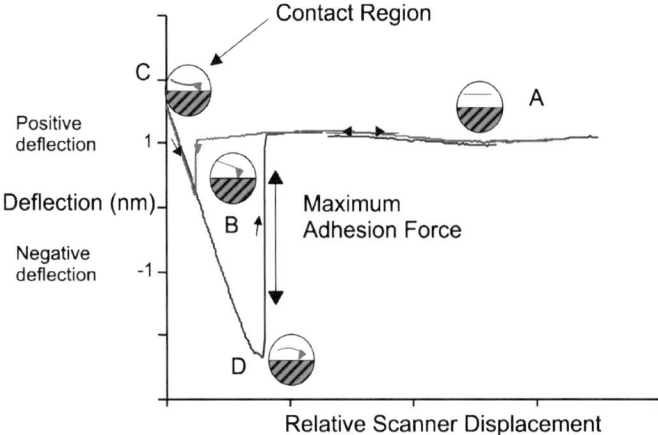

Fig. 2 Representation of a force–distance curve (FDC). At the beginning of the cycle (**A**) the probe is moved towards the surface (the approach trace, *red line*) until contact is made (**B**) and finally a predefined point of maximal load is reached (**C**). The tip is then removed from the surface (retraction trace, *blue line*). As the cantilever withdraws, it remains in contact with the surface until the maximum adhesion force is overcome (**D**). Reproduced with permission from [230]

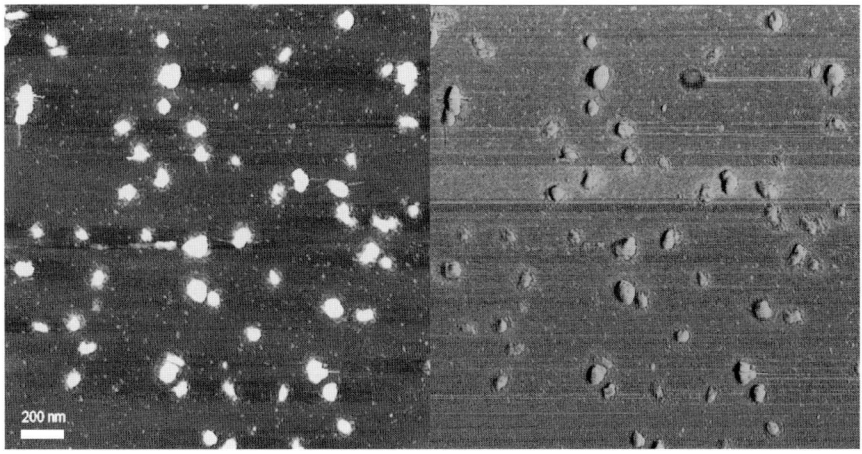

Fig. 3 Comparison between topography (*left*) and phase data (*right*), $Z = 10$ nm and $60°$ respectively. Scale bar = 200 nm. Figure depicts condensed DNA in the presence of 25 kDa polyethylenimine (PEI)

length [72]. When the pre-set maximum load of the cantilever is exceeded, the direction of motion of the cantilever is reversed, and the tip begins to retract from the surface [84]. The retraction curve may follow the approach curve; however, hysteresis often changes the curve's shape [34]. Hysteresis is often experienced in air due to the capillary forces arising from the fluid layer on the surface [34]; such forces may be such that they mask the fundamental force of interaction under investigation [84].

If pick-up of individual polymer molecules is achieved by the AFM probe, the retraction curve will depict this, by displaying a negative deflection (see Fig. 2) of the cantilever until the polymer becomes detached from the tip and/or the surface. Here specific events associated with biopolymer stretching and conformational changes can be interpreted based on the specific shape of the curve [34].

2
Advances in Single Molecule Biopolymer Investigations Using AFM

2.1
DNA

DNA is one of the most frequently studied biopolymers, not least because of its biological importance, but also because of the ease at which it can be synthesized and prepared for imaging [2, 17]. Early images of DNA were obtained in air using contact mode [17], soon after the ability to resolve DNA in liquid conditions was realized, enabling a more natural confirmation of DNA to be achieved [85]. The number of different applications of AFM to DNA are numerous in the literature. Here we discuss specific areas of interest, including the mechanisms of immobilization of DNA to mica, and the application of AFM to the study of gene therapy DNA systems. For gene therapy, DNA may be condensed with a cationic polymer and this is then used to deliver DNA to a cell in order to replace damaged or faulty parts of a genome [86]. AFM can be used to study the process of DNA condensation [10, 30] and to characterise different morphologies of condensed DNA [87]. The circumvention of DNA degradation in the endosome is an important task for gene therapy [88]; AFM has been used to investigate the process of DNA degradation with enzymes [9, 18, 89], therefore enabling the protection conferred by polymers to be predicted. In addition to gene therapy research, a lot of interest has been shown in the use of AFM to investigate both drug–DNA and protein–DNA interactions [30, 90]; these interactions are important pharmacologically and provide interesting studies in which AFM can be used to explore the physical and chemical processes that occur.

2.1.1
DNA Immobilisation

Muscovite mica is often the preferred substrate to image DNA, but the natural negative surface charge of mica needs to be overcome to enable immobilisation. There are two paths to immobilisation; firstly by treating the surface of the mica with cationic chemicals such as silanes [24, 91], and secondly by using both monovalent and divalent cations [5, 9, 28, 92] to attach DNA to mica. The mechanical properties of DNA differ according to the method of immobilisation [30]. The process of adsorption of DNA onto a mica surface has been described as a transition from three dimensions (solution) to two dimensions (surface), reducing the number of conformational possibilities of the molecule [93]. Adsorption follows two possible routes, (i) where DNA is free to change position (equilibrate) on the surface, and (ii) where DNA becomes kinetically trapped to the surface; the conformation it adopts on the surface represents its conformation upon adsorption. The latter case does not allow us to determine the effect of the substrate on DNA conformation [93].

Surface Pretreatment of Mica Using Silanes

The silane aminopropyltriethoxysilane (APTES) is commonly used to functionalise the mica surface with positively charged amino groups [24, 91], producing AP-mica [94]. Here, surface functionalisation enables effective reproducible imaging of DNA [95]; however, the persistence length of DNA may be lowered to approximately half the original value in doing so [56], and may initiate DNA condensation in some circumstances [23]. It is however, still possible to image dynamically with AP-mica [96]. In addition, as with divalent cations [97], it is possible to modify the number of adsorbed DNA molecules by adjusting the silane concentration, allowing for the visualisation of single DNA molecules [98].

Immobilisation by Divalent Cations

Divalent cations are frequently used to immobilise DNA to mica before imaging in air or liquid. Here, dilute DNA solutions containing between 1 and 10 mM of divalent ions are deposited onto mica, and either imaged immediately, or dried under nitrogen [3, 5, 92]. Both nickel and zinc ions are commonly employed where it is believed DNA is secured to mica by kinetic trapping processes [99] caused by the action of counter-ion correlations [100–102], coulombic repulsions [3, 5] and ion–base pair interactions [92]. It was originally proposed that both the ionic radius and the enthalpy of hydration of the immobilising ion affected the strength of binding of DNA to mica, where a greater binding force was achieved with the smaller ions such as Ni^{2+}, Co^{2+} and Zn^{2+}, with maximal deposition at concentrations of 1 mM [97]. Interestingly, the binding efficacies of both Mg^{2+} and Ca^{2+} were believed insufficient to mediate immobilisation [97, 99].

There is an increasing need for DNA to have a greater degree of conformational freedom on a surface. Thomson et al. [64] found that both pH and ionic strength influenced the degree of DNA immobilisation, allowing us to adjust the extent of immobilisation of DNA. Subsequent work utilised the competition between both monovalent and divalent ions to modify immobilisation [63], enabling the study of DNA-protein interactions at physiological ion concentrations.

Immobilisation of DNA in the Presence of Monovalent Ions

The first discussion of DNA immobilisation by monovalent ions was published in 1991 [103], where both monovalent and divalent ions were used to immobilise DNA to quartz sand. A recent publication [3] cited, for the first time, direct AFM evidence of monovalent ion immobilisation of DNA. Here, plasmid DNA was diluted in 10% PBS and incubated for either 10 min or 24 h at room temperature. After incubation, DNA was imaged on untreated, freshly cleaved mica using TM-AFM. The mobility of DNA could be observed with both immobilisation times. However, the immobilisation

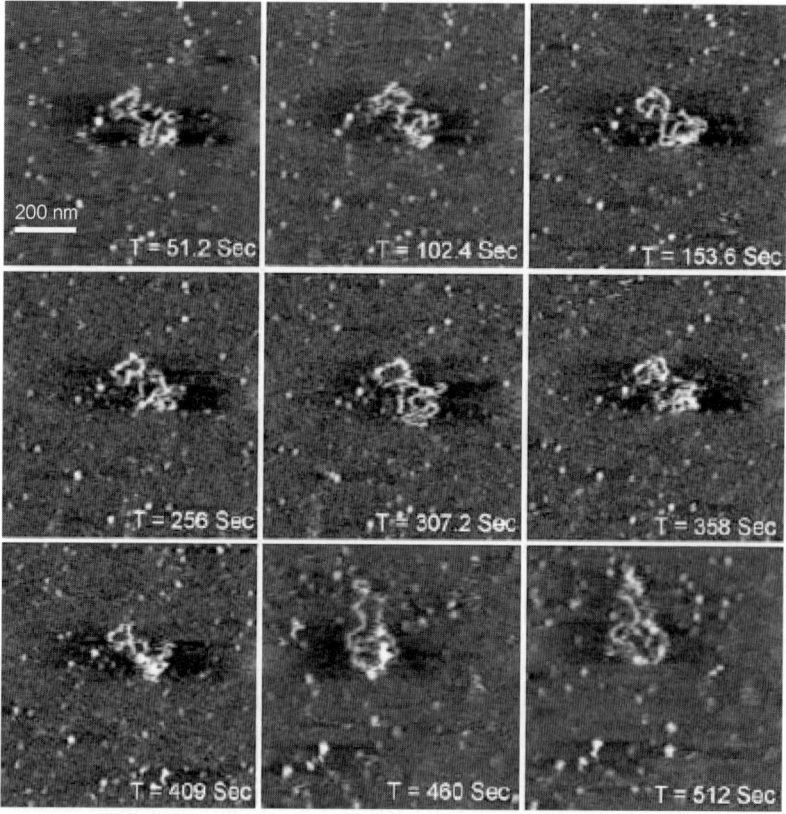

Fig. 4 DNA plasmid pBR322 mobile on mica after 10 minutes incubation with 10% PBS. Reproduced from [3] with permission from Blackwell Publishing

was greater after 24 h of incubation than after 10 min, suggesting a greater degree of immobilisation after 24 h (Fig. 4). However, this approach is less reliable in comparison to other immobilisation methods. At this stage, the use of divalent cations or pretreated ionised mica to image DNA is preferred.

2.2
AFM Investigations of DNA Condensation

2.2.1
Visualisation of DNA Condensation by Polycations

AFM has been used extensively in the study of DNA condensation and systems related to gene therapy [23, 24, 30, 104–106]. Both Hansma et al. [107] and Golan et al. [26] used TM-AFM to image various stages of DNA condensation with both poly(L-lysine) (pLL) and pLL-asialoorosomucoid (pLL-AsOR).

Here DNA was shown to collapse upon the addition of pLL, where small rods and toroids were observed as endproducts of condensation [26]. The presence of intermediary structures suggested a state of equilibrium between these rods and toroids. These data supported earlier findings suggesting that an equilibrium exists between rods and toroids [105]. The AFM data presented by Golan et al. [26] suggested incomplete DNA condensation with pLL compared with pLL-AsOR, which showed effective DNA condensation. In addition, the polymeric chain length had little effect on the DNA condensation with pLL, but had a significant effect where pLL-AsOR was concerned. The high charge density surrounding the AsOR moiety was believed to be responsible for the increased condensation, increasing the interactions between the anionic DNA and the cationic polymer [26, 30]. Time-lapsed AFM may also been used [19] to identify the pathways of DNA condensation. Fig. 5 depicts a series of images showing the formation of toroidal DNA.

AFM has also been used to investigate the condensation of DNA by the cationic polymer poly(ethylenimine) (PEI) [108, 109]. Kleeman et al. [110], for example, used AFM to investigate the effect of both air-jet and ultrasonic nebulisation on PEI-condensed DNA. Smaller nonspherical DNA condensates were observed with air-jet nebulisation in comparison to untreated control samples. In contrast, ultrasonic nebulisation appeared to have little effect on the morphology and size of condensed DNA [110]. An additional study [111] used AFM to investigate the morphology of PEI-condensed DNA both before and after freeze-drying, where freeze-drying was shown to have little effect on either the physical stability or the biological activity of the condensed DNA.

Ambient TM-AFM has also been used to investigate the effect of PEI : DNA ratios on particle morphology [105]. An increase in polymer : DNA ratio from

0.08 : 1 to 1.6 : 1 resulted in a change from partial condensation to complete condensation; here compact structures of 20–40 nm were observed. DNA characterisation was aided by the pretreatment of mica with poly-L-ornithine to create a uniform negative charge over the mica surface. In this study the ability of AFM to resolve single DNA strands within condensed DNA enabled the visualisation of looping and packing of DNA into the condensates [105].

In addition to PEI, a recent study by Wittmar et al. [87] investigated the use of amine-modified poly(vinyl alcohol) (amine-PVAL) as gene therapy agents using AFM amongst other techniques. Using TM-AFM, Wittmar et al. used TM-AFM in liquid to investigate the effects of various polymer properties (amine spacer length, the degree of amine substitution, and the polymer : DNA ratio) on the degree of DNA condensation. Both spacer length and polymer : DNA ratio were shown to have a significant effect on the size and morphology of condensates. In addition, distinct differences were seen between low (2%) and high (35%) amine substitutions, where fewer amine substitutions showed less compact condensation with very few immobilised molecules, and more amine substitutions produced more compact, occasionally aggregated condensates which demonstrated a greater surface coverage.

2.3
Direct Visualisation of DNA–Protein Interactions

A great deal of interest has been shown in the study of DNA–protein interactions [30]. As such, many reviews cite the application of AFM to these interactions [30, 104, 112, 113]. AFM has been used to provide structural information on DNA binding sites and the stoichiometry of proteins that bind to the DNA [62]. The regulation of bacterial transcription [104], and the mechanisms involved in its initialisation are of considerable interest [114, 115], as is the application of AFM in order to elucidate mechanisms of DNA–protein interactions that are involved in the repair [62] and replication [90, 116] of DNA.

In relation to DNA repair, van Noort et al. [62] used a home-made AFM, offering increased cantilever deflection sensitivity and a modified liquid cell, to investigate the action of the photolyase enzyme (obtained from *Anacystis nidulans*) on a restriction fragment of DNA. Photolysase repairs lesions in DNA and is specific to pyrimidine dimers resulting from UV damage [112]. AFM data suggested the photolyase enzyme showed mobility on loose (immobilised) sections of the otherwise immobilised DNA molecules; this enzyme mobility was limited to the sections of DNA that were not immobilised to mica.

A subsequent study by van Noort [117] used the *Nco*1 enzyme to digest DNA plasmid pET-XPB to produce both 1150 and 735 bp fragments. Using persistence length measurements, van Noort identified UV-damaged DNA fragments using AFM, where undamaged DNA molecules showed contour

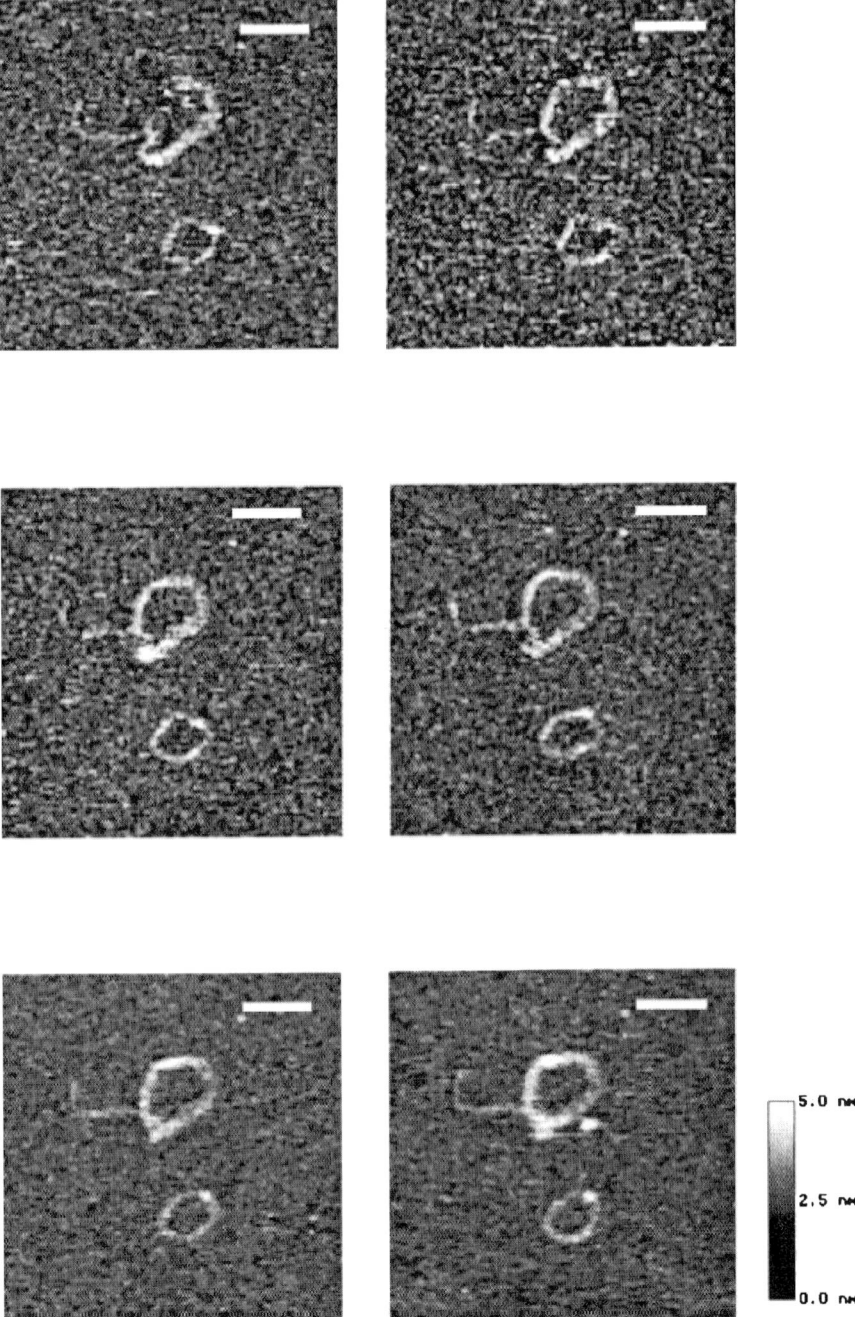

Fig. 5 Depicts a series of real-time images of the condensation of DNA plasmid pBR322 with cationic polymer. Each image depicts a 5 minute scan. Scale bars are equivalent to 200 nm. Reproduced with permission from [19]

lengths of < 300 nm, whereas damaged DNA molecules suggested greater contour lengths (> 300 nm). The photolyase enzyme adopted a globular structure, measuring some 3 nm in height, causing a $36 \pm 30°$ bend in DNA at the site of binding [117].

In addition to measuring protein-induced DNA bending [62, 117], AFM can be used to map protein binding sites on DNA, as demonstrated by Yokota et al. [118], where the interactions between the GAL4 protein and *Xba*I with linearised DNA (ratio of 100 : 1 protein : DNA) were monitored using a sample preparation technique which prepared samples with straightened DNA, facilitating the creation of a large-scale map of protein binding sites. Using this technique, protein molecules were mapped at the apexes of bends of DNA, in addition, the binding sites of both *Xba*I and GAL4 were determined to be at 23 and 35% of the DNA's total length.

RecA is an additional protein that is important for the recombination of two DNA strands. The interaction between DNA and RecA has been investigated using TM-AFM operating with CNT tips [119]. The morphology of the RecA–DNA complex was investigated at a 1 : 3 (protein : DNA) molecular ratio, suggesting that RecA was bound to DNA as a monomer, and not a hexamer as previously believed. However, as tapping mode in air was used, the mobility of the molecules was limited [4]. The formation of RecA–DNA complexes was studied over a timescale of one hour, suggesting an initial "lag phase" where nucleation was slow which was then followed by a faster growth phase.

An additional investigation [120] studied the interaction between the tumour suppressor protein (p53) and partially immobilised DNA. A series of time-lapsed images displayed the movement of p53 across the DNA, where it was observed that $\sim$ 25% of p53 stayed stationary, $\sim$ 25% moved less than 100 nm and $\sim$ 50% moved by as much as 400 nm. As a further control study, p53–DNA complexes were imaged in air using TM-AFM to ensure that they were capable of forming complexes. Unlike other protein-DNA experiments, the positions of the p53 molecules on DNA appeared to be random. This finding is compatible with the two known modes of binding of p53, namely direct and initial nonspecific binding with one-dimensional diffusion to a specific site on DNA [120].

DNA–enzyme interactions have also been investigated using AFM [9, 30, 121]. Here DNA–enzyme activity has particular relevance to gene therapy, where enzymatic degradation of DNA in the endosome is a major hindrance to successful transfection [88]. TM-AFM in air has been used to investigate the protection that chitosan confers to DNA for gene delivery [121]. Here the appearance of DNA fragments with naked DNA suggested no protection from enzymatic assault, but when chitosan was present, little evidence of DNA degradation was seen [121].

The use of time-lapse AFM to image enzyme action on DNA was explored by Bezanilla [18], where the process of DNA degradation in the presence of DNaseI was observed. The rate at which images were recorded (one every

30–60 seconds) suggests that key stages in the metabolism of DNA were not observed. Subsequent work showed the endonuclease enzyme EcoKI cleaving DNA following the addition of ATP to the DNA-EcoKI complex [89].

An early attempt to determine the rate of degradation of DNA was described by Hansma [30], where a small prototype cantilever capable of capturing one image every two seconds was employed. Here, the rates of degradation of DNA were observed with two DNase concentrations. Interestingly, a ten-fold difference in the rate of degradation was observed. A kinetic study of the degradation process using a specialised imaging programme (Scion Image Software) suggested a reaction rate of 16 particles per second (the number of additional particles observed on the mica surface).

Finally, building on the previous success of both Ellis et al. [89] and Bezanilla et al. [18], Abdelhady et al. [9] used AFM to depict a series of time-lapse images (Fig. 6) of the degradation of DNA–polymer complexes by DNaseI. Here DNA was exposed to the enzyme either in its condensed state (condensed by poly(amidoamine) (PAMAM) dendrimers) or its naked state (in the absence of polymer). The protection conferred to DNA by the PAMAM

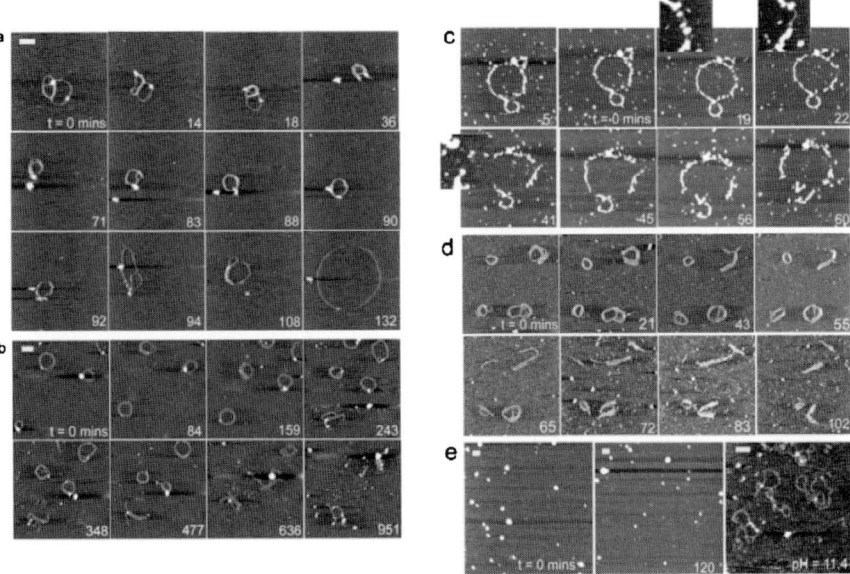

Fig. 6 A series of images depicting the degradation of a 1 : 1 ratio of G4 PAMAM dendrimers: DNA where the dendrimers were incubated with DNA for 15 minutes (**a**) and 2 hours (**b**) prior to deposition onto mica. Scale bars show 70 and 100 nm for **a** and **b** respectively, $Z = 3$ and 5 nm respectively. Figures **c,d,e** depict G4-DNA complexes at ratios of 0.5, 1 and 5 : 1, respectively. No degradation is observed in **e**, so the pH was raised to 11.4 to collapse the complexes. Reproduced from [9] with permission from Oxford University Press

dendrimers was investigated, where the polymer : DNA ratios were seen to affect the breakdown of DNA. A lower ratio suggested the formation of "open-nicked DNA" which then degrades into smaller dinucleotides. In addition, it was possible to observe a globular feature, believed to be an individual DNase enzyme, interact with the DNA complex, resulting in the degradation and opening-out of complexed DNA.

2.4
DNA–Drug Interactions

Many medicines and chemicals can interact with DNA by a series of mechanisms, including intercalation, bis-intercalation, and minor and major groove binding [8, 27, 122–126]. Using these interactions, AFM, and in particular single molecule force spectroscopy (SMFS), has been used to detect changes in DNA contour length in both the presence and absence of DNA-active drugs [127]. SMFS can provide a wide range of information on the mechanical properties of a polymer. Apart from AFM [123–129], other techniques may also be used to investigate the effects of force on DNA, including optical tweezers, magnetic beads and glass needles [75]. Although the force associated with these techniques differs, for AFM SMFS, the forces are typically in the range of 10 to 1000 pN [75].

Many studies [122–126, 130] have used SMFS to investigate the elastic properties of DNA; often to study the binding properties of small molecules. As DNA yields a unique fingerprint when stretched [124], SMFS can reveal specific force-extension characteristics dependent on the concentration and type of molecule bound to it [123]. There are several mathematical models that describe the force–extension characteristics of DNA when stretched, including the worm-like chain (WLC) model [75] which can be used to describe DNA stretching up to a force of 50 pN [131]. The WLC model cannot be applied after this force [123], so an alternate model must be used. At higher forces (between 65 and 70 pN), a plateau is usually seen in the force curve, depicting the overstretching of B-DNA [125, 126], where it may stretch by up to 1.8 times its original contour length.

Higher forces are believed to disrupt the double helix, breaking hydrogen bonds and leading to the loss of DNA base stacking interactions [131]. When forces in the range of 150 pN are encountered, strand separation (melting) occurs [126]. Relaxation enables the recombination of single DNA strands, enabling the reformation of the DNA double helix.

2.4.1
DNA Crosslinkers

Cisplatin is an example of a crosslinker commonly used for the combined treatment of neuroendocrine tumours [132]. It exerts its pharmacological action by crosslinking two guanine bases at their N7 positions [126]. Krautbauer et al. [126] used SMFS to show how cisplatin affects the shape of the FDC using linear fragments of λ-phage DNA stretched between a gold tip and a gold substrate. Upon the introduction of cisplatin, the FDC was found to be significantly different to that of untreated DNA, where, even with forces as high as 318 ± 22 pN, no B-S transition was observed. In addition, the relaxed curve generally followed the extension curve, even up to forces as high as 500 pN. The effect of incubation time (DNA with cisplatin) on the FDC fingerprint was also discussed, where the DNA molecule appeared shorter after 1 hour of incubation, with a similar FDC to that of untreated DNA. Here the presence of the crosslinker stabilised the DNA double-helix and prevented strand separation (melting), so no melting transition was observed. A more pronounced effect was observed after 24 hours incubation with DNA, where a decreased overstretching force was observed at the start of the plateau. These data agree with solution experiments that show decreased melting temperatures of DNA with cisplatin.

2.4.2
Intercalators

The phenanthridine dye ethidium bromide (EtBr), whilst not a drug, is commonly used in biochemistry as a UV marker for gel retardation assays [9]. A single EtBr molecule is known to insert between base pairs of DNA, causing both an increase in the base pair rise and an unwinding of the double helix by $26°$ [133]. Krautbauer et al. [123] demonstrated the effect of EtBr on DNA using the same SMFS set-up as for the cisplatin experiments. It was noted that only a very small ratio (one molecule of EtBr to ten DNA) was sufficient to significantly change the FDC, where the B-S transition with 1 : 10 EtBr : DNA base pairs occurred at a reduced force, showing a steeper gradient during the B-S transition. A high force melting transition and hysteresis were observed with the 1 : 10 ratio. The 1 : 2 ratio showed no hysteresis, with no apparent B-S plateau. The curve shows a gradual, almost exponential increase in force. At this high ratio it was suggested that EtBr prevents the separation of the DNA strands; in addition, EtBr induces the unwinding of the double helix at high concentrations, so it reduces the likelihood of a B-S transition and increases the contour length of DNA.

Proflavine is another example of an intercalating dye, which has its uses in the treatment of warts. Its mode of action is similar to EtBr, where it inserts itself between two base pairs, although proflavine shows little sequence selectivity. Once incorporated into the DNA double helix, proflavine induces unwinding by approximately $11°$ per bound molecule [124]. Single molecule studies [124] suggest a shortening of the B-S plateau with increasing proflavine concentration, where ratios of 1 : 10, 1 : 4 and 1 : 1 (proflavine molecules : base pairs) are concerned. As with the EtBr study by Krautbauer [123], the melting hysteresis was dramatically reduced with increasing proflavine concentration. In addition, the extension curve differs very little from the relaxation curve.

A follow-up study by Eckel et al. [130] used a synthetic dsDNA molecule (724 bp) containing poly(dG-dC) to investigate the effects of different chemicals on DNA's FDC. Here DNA was immobilised onto a gold surface overnight at ambient conditions, and then a typical AFM set-up was used to investigate the effects of the intercalators, daunomycin, EtBr, oxazole yellow (YO) and dimerised YO (YOYO). YO and YOYO both bind to the major groove and intercalate DNA. As predicted, the DNA showed considerable resistance to pulling in the presence of intercalators, with no evident plateau suggesting a B-S transition. The FDC from the bis-intercalators YO and YOYO differed slightly in terms of their overstretching behaviour [130]. It appears that intercalators stabilise the DNA helix, giving some resistance to melting and preventing strand breakage. This is reflected in the various FDCs by virtue of the lack of a B-S transition, and similar extend–retract curves.

2.4.3
Minor Groove Binders

Many minor groove binders are secreted by viruses or fungi and they interact selectively with the minor groove of DNA [134], requiring only slight adaptations in the conformation of the double helix of DNA [122]. Distamycin A is an example of a protein-based antibacterial and antiviral minor groove binder which is secreted by the actinomycete *Streptomyces distallicus*. Distamycin A exerts its action by reversibly binding to the minor groove of DNA by a mixture of hydrogen bonds, van der Waals forces and electrostatic interactions, demonstrating a strong preference for DNA sequences rich in adenine and thiamine [134]. Using the same synthetic dsDNA, Eckel et al. [130] used SMFS to study the effect of distamycin A on the FDC of DNA, where little difference was observed between DNA in the presence of and in the absence of this groove binder.

Another minor groove binder, netropsin, was investigated by Krautbauer et al. [124] with a ratio of one netropsin molecule per 0.4 base pairs of λ-phage DNA (which contains a random sequence of bases). Operating in a similar way to distamycin A, each end of netropsin forms three hydrogen bonds with DNA, stabilising the structure. As observed by Eckel et al. [130], initially the FDCs for DNA both with and without netropsin appeared similar, but netropsin appeared to increase the force (75–90 pN) required to overstretch DNA. Other groove binders (Berenil and Hoechst 33 258) were also investigated, again showing an increase in the B-S transition from $\sim$ 65 pN to 70–90 pN. However, as Berenil behaves as both a groove binder and an intercalator, data analysis has hidden complexities [123].

2.5
Direct Imaging of DNA–Drug Interactions

2.5.1
Intercalators

An initial study by Pope et al. [8] demonstrated the effect of the addition of EtBr (8 : 1 EtBr : base pair stoichiometry) to previously immobilised DNA. Ambient TM-AFM data depicts changes in the morphology of DNA, where DNA appears to change from a relaxed open loop to a more tightly wound plasmid with supercoiled (plectonomic) loops. Similar events were observed in liquid 12 minutes after the addition of EtBr; again DNA is seen to become progressively more supercoiled until plectonomic morphologies are observed.

A subsequent study [27] used both AP-mica and a magnesium-containing buffer to immobilise DNA to mica. Following DNA immobilisation a range of EtBr concentrations (1 : 50 to 8 : 1 EtBr to DNA bp) were used to investigate the effect of EtBr on DNA morphology. At low (1 : 50 to 1 : 10) EtBr : DNA ratios intercalation seemed to have minimal effects on DNA, but when the EtBr concentration was increased to 1 : 5 and 1 : 2.5, toroidal DNA was evident [27]. As the concentration of EtBr increased, so too did the degree of supercoiling until a complete plectonomic-like structure was formed again.

A later study [135] describes the structural changes pBR322 underwent following the addition of EtBr, where the DNA molecule was first relaxed by a topoisomerase I enzyme prior to the addition of EtBr. Following the addition of EtBr, some DNA molecules presented a rod-like morphology and were described as having condensed plectonomic forms [135]. The degree of compaction was believed to increase with increased EtBr concentration. The three studies (by Pope et al. [8, 27] and Utsuno et al. [135]) have provided a new perspective on the ability of AFM to observe the impact of drug molecules on the macromolecule conformation of DNA.

An interesting study by Kaji et al. [136] measured the changes in persistence length (stiffness of DNA), contour length and end-to-end distance of DNA in the presence of the intercalating dye YO-PRO-1. Assuming that the 500 and 1000 bp DNA fit the worm like-chain model, Kaji et al. plotted the persistence length versus YO-PRO-1 concentration [136]. YO-PRO1 appeared to increase the persistence length from 33.0 nm (no YO-PRO1) to as high as 80.7 nm (1000 bp DNA with a relatively low YO-PRO1 concentration). Where the effect of EtBr increased with concentration, this was not the case for YO-PRO1. The persistence length was at its greatest (for both 500 bp and 1000 bp DNA molecules) at medium concentrations. With these data in mind, Kaji et al. suggested that the intercalation of one molecule causes a deformation of three other sites, rather like the reverse of allosterism. They went on to suggest that over half of all binding sites will be deformed at a ratio of 0.2, and as a result the intercalators will not be able to bind [136]. In addition, it was suggested [136] that YO-PRO1 changes the persistence length by two mechanisms, namely electric effects and nonelectric effects. The electric effects (charge-related) decrease the persistence length (make DNA less flexible), whereas the nonelectric effects counteract these to increase the persistence length.

3
AFM Studies of RNA

RNA is an important molecule that is involved in almost all key processes of cellular metabolism [137]. Despite such importance, comparatively few studies have been conducted on it using AFM. For this reason, RNA has been described as the "Cinderella" of the nucleic acids [39]. The lability of RNA [94, 138, 139] is believed to account for the apparent lack of studies, since it makes reliable imaging and immobilisation difficult [39]. In addition, the natural tertiary and secondary conformations of RNA required for biological applications are highly dependent on environmental conditions [94, 140]. For example, ionic strength was shown to have a dramatic effect on the stability of dsRNA from *Penicillium chrysogenum* [141]. An early method for immobilising RNA to mica using silanes was described by Lyubchenko et al. [142]. The AFM images obtained in air depict rather convoluted packed morphologies, but the resolution was believed to be on par with electron microscopy [142].

Interesting observations have also been made by Shao and Zhang [31], who used cryo-AFM to investigate the effect of osmotic stress on the influenza virus, where the virus was first adsorbed onto mica, where excess particles were removed. The remaining surface particles were then immersed in deionised water, producing osmotic shock, which caused the capsid envelope to burst and release RNA. Here the rupture pore was also clearly visible, allowing the internal structure to be analysed.

Further advances were made by Kienberger et al. [143], who used a magnetically oscillating cantilever to visualise the release of RNA from a human rhinovirus (HRV2). RNA release was promoted by the rapid reduction of pH to 4.1; the pH was then increased to pH 7.6 after two hours. These RNA molecules were observed both connected to HRV2 and on mica, where the height of RNA varied between 1 and 1.5 nm. Average diameters were in the region of ~ 10 nm, suggesting tip-induced broadening. The presence of RNA was confirmed by the addition of RNaseA, which was injected into the liquid cell, and images obtained both before and after RNA degradation. Whilst RNAs are commonly found to have interesting three-dimensional structures, the molecules observed by Kienberger et al. were either straight or bent, with no secondary structure. Fork-like RNA structures were classified as RNA molecules that had fully dissociated themselves from the viral capsid. However, it was believed that complete genomes was observed, as the measured contour lengths were an order of magnitude lower than the predicted value based on 7100 nucleotides.

One issue that came from the work of Kienberger et al. was that the lack of use of RNAse inhibitor will allow partial degradation of RNA by endogenous enzymes or contaminants. It was claimed that where RNA immobilisation on mica was dense, the tight packing inhibited the action of the RNAse, so little degradation was observed [143].

3.1
Observation of In Situ RNA Synthesis

Kasas et al. [144] visualised the synthesis of RNA using a 1 : 1 molar ratio of DNA to RNA polymerase (RNAP) with DNA templates, including a "rolling circle plasmid" which is capable of producing transcripts as large as 9000 nucleotides (2.7 µm) [145]. Here, RNA molecules were observed on the surface of mica (where the sample was dried down as these molecules were not visible in liquid); in addition enzyme molecules were observed attached to transcribed RNA. Rates of RNA synthesis were determined based on the length of the transcribed RNA and the time given to transcribe, where rates between 0.4 and 1 bases per second were suggested. A time-lapsed (liquid) investigation was also conducted [144], using 373 and 1047 bp ds-DNA templates that were originally added to mica in a low salt Ni^{2+} buffer. Here, both DNA and RNAP were observed on the surface of mica, and once mobility of DNA was observed, a transcription buffer containing four nucleotides was added. The DNA was then seen to dissociate from three RNAP molecules.

An additional study by Hansma et al. [146] used Ni-mica (1 mM nickel(II) chloride) to immobilise Φ X174 DNA and a purine-rich plasmid (5'-GATTCCTTTCTTCTTTCCTTC) (the DNA templates) onto mica, where again RNAP was employed to synthesise RNA in situ. Here both ribozyme (Φ X174) and purine-rich RNA (transcript product) were observed, for which three main differences were observed. Finger-like projections (believed to be formed by non-Watson-Crick base pairing) were seen to protrude from purine-rich RNAs, but not from ribozymal RNA. However, ribozymal RNA possessed flatter regions (0.3 nm), presumably due to the stacking energies of this RNA. In addition, uniformly thick ribosomal RNAs were occasionally observed; this was not the case for purine-rich RNAs.

3.2
RNA Tectonics

Tectonics describes the ability to break down large molecules of RNA into smaller subunits with additive behaviour and reassemble them [147]. Here, three-dimensional structures can be built from nanoscopic building blocks to create "jigsaw puzzle" pieces termed tectosquares [147]. Here we briefly discuss the application of AFM to the science of RNA tectonics.

Hansma et al. [139] described the first use of AFM to characterise two forms of RNA self-assembly, namely dimerisation of two RNA molecules by the formation of a "kissing loop" and the formation of a supramolecular fibre from tectoRNA. The kissing loop RNA was prepared from a 230 nucleotide Moloney murine leukaemia virus (MoMLV), using polymer chain reactions to generate the desired sequence.

An additional fascinating study recently published by Chworos et al. [147] described AFM evidence for the formation of programmable jigsaw puzzles using tectoRNA [147]. Here two sets of tectoRNA were used for the assembly of both large and small tectosquares, with side lengths typically measuring 10 nm and 13 nm with hairpin stems of 9 and 15 base pairs respectively [147]. Large tectosquares (LT) were analysed in both air and liquid using TM-AFM. Here it was suggested that the LT folds into a stiff-sided square morphology

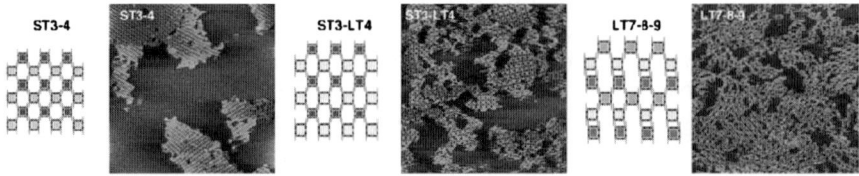

Fig. 7 Various AFM images of different nanopatterns, including fishnet (ST3-4), striped velvet (ST3-LT4) and basket weave LT7-8-9 patterns. Reprinted with permission from [147]. Copyright (2004) AAAS

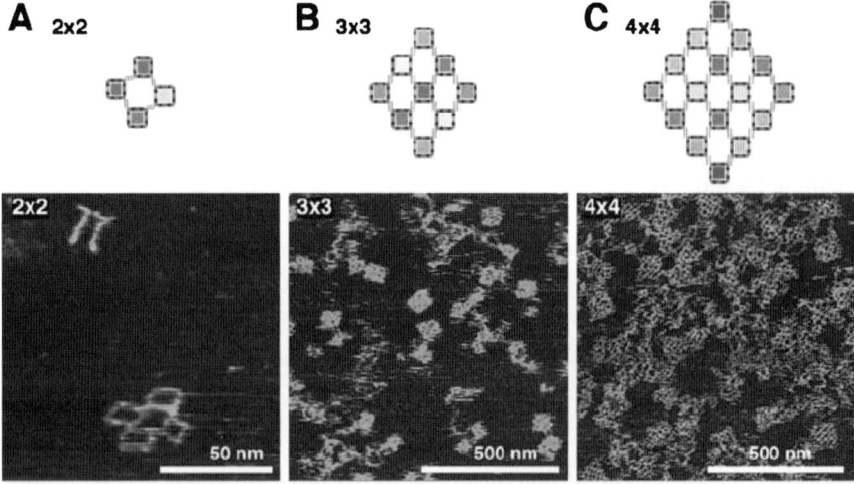

Fig. 8 Various different nanogrids made from RNA tectosquares. Reprinted with permission from [147]. Copyright (2004) AAAS

with sides measuring some 13 nm and corner angles of 70 and 110°. The width of the helical region was determined to be 3 ± 1 nm based on width at half height measurements as described by Golan et al. [26]. The patterns were constructed from a total of 22 tectosquares, which were originally synthesised from 49 tectoRNAs. Patterns were formed in 15 mM magnesium acetate by cooling the tectoRNA slowly from 50 °C to 4 °C, and monitored using liquid AFM (Fig. 7). In addition to these patterns, tectoRNAs were designed to assemble into nanogrids, where 2 by 2, 3 by 3 and 4 by 4 grids were assembled by controlling the directionality of the 3' tail (see Fig. 8). In some cases, edges were formed by removing the 3' end at specified corners. Here AFM was used to characterise a 4 by 4 grid showing how 27 different tectoRNAs can be reproducibly arranged and controlled to form nanoscopic jigsaw puzzles [147].

3.3
RNA Crystallization

Many aspects of RNA crystallisation remain unclear [39], and very few AFM studies have been conducted on RNA crystallisation. The first in situ (AFM) study of RNA crystallisation was presented by Ng et al. [148], where phenylalanine transfer RNA (tRNA) was investigated on glass substrates using a vapour diffusion method. The glass substrate was transferred into a sealed liquid cell with a thermoelectric cooler, where supersaturation conditions were controlled by modulating the temperature. It was possible to observe the effects of temperature on crystal habit; in the temperature range of 16–11 °C, a transition from one growth mechanism to another was observed at a single locus. The shape of the crystal became increasingly isotropic, and two-dimensional nuclei appeared. At 12 °C, three-dimensional nuclei began forming on the surface, appearing as large stacks of steps. This effect of crystal growth was completely reversible; a rise in temperature above 16 °C promoted crystal dissolution. The crystal formation and subsequent dissolution could be repeated consistently.

3.4
Force Investigations of RNA

In common with DNA, force investigations can reveal important information on RNA properties, as demonstrated by the stretching of both dsRNA and ssRNA in comparison to dsDNA [140]. Here both DNA and RNA were stretched between a silicon nitride tip and a mica surface. First, identical sequences of dsDNA and dsRNA were compared. DNA molecules adsorbed to a tip were extended for approximately 200 nm and they showed the typical B-S transition at approximately 65 pN. The force curve was similar to those described for DNA earlier. After applying forces in excess of 150 pN, DNA melting occurred with subsequent detachment of DNA at approximately 350 pN. In comparison, RNA showed a plateau force of approximately 60 pN which measured 120 nm in length with no observed melting transition. RNA detachment occurred at a much lower force of approximately 200 pN. After more than 300 force measurements with RNA, the plateau force was found to be in good agreement with the predicted value of 65 pN [149]. Here, a greater stacking force between RNA nucleotides was suggested as the cause of the difference between DNA and RNA. In addition it was suggested that RNA hybrids may account for the greater melting temperatures.

4
Polysaccharides by AFM

Polysaccharides are abundant throughout nature with numerous applications in both the pharmaceutical and food industries, and so they have been the subject of considerable AFM investigation [150–153]. AFM investigations of polysaccharides are important not only for our understanding of food sources (as reviewed by Morris et al. [154]) but also for our understanding of the mechanical properties of polysaccharides and how these affect their host cells.

4.1
Cellulose

High-resolution images of cellulose I microcrystals [155] demonstrated the application of CM-AFM to the study of large biopolymers. Here, the 0.52 nm repeat along the chains of the cellulose from *Valonia ventricosa* (a dark-green balloon-like marine alga) was observed. AFM has also been used to study the polysaccharides present in wheat straw cell walls [156], where clear differences can be seen both before and after de-waxing. Cellulose appeared to form microfibrils which were orientated in one direction. These fibrils measured some 20 nm in diameter and are believed to contain as many as 60–80 cellulose molecular chains.

4.2
AFM of Starch Grains

An early study by Baldwin et al. [157] highlighted the application of AFM to the characterisation of starch granules. This time-dependent study using TM-AFM in combination with low voltage scanning electron microscopy investigated both wheat and potato starch. High-resolution images were obtained, enabling the determination of topographical differences between starches of different species of plant. The surface of potato starch appeared much rougher than that of pea starch, which was believed to be due to the presence of "blocklet structures", as previously described [158].

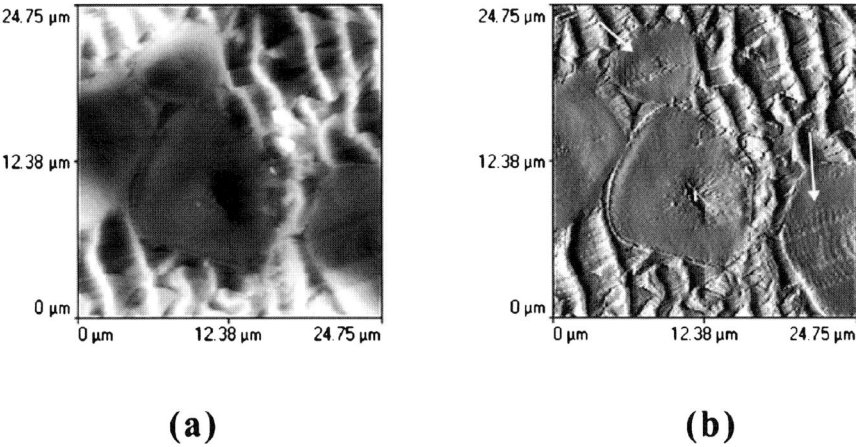

Fig. 9 A depiction of the topography (**a**) and the error signal (**b**) for a starch granule embedded in resin. The arrow points to damage scars caused by cutting. Reproduced from [159], with permission from Elsevier. Copyright (2002) Elsevier

A subsequent investigation by Ridout et al. [159] investigated the structure of starch granules using CM-AFM in both air (Fig. 9) and liquid. Topographical images were presented showing the typical morphology of starch, where the helium (air pocket in starch) is observable as a black hole. No growth rings are observable with these starch granules, which have previously been observed using AFM [160]. Higher magnification was utilised, and objects measuring some 50–80 nm could be observed, displaying globular morphologies. These structures were observed on both potato and maize starch and deemed to be of similar sizes. An early study [161] used AFM to investigate the in situ degradation of starch granules by alpha-amylase enzymes, where the creation of a "pin hole" in the centre of the granule was observed. Subsequent studies by Ridout et al. [160, 162] used similar techniques to investigate the surface characteristics of pea starch imaged in air using CM-AFM, where growth rings were present, allowing the spacing between the rings to be measured. High-resolution images again suggested the appearance of "blockets", adding further evidence to the "blocket model" of starch structure proposed by Gallant et al. [158]. Latterly, a study by Ridout et al. [162] investigated the effect of hydration on the structure of starch using CM-AFM. A change in the surface topography suggested swelling of the granules, which was reversible after three months of drying.

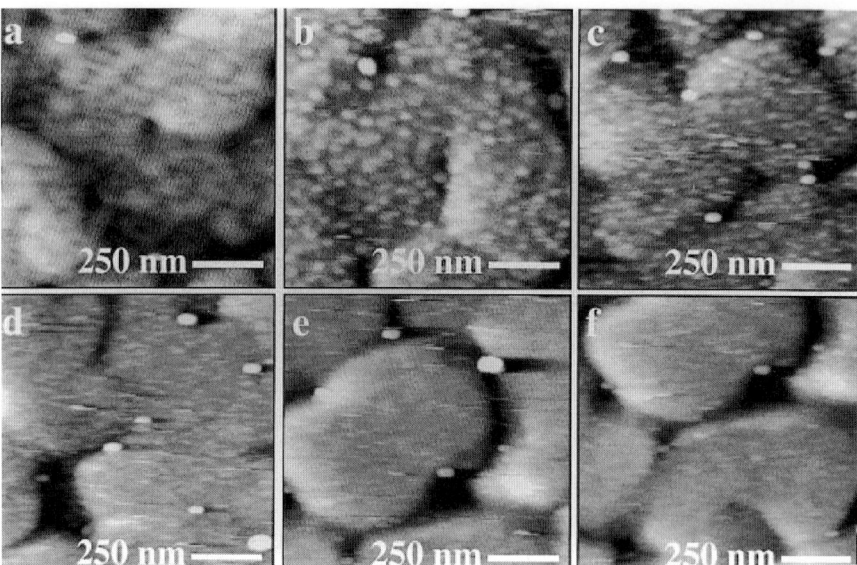

Fig. 10 A series of images depicting thiolated dextran on a gold surface before (**a**) and after (**b–f**) the addition of water. Initial surface changes, showing spheroidal particles, are seen after 5 minutes incubation in water (**b**). These swell on the surface until 70 minutes (**d**), when a continuous film is formed. Reproduced with permission from [164]. Copyright (1997) American Chemical Society

4.3
Dextran

Dextran is a branched (α-D-glucan) polysaccharide that is often used as a blood-plasma expander [163]. AFM has been applied to the study of dextran [163–165], and has been used to investigate the effect of polymer molecular weight [163] and hydration [164] on dextran polymer morphology. It was suggested [164] that the degree of hydration would affect the particle morphology, as intramolecularly bound water provides structural support, where it forms a bridge between two OH^- groups in the polymer. Fully hydrated dextran molecules are believed to adopt a spherical conformation, whereas dehydrated molecules display an ellipsoidal morphology [163]. Subsequent investigations by Frazier et al. [164], immobilised a 70 kDa thiolated dextran polymer to a gold substrate (Fig. 10), where differences in morphology were observed after the addition of water. The initial air image depicts a dense coverage of globular structures, believed to be single molecules; after the addition of water (5 min) spherical morphologies were observed, and as time progresses these particles are observed to swell and coalesce into a continuous film. After one hour a continuous layer of dextran (no individual spheroids) was observed, following the contours of the gold islands on the substrate. FDCs were used to study the tip–surface deformation, suggesting a decrease in the elastic modulus of dextran as time increased (after addition of water). This is one of the first examples of a direct observation of the collapse of a steric barrier. The subsequent addition of propanol caused dehydration of the dextran, resulting in a FDC that is similar to that in air and a decrease in stiffness. In addition, the AFM images show the reversal of hydration after the addition of propanol to reform the globular structures. Subsequent investigation by Frazier et al. [165] suggested the benefits of modifying dextran with thiol, where surface coverage increases with increased degree of thiol substitution; this appeared irrespective of substrate (silver or gold). Conversely, an increase in molecular weight was seen to produce a decrease in the surface coverage of dextran, which was believed to be due to an increase in conformational freedom and steric hindrance, resulting in fewer binding sites being available for the thiol groups.

4.4
Polysaccharide–Enzyme Interactions

Polysaccharide–enzyme interactions have not been extensively studied by AFM. However Lee et al. [166] have used AFM to study the effect of two different cellulases from the fungus *Trichoderma reesei*, which produce an exogluanase (CBH I) and an endogluconase (EGII), on cotton fibres. Here, TM-AFM was used under ultrapure nitrogen (< 5% RH), where native CBH I molecules were imaged on a chemically treated gold substrate. Additional images were obtained of inactivated (hexachloropalladate) CBH I molecules on cellulose, where it is believed to be capable of binding to cellulose fibres. Size analysis suggested that native CBH I had a width of 6.5 nm, while the inactivated enzyme appeared to have a larger width of 13.1 nm; both enzymes appeared as spherical particles.

The appearance of pores (13.1 nm), in the cotton fibres (after incubation with inactivated CBH I) was believed to be due to the insertion of the enzymes' tails into the cellulose fibres at the binding sites. The subsequent addition of active enzyme (CBH I) produced slight changes in the structures of the fibres, where high-resolution images suggest surface indentations that were not evident in the control studies. It was suggested that CBH I disrupts the cotton surface by "tracking" along the fibres. It was also suggested that the mechanisms of action of EGII and CBH I differ, although they appeared to act in synergism when used together.

5
AFM of Proteins and Viral Capsids

Many classes of individual proteins have been successfully imaged by AFM, sometimes with submolecular resolution [167]. Resolution is typically greatest when the protein is imaged in two-dimensional arrays [104]. As with other biopolymers, the application of AFM to protein imaging and pulling has been the subject of many reviews [6, 28, 56, 104, 168]. Depending on the protein under study, immobilisation strategies may differ compared to those used to immobilise nucleic acids. The methodologies used to immobilise proteins differ depending upon the nature of the protein, the pH and ionic conditions in which the study is conducted, and more importantly they are affected by the chemistry of the protein. Here we will discuss some of the key AFM studies that have investigated proteins.

5.1
Membrane Proteins

The purple membrane was one of the very first membranes to be characterised by AFM [169]; here trimers of bacteriorhodopsin (BR) were observed on membranes of *Halobacterium halobium*. BR serves as a light-driven proton (H^+) pump. It contains eight typtophan residues and is composed of seven helices [170]. Both the membrane and BR have been the subject of a great deal of research [83, 171–177] into the detailed structure of the membrane, which consists of approximately 75% BR and 25% lipids [178]. Worcester et al. [179] presented a study of the purple membrane, using platinum/palladium (80/20) wire cantilevers to study the membrane in air at a controlled relative humidity (55–75%). This study enabled the observation of unidirectional parallel rows, spaced approximately 5 nm apart from each other. It was suggested that they directly represented the surface structure of the purple membrane.

Following the adsorption of the purple membrane onto mica, Müller et al. [175] introduced a BR-specific antibody (which recognises the c-terminus of BR) to label the purple membrane. Here AFM studies suggested that some membranes remained unlabelled (no change in topography or roughness), whereas the roughness of other membranes was seen to increase due to antibody attachment. The subsequent removal of the c-terminus (using Papain enzyme) of BR and the subsequent addition of antibody resulted in the observation of a smooth membrane, suggesting no interaction between membrane and antibody.

A later study by Müller et al. [173] revealed important structural information on the purple membrane, where trimeric structure of the BR was observed (showing a side length of 6.2 nm). Each subunit boasted protrusions extending some 0.83 nm above the membrane surface; these were believed to be a loop that connects the α-helices E and F on BR [173]. This loop was believed to be quite flexible, as it was able to sustain forces of up to 200 pN, returning to its original state after the reduction in force [173, 176]. The application of imaging forces greater than 200 pN forced this loop into a conformational change of approximately 0.2 nm, which Müller et al. suggested corresponded to an energy difference of 4 kJ mol^{-1}.

Persike et al. [180] published an interesting study on the effect of illumination on the purple membrane using *Halobacterium salinarum* membrane. It was suggested that proteins on the extracellular side appeared more compact than those on the cytoplasmic side, possibly due to the proteins' preference for the extracellular side to be on the mica surface.

As Müller et al. [173] suggested, the morphology of BR depends on the force at which it is imaged. Persike et al. used an average force of 100 pN to image BR, and the E-F loop was observed, which was compressed after forces of 1000 pN were applied [173]. Further interesting observations were made when BR was imaged in liquid in both the presence and absence of (white) light. Here a force of 200 pN was applied and maintained on the sample (light was observed to decrease the imaging force due to the effect of temperature on the bending of the cantilever). It was possible to observe clear structural changes after the sample had been irradiated. Their averaged data suggested that changes in the crystal lattice occurred and that the protein trimers appeared to be more condensed. This process is believed to be completely reversible [173].

In addition to natural membranes, artificial membranes have been constructed for the purpose of monitoring protein movement [181]; one such membrane was constructed from palmitoyl-oleoyl phosphatidylcholine (POPC), into which ion-driven rotors of *Ileobacter tartaricus* ATP synthase were incorporated. AFM showed the formation of assemblies of proteins, while some proteins failed to make contact with each other. Random movements of single proteins were observed, although with diffusion coefficients that were lower than predicted. Müller et al. [181] gave two possible explanations for this. First, the mobility coefficient through an adsorbed (supported) membrane (to mica) is an order of a magnitude lower than through an unsupported membrane. Secondly, when adsorbed onto mica, a layer of water produces a 1 nm separation distance between the mica and the membrane, which increases the friction on the protein whilst maintaining a degree of mobility.

The application of TM-AFM to the study of protein and membrane surfaces was discussed by Möller et al. [54]; TM-AFM affords the ability to study loosely immobilised membranes. In their study, Möller et al. used both the hexagonally packed intermediate layer (HPI) from *Deinococcis radiodurans* and the purple membrane from *Halobacterium salinarum*. TM-AFM clearly identified the hexamer arrangement of single proteins found on the HPI layer. When the dimensions were compared with CM-AFM (from [182]), good agreement was found with TM-AFM.

5.2
Viral Proteins

A range of substrate preparation methods have been employed to study viral proteins. Poly-L-lysine, for example, has been used to prevent viral particles form being dislodged from the surface [39]; in addition, three different antibodies (anti-A5 anti-ELd and anti-E2d) have be used to immobilise Semliki Forest virus particles to a silanised silicon wafer [183]. Dimension and height measurements suggested collapse of the viral particle onto silicon. Interestingly, the presence of an envelope on the viral capsid did not stop collapse, although the collapse was less significant where the envelope was present.

An alternate immobilisation method was suggested by Kuznetsov et al. [184] where 0.05% glutaraldehyde was used to fix the cells and virons to the substrate. This was followed by subsequent fixation with osmium tetrox-

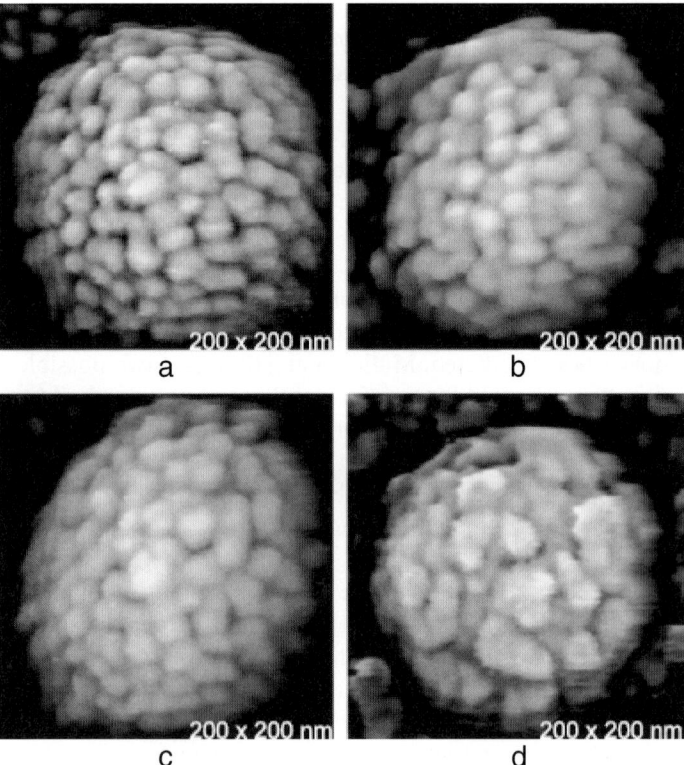

Fig. 11 Images **a–c** depict wild-type MoMLV exhibiting protein bumps on the surface of the capsid; **d** depicts a capsid with an unnaturally low density of proteins on the surface. Approximately 100 protein bumps per virus can be observed, extending the girth of the virus by 6–10 nm. Reprinted from [184] with permission from Elsevier. Copyright (2004) Elsevier

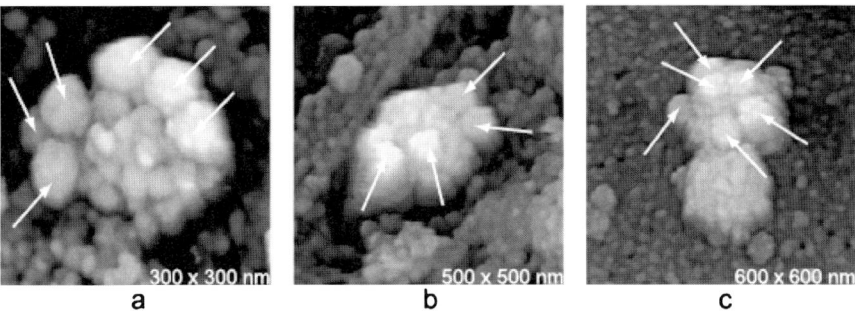

Fig. 12 Two MoMLV virons emerging from its host cell. The positions of the antibodies are highlighted by the arrows (**a**–**c**). **c** depicts two virons adjacent to each other; one has been labelled by antibody whilst the other has not. Reprinted from [184] with permission from Elsevier. Copyright (2004) Elsevier

ide, dehydration, and then imaging under ethanol (Fig. 11). High-resolution AFM images showed a wild-type Moloney murine leukaemia virus (MoMLV) emerging from an infected 43-D cell (Fig. 12). Here two virons were observed escaping from a host cell. Soon after this was observed, monoclonal antibodies against SU proteins were administered; soon after administration, 30 nm gold-clustered secondary antibodies were added and were observed attached to the primary antibody. Figure 12a–b depicts tagged virus particles leaving the cell that have been marked by the antigens, and also depicts two virons in close proximity, one of which is marked, while the other, which has the characteristic protein blobs or tufts, remains unmarked by the antibody.

A similar study investigated both the ingress and release of Theiler's murine encephalomyelitis virus from baby hamster kidney cells [185]. Here, it was concluded that there was no difference in the effect on the cell surface between both enveloped and nonenveloped cells upon both entrance and departure from the cell. Although line analysis suggests a difference between the cells, it is hypothesised that this could be due to different mechanisms for enveloped and nonenveloped cells.

5.3
Imaging Proteins on Substrates

AFM enables the observation of surface-immobilised proteins with a lateral resolution of 0.2 [50] to 0.6 nm [186] (depending on the apex of the tip amongst other factors) and a vertical (z) resolution of 0.1 nm. Whilst it has been claimed that high resolution of proteins is possible irrespective of their arrangement [186], Hansma et al. [104] suggest that resolution is highest when the protein is packed into two-dimensional arrays, as has been shown for ferritin [187]. Irrespective of resolution, the globular nature of many proteins is believed to make their identification more difficult, especially if this is to be based solely on their morphology [167]. An early investigation by Schneider et al. [167] suggested a correlation between protein molecular weight and its volume as measured by AFM, where various proteins (molecular weights 38 to 900 kDa) were imaged on mica. Optimal resolution was achieved not by simple fixation of proteins to mica, but by the use of a detergent-spreading method [188]. This technique was particularly useful when imaging proteins that possess few positively charged residues (such as the immunoglobins), which tended to immobilise very loosely to mica, and would therefore become loose during imaging without this method.

Proteins possessing a large number of positively charged residues require little pretreatment of mica to facilitate immobilisation [167, 189]. Avidin is an example of such a protein [189], and it has been investigated by Allen et al. [190], who presented an early study investigating the adsorption of streptavidin onto the surface of a polystyrene microtitre plate (as used in enzyme-linked immunosorbant assays). Here, streptavidin was imaged on the surface of a well, suggesting the formation of supramolecular networks on the polystyrene surface, with random pores ranging between 20 and 50 nm. The subsequent characterisation of a biotinylated antiferritin antibody on the plate revealed similar supramolecular networks. Globular features were also observed, measuring some 15 to 17 nm, which were believed to be individual IgG molecules [190]. After the same surface was exposed to ferritin, a continuous layer of spherical features was observed, this time measuring some 15 to 25 nm.

A subsequent study by Chen et al. [191] investigated the interactions between a protein (albumin)-coated tip and protein-coated polystyrene surfaces using FDCs. Here it was possible to recognise the interaction between the tip and the proteins adsorbed on the polystyrene surfaces, irrespective of how they were adsorbed. In addition it was possible to image the proteins on the surface of the polystyrene without any reduction in resolution.

An interesting immobilisation technique was also described by Li et al, where uniform distributions of immuno reagents were observed on gold immobilised with self-assembled monolayers (SAMs) of both carboxylic acid and hydroxyl-terminated thiols (Fig. 13) [85]. These SAMs were activated using N-hydroxysulfosuccinamide (NHS) prior to the addition of protein. This study suggested that less deposition of protein occurred when the SAM was composed solely of carboxylated groups, in comparison to a mixed system (of carboxylated and hydroxylated SAMs). Using a sample relocation device [192–194], Li et al. investigated the binding of antibodies to antigens using TM-AFM, enabling specific and nonspecific binding to be differentiated. A similar investigation [195] studied the immobilisation of the protein catalase on gold using two different SAMs based on either 3-mercaptopropanoic acid (3-MPA), 11-mercaptoundecanoic acid (11-MUA), or a mixture of the two. Both NHS and 1-ethyl-3-(3-dimethylaminopropyl)-carbodiimide (EDC) were also used to facilitate protein coupling to the SAMs, and hence promote covalent immobilisation. Differing degrees of adsorption of catalase were observed using AFM when the surface coverage was calcu-

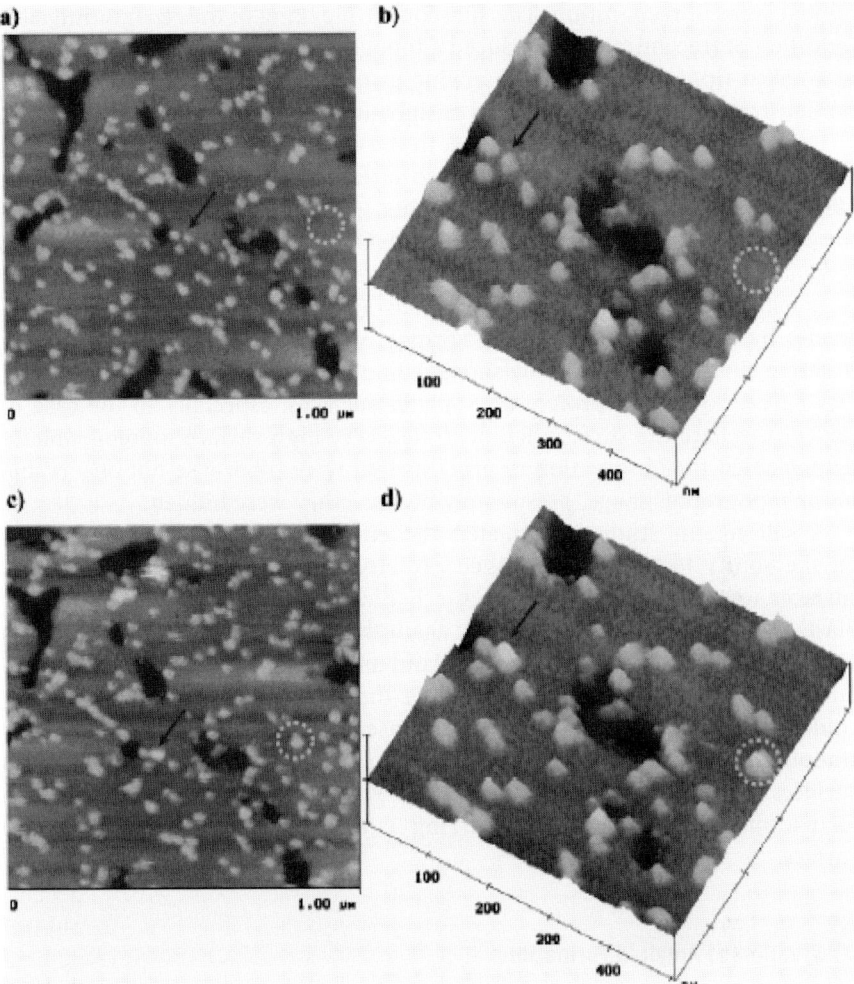

Fig. 13 TM-AFM images of immobilised anti-goat IgG on a mixed SAM before (**a** and **b**) and after (**c** and **d**) the addition of goat IgG. The *dashed circle* refers to an area of non-specific binding, whilst the *arrow* refers to an area of specific binding. Reproduced with permission from [85]. Copyright (2002) American Chemical Society

lated [196]. Here it was suggested that, with both physical adsorption and covalent immobilisation, the degree of deposition of catylase was greatest when both 3-MPA and 11-MUA were present on the surface, while the least deposition was observed with 3-MPA alone.

An investigation by Raab et al. [197] also used AFM to aid recognition of antibodies, but this time using a magnetically oscillated cantilever, tethered to an antibody. This cantilever was scanned across a lysozyme-bound surface, where changes in the amplitude (in relation to the cantilever) highlighted areas of recognition. The subsequent blocking of the recognition sites (addition of free antibody) produced a topography that was equivalent to that of a naked tip. In addition to observing proteins at the single molecule level on substrates in order to gain an insight into their gross macromolecular structures, there is a growing interest in the ability to image the distributions and adsorptions of proteins on modified surfaces [198] and biomaterials [165, 199–206]. AFM may provide key insights into protein–surface interactions for medical devices such as implants [200]. Unlike AFM substrates, these surfaces are often rough, making it difficult to observe individual proteins with reasonable clarity. Holland and Marchant (2000) [200] overcame such obstacles by exploring both the topography and phase data of fibrinogen on different surfaces (polydimethylsiloxane and polyethylene). As anticipated, it was harder to resolve protein on rougher surfaces, especially in the case of (ePTFE). Where the surface demonstrated very high surface topology, phase data enabled the observation of proteins that were otherwise not observable in the topography image, presumably because they were hidden.

A subsequent study investigated the effect of titanium (Ti) roughness on the ability to resolve fibrinogen proteins [201]. Here seven different Ti surfaces were used, ranging from mechanically polished (smooth) to sandblasted and etched surfaces (rough); again the plasma protein fibrinogen was used as a model protein. As with Holland and Marchant [200], protein resolution was affected by the roughness of the sample surface, where protein could only be visualised (with height, phase and amplitude data) on the two smoothest substrates. The dimensions of proteins adsorbed on the polished surfaces appeared larger than those on the ground surfaces, suggesting a conformational change, possibly occurring as a result of the roughness of the surface.

5.4
Observation of Amyloid Formation

The formation of fibrillar deposits (also termed amyloids) due to the misfolding and subsequent aggregation of polypeptides has been of recent inter-

Table 1 Typical diseases associated with amyloid fibrils. Adapted from [207]

Disease	Protein
Alzheimer's disease	$A\beta$ peptide
Spongiform encephalopathy	Prions
Type II diabetes	Fragment of islet-associated polypeptide
Haemodialysis-related amyloid angiopathy	$\beta 2$-microglobin
Atrial amyloidosis	Atrial natriuretic factor

est [207, 208]. Some diseases commonly associated with amyloid formation are listed in Table 1. The severity of many diseases associated with amyloid formation has prompted research into the nature of these fibrils under the AFM [22, 208–213].

The 39–43 amino acid residue β-amyloid associated with Alzheimer's disease was initially studied in air [209] but later in liquid [210] using TM-AFM. Here the formation of early fibrillar structures (nascent fibrils) from smaller oligomeric units (constructed from five spherical units) was observed, where the fibrils were seen to grow by the addition of further monomers or oligomers [209]. Such growth has also been observed directly and in situ [210], where bidirectional growth of premature fibrils (protofibrils) was identified.

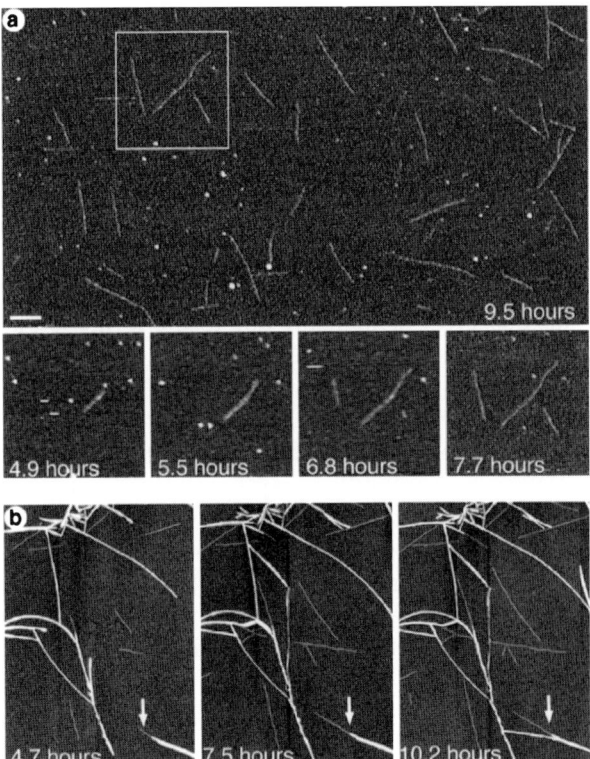

Fig. 14 Time-lapse images of amylin fibrils forming on a mica surface. Image **a** depicts the formation of protofibrils, showing evidence of bidirectional growth. The *boxed* section depicts the area in which the four previous images were taken (4.9 to 7.7 hours). Image **b** depicts another experiment showing the growth of a protofibril (*arrow*) from a higher order fibril. Scale bar 200 nm. Reprinted from [22] with permission from Elsevier. Copyright (1999) Elsevier

In addition to the studies conducted by Blackley et al, Wang et al. [213] utilised both TM-AFM (in situ) and STM (in vacuo) to study folded protein structure on freshly cleaved HOPG. Self-assembled sheets of the protein with a height of approximately 1 nm and narrow tubes were observed with ridges exhibiting a right-handed axial periodicity using AFM. It was concluded that fibrilisation followed two paths; one where the peptides aggregate into beads, and the other being where the proteins self-assemble into filaments which subsequently twist into fibrils. The data obtained in air suggested the appearance of protofibrils, intermediary between protein monomer and amyloid. This was taken as support for the theory that $A\beta$-amyloids are formed via the association of these protofibrils with monomeric units of $A\beta$-protein [213].

The application of real-time AFM to study the dynamic nature of fibril amyloid formation has been briefly reviewed by Stolz et al. [214]. An example of the application of time-lapse AFM to the study of amyloid fibrils was presented by Goldsbury et al. [22], where the 37 amino acid hormone amylin was observed undergoing fibrilization on a mica surface (Fig. 14). The TM-AFM data shows the formation of protofibrils, which then formed complete fibrils. It was established that fibrils form from each of their free ends, whilst growth rates were deemed to be similar at both ends. A similar investigation used TM-AFM in air to image a 20–29 amino acid believed to be a major constituent of the amyloid deposits formed in type II diabetes mellitus [215]. Here aliquots were removed and imaged every two hours, allowing for the observation of fibril formation. The interaction between amyloid proteins and lipid bilayers has also been investigated. Rupture of the lipid bilayer was observed, presumably due to the formation and spread of defects [211, 216].

5.5
Single-Molecular Force Interactions of Protein Molecules

The application of force spectroscopy to the study of proteins has been cited by many reviews [36, 72, 217–220] including Best et al. [36], which focuses on the application of AFM to the study of protein folding. It has been suggested that AFM can give new insights into protein folding and the effect of force on protein unfolding, and can provide information on folding energy landscapes.

5.6
SMFS Investigations of the Mechanical Properties of Proteins

The 3 MDa protein titin was the focus of many of the first SMFS investigations into protein unfolding [221]. Titin is composed of multiple-tandem repeats of Ig and fibronectin III-like domains [218]. Figure 15 depicts a typical force-distance curve obtained from stretching titin I27 using SMFS. Each individual

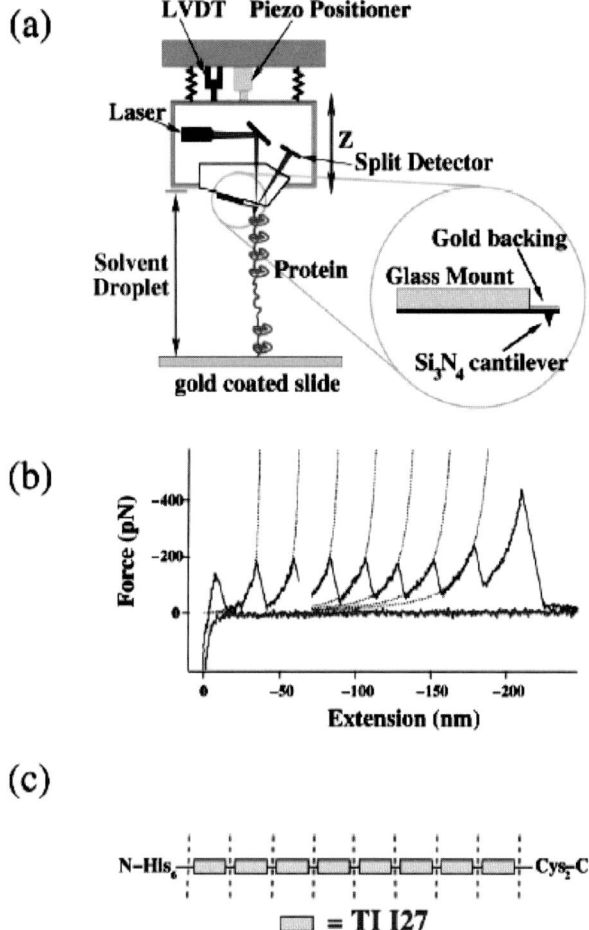

Fig. 15 The application of force mode AFM to the pulling of titin 127 (**c**). (**a**) The AFM set-up; (**b**) the force curve generated from titin, showing the unfolding of each different domain. Reproduced from [36] with permission from the Royal Society of Chemistry

peak (or spike [218]) in the FDC of titin has a characteristic curvature. Best et al. suggested that this is the soft elastic extension of the already unfolded titin molecule [36]. Further stretching results in a drop in the force, resulting from the unfolding of each domain, giving the characteristic "saw-tooth" effect [220]. The final peak observed in Fig. 15 shows the release of the protein molecule from the tip. In this example of titin stretching, only seven out of the eight domains (Fig. 15c) were observed using SMFS. It has been suggested that the first domain was involved in the binding of the protein to the tip, and so it can't be distinguished from the first peak, which usually accounts for nonspecific surface–tip interactions [36]. Lack of homogeneity between the

Ig domains within wild-type titins may produce variations in the FDC [220], so proteins were synthesised with the aim of producing more homogenous FDCs [222]. One such protein was investigated by Williams et al. [222] and compared with wild-type titin I27 using SMFS. Where the process of I27 unfolding was believed to occur via the formation of an intermediary structure (in which the A-strand becomes detached) at around 100 pN [223], Williams et al. further demonstrated that the force transitions of I27 were dependent on the loading rates used. This has also been described using both optical tweezers (OT) and AFM [224].

5.7
The Application of SMFS to Ligand–Receptor Interactions

The unusually high affinity of the receptor–ligand pair biotin–streptavidin has lead to its frequent use as a model system for the study of receptor–ligand interactions [72]. Such interactions have been described in detail in the reviews by both Zlatanova et al. [72] and Willemsen et al. [225]. We shall briefly discuss the application of SMFS to understanding protein–ligand interactions. Early applications of SMFS to streptavidin–biotin interactions were first published by Lee et al. [226]. Here AFM was used to investigate the interaction between biotin and streptavidin using both glass microspheres and mica

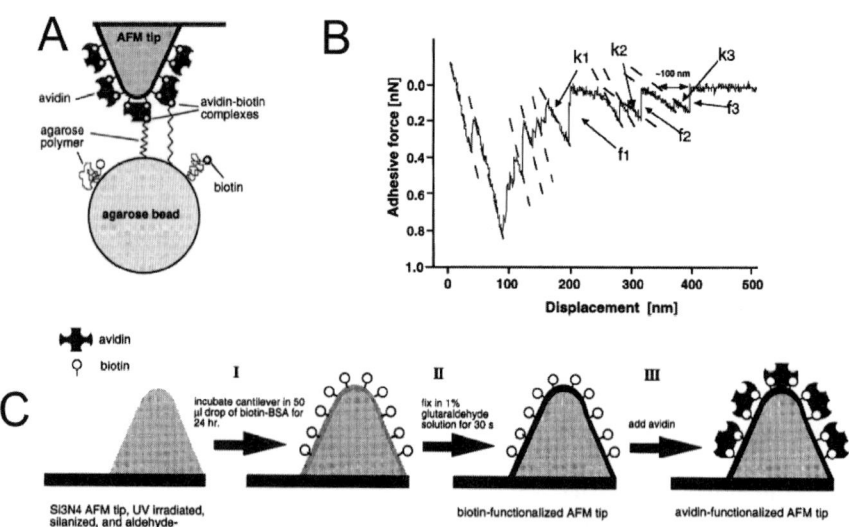

Fig. 16 The avidin–biotin interaction (**A**) and the process of tip coating (**C**). (**B**) shows a FDC for the breakage of a small number of avidin–biotin bonds. Here it is suggested that between one and six complexes are ruptured, where $f3$ refers to the force required to rupture the last complex. Reprinted from [227], with permission from Elsevier. Copyright (1999) Elsevier

surfaces, and thus it was possible to investigate individual streptavidin–biotin interactions. An additional study by Wong et al. [227] used both surface force apparatus and AFM to investigate the interaction between avidin and biotin (Fig. 16). It was suggested that the surface force apparatus permitted the determination of long-range attractive forces, whereas the AFM allowed short-range forces to be determined.

A more recent study by the Gaub group [228] used SMFS to investigate cell–cell mediated interactions using the eukaryotic cell *Dictyostelium discoideum*, where the cell glycoprotein contact site AS (csA) participates in cell aggregation. A tipless cantilever functionalised with lectin was used to pick up a cell from *Dictyostelium discoideum*. A target cell was positioned at the bottom of a dish, and was approached until a contact force was felt. It was possible to measure the desorption between the cells, which suggested that the unbinding force was approximately 1 nN. However, it was recognised that it was not possible to resolve the nature of these interactions (whether they were specific to csA or to another protein) [228]. As adhesion was known to be calcium-dependent, the chelating agent EDTA was used to remove calcium ions, which lead to a reduction in the number of detectable adhesion interactions. Here both calcium-dependent de-adhesion forces, and calcium-independent (EDTA had no effect) interactions were investigated. It was suggested that whilst the de-adhesive forces were within the same range (piconewtons), differences existed between the two reactions: calcium-dependent de-adhesion showed a broad range of forces, whereas calcium-independent force showed a peak at 23 pN. It is believed [228] that calcium-dependent adhesion utilises several molecules, each with different adhesion characteristics. On the other hand, EDTA-resistant adhesion may involve a single molecule, such as csA. This provides an interesting study of the application of AFM to living cells in order to identify important mechanisms of adhesion; however, due to the dynamic nature of living cells, it was not possible to identify single molecule interactions using this method.

Finally, Kaur et al. [229] discussed the use of SMFS to investigate the interaction between both antibodies and pesticides. Here a gold surface (gold-coated glass) was used to immobilise the antihapten antibody using protein A on the gold surface. AFM tips (silicon nitride) were cleaned in sodium hydroxide, and then in piranha solution, where they were functionalised with APTES and glutaraldehyde. The probes were then further functionalised with either BSA-MPAD (a mercaptopropanoic acid derivative of atarazine) or 2,4-conjugated BSA. The substrate was blocked using 2% casein, and force curves were recorded for each tip BSA-MPAD or 2,4-conjugated BSA. The method used by Kaur et al. ensured the Fab site of the antibody remained free to access the pesticides. Protein A has a high affinity for the Fc region of antibodies [229], ensuring that the Fab region was unaffected. Here it was possible to detect the antigen binding activities of two different pesticides and determine the adhesive forces involved in the interactions between the pesticide and antibody (240–940 and 80–400 pN for MPAD and 2,4-conjugated BSA respectively).

6
Conclusions

AFM is an extremely versatile area of microscopy, with numerous applications to the study of biopolymers. Here we discussed the application of AFM to the study of DNA, proteins, RNA and polysaccharides, where both SMFS and imaging have been used to further our understanding of the properties and interactions of these biopolymers. Recent publications have proved that AFM offers the potential to obtain a wide range of both quantitative and qualitative information on biopolymers. The future of AFM should see some further new and exciting information gather on both old and new biopolymers. As major improvements and development of the instrument is currently underway, and with the advances in computer technology (allowing an increase in the rate of data collection), the future applications of AFM to biopolymer research are likely to grow.

Acknowledgements The authors would like to thank Omar Wahab and Cătălin Fotea (University of Nottingham) for their scientific input. JE and YTAC would like to thank the BBSRC for their studentships.

References

1. Rippe K, Mucke N, Langowski J (1997) Bioforum Int 1:42
2. Hansma PK, Elings VB, Marti O, Bracker CE (1988) Science 242:209
3. Ellis JS, Abdelhady HG, Allen S, Davies MC, Roberts CJ, Tendler SJB, Williams PM (2004) J Microsc 215:297
4. Rivetti C, Guthold M, Bustamante C (1996) J Mol Biol 264:919
5. Pastre D, Pietrement O, Fusil P, Landousy F, Jeusset J, David MO, Hamon C, Le Cam E, Zozime A (2003) Biophys J 85:2507
6. Hansma HG, Hoh JH (1994) Annu Rev Biophys Biomol Struct 23:115
7. Lyubchenko YL, Shlyakhtenko LS, Harrington RE, Oden PI, Lindsay SM (1993) Proc Natl Acad Sci USA 90:2137
8. Pope LH, Davies MC, Laughton CA, Roberts CJ, Tendler SJB, Williams PM (1999) Anal Chim Acta 400:27
9. Abdelhady HG, Allen S, Davies MC, Roberts CJ, Tendler SJB, Williams PM (2003) Nucleic Acids Res 31:4001
10. Martin AL, Davies MC, Rackstraw BJ, Roberts CJ, Stolnik S, Tendler SJB, Williams PM (2000) Fed Eur Biochem Soc Lett 480:106
11. Putman CAJ, Van der Werf KO, De Grooth BG, Van Hulst NF, Greve J (1994) Appl Phys Lett 64:2454
12. Binnig G, Rohrer H (1982) Helv Phys Acta 55:726

13. Yang GZ, Ling LS, Wan LJ, Fang XH, Bai CL (2004) J Nanosci Nanotechnol 4:561
14. Giessibl FJ (2003) Rev Modern Phys 75:949
15. Binnig G, Quate CF, Gerber C (1986) Phys Rev Lett 56:930
16. Grange W, Strunz T, Schumakovitch I, Guntherodt H-J, Hegner M (2001) Single Mol 2:75
17. Bustamante C, Vesenka J, Tang CL, Rees W, Guthold M, Keller R (1992) Biochemistry 31:22
18. Bezanilla M, Drake B, Nudler E, Kashlev M, Hansma PK, Hansma HG (1994) Biophys J 67:2454
19. Martin A, Stolnik S, Davies MC, Laughton CA, Roberts CJ, Tendler SJB, Williams PM (2000) Fed Eur Biochem Soc Lett 480:106
20. Jiao Y, Cherny DI, Heim G, Jovin TM, Schaffer TE (2001) J Mol Biol 314
21. Steele A, Goddard D, Beech IB, Tapper RC, Stapleton D, Smith JR (1997) J Microsc 189:2
22. Goldsbury C, Kistler J, Aebi U, Arvinte T, Cooper GJS (1999) J Mol Biol 285:33
23. Fang Y, Hoh JH (1998) Nucleic Acids Res 26:588
24. Fang Y, Hoh J (1999) Fed Eur Biochem Soc Lett 459:173
25. He SQ, Arscott PG, Bloomfield VA (2000) Biopolymers 53:329
26. Golan R, Pietrasanta LI, Hsieh W, Hansma HG (1999) Biochemistry 38:14069
27. Pope LH, Davies MC, Laughton CA, Roberts CJ, Tendler SJB, Williams PM (2000) J Microsc-Oxford 199:68
28. Hansma HG (1996) J Vac Sci Technol B 14:1390
29. Czajkowsky DM, Iwamoto H, Shao ZF (2000) J Electron Microsc 49:395
30. Hansma HG (2001) Annu Rev Phys Chem 52:71
31. Shao Z, Mou J, Czajkowsky DM, Yang J, Yuan JY (1996) Adv Phys 45:1
32. Florin E-L, Rief M, Lehmann H, Ludwig M, Dornmair C, Moy VT, Gaub HE (1995) Biosens Bioelectron 10:895
33. Chen Y, Cai J, Xu Q, Chen ZW (2004) Mol Immunol 41:1247
34. Heinz WF, Hoh JH (1999) Trends Biotechnol 17:143
35. Burnham NA, Chen X, Hodges CS, Matei GA, Thoreson EJ, Roberts CJ, Davies MC, Tendler SJB (2003) Nanotechnology 14:1
36. Best RB, Clarke J (2002) Chemical Commun 183
37. Lu JH (2004) Colloids Surf B 39:177
38. Costa LT, Pinto JR, Moraes MB, de Souza GGB, Sorenson MM, Bisch PM, Weissmuller G (2004) Biophys Chem 109:63
39. Morris VJ, Kirby AR, Gunning AP (2001) Atomic force microscopy for biologists. Imperial College Press, London
40. Howland R, Benatar L (1996) A practical guide to scanning probe microscopy. Park Scientific Instruments, Sunnyvale, CA
41. Bonnell DA (2001) Wiley-VCH, New York
42. Henderson RM, Oberleithner H (2000) Am J Physiol Renal Physiol 278:F689
43. Ikai A (1996) Surf Sci Rep 26:261
44. Anczykowski B, Krüger D, Babcock KL, Fuchs H (1996) Ultramicroscopy 66:251
45. Martin Y, Williams CC, Wickramasinghe HK (1987) J Appl Phys 61:4723
46. Sulcheck T, Hsieh R, Adams JD, Minne SC, Quate CF, Adderton DM (2000) Rev Sci Instrum 71:2097
47. Samori P (2004) J Mater Chem 14:1353
48. Lillehei PT, Bottomley LA (2001) Methods Enzymol 320:234
49. Chen GY, Warmack RJ, Thundat T, Allison DP, Huang A (1994) Rev Sci Instrum 65:2532

50. Fotiadis D, Scheuring S, Muller SA, Engel A, Muller DJ (2002) Micron 33:385
51. Hansma PK, Cleveland JP, Radmacher M, Walters DA, Hillner PE, Bezanilla M, Fritz M, Vie D, Hansma HG, Prater CB, Massie J, Fukunaga L, Gurley J, Elings V (1994) Appl Phys Lett 64:1738
52. Möller C, Allen M, Elings V, Engel A, Muller DJ (1999) Biophys J 77:1150
53. Jalili N, Laxminarayana K (2004) Mechatronics 14:907
54. Möller C, Allen M, Elings V, Engel A, Muller DJ (1999) Biophys J 77:1150
55. Moreno-Herrero F, Colchero J, Baro AM (2003) Ultramicroscopy 96:167
56. Hansma HG, Kim KJ, Laney DE, Garcia RA, Argaman M, Allen MJ, Parsons SM (1997) J Struct Biol 119:99
57. Argaman M, Golan R, Thomson NH, Hansma HG (1997) Nucleic Acids Res 25:4379
58. Santos NC, Castanho M (2004) Biophys Chem 107:133
59. Sheng S, Shao Z (1998) Jpn J Appl Phys 37:3828
60. Jalili N, Laxminarayana K (2004) Mechatronics 14:907
61. Gao C (1997) Appl Phys Lett 71:1801
62. van Noort SJT, van der Werf KO, Eker APM, Wyman C, de Grooth BG, van Hulst NF, Greve J (1998) Biophys J 74:2840
63. Pietrement O, Pastre D, Fusil S, Jeusset J, David MO, Landousy F, Hamon L, Zozime A, Le Cam E (2003) Langmuir 19:2536
64. Thomson NH, Kasas S, Smith B, Hansma HG, Hansma PK (1996) Langmuir 12:5905
65. Weigert S, Dreier M, Hegner M (1996) Appl Phys Lett 69:2834
66. Humphris ADL, Round AN, Miles MJ (2001) Surf Sci 491:468
67. Santos NC, Castanho MARB (2004) Biophys Chem 107:133
68. Kasas S, Thomson NH, Smith BL, Hansma PK, Miklossy J, Hansma HG (1997) Int J Imag Syst Technol 8:151
69. Chen Y, Cai JY, Liu ML, Zeng GC, Feng Q, Chen ZW (2004) Scanning 26:155
70. Marcus RB, Ravi TS, Gmitter T, Chin K, Liu D, Orvis WJ, Ciarlo DR, Hunt CE, Trujillo J (1990) Appl Phys Lett 56:236
71. Burnham NA, Colton RJ, Pollock HM (1993) Nanotechnology 4:64
72. Zlatanova J, Lindsay SM, Leuba SF (2000) Prog Biophys Mol Biol 74:37
73. Williams MC, Rouzina L (2002) Curr Opin Struct Biol 12:330
74. Hoh J, Engel A (1993) Langmuir 9:3310
75. Bustamante C, Smith SB, Liphardt J, Smith D (2000) Curr Opin Struct Biol 10:279
76. Wang DW, Yin H, Landick R, Gelles J, Block SM (1997) Biophys J 71:1335
77. Baumenn CG, B SS, A BV, Bustamante C (1997) Proc Natl Acad Sci USA 94:6185
78. Han T, Williams JM, Beebe TP (1995) Anal Chim Acta 307:365
79. Noy A, Sanders CH, Vezenov DV, Wong S, Lieber CM (1998) Langmuir 14:1508
80. Noy A, Frisbie CD, Rozsnyai LF, Wrighton MS, Lieber CM (1995) J Am Chem Soc 117:7943
81. Lieber CM, Noy A, Vezenov DV, Wong SS, Woolley AT, Joselevich E (1998) Abstracts of Papers of the American Chemical Society 216:314
82. Smith DA, Connell SD, Robinson C, Kirkham J (2003) Anal Chim Acta 479:39
83. Muller DJ, Engel A (1997) Biophys J 73:1633
84. Green NH, Allen S, Davies MC, Roberts CJ, Tendler SJB, Williams PM (2002) Trends Anal Chem 21:65
85. Li LY, Chen SF, Oh SJ, Jiang SY (2002) Anal Chem 74:6017
86. Merdan T, Kopecek J, Kissel T (2002) Adv Drug Deliv 54:715
87. Wittmar M, Ellis JS, Morell F, Unger F, Schumacher JC, Roberts CJ, Tendler SJB, Davies MC, Kissel T (2005) Bioconjug Chem (accepted)
88. Merdan T, Kopecek J, Kissel T (2002) Adv Drug Deliv Rev 54:715

89. Ellis DJ, Dryden DTF, Berge T, Edwardson IM, Henderson RM (1999) Nat Struct Biol 6:15
90. Thirlway J, Turner IJ, Gibson CT, Gardiner L, Brady K, Allen S, Roberts CJ, Soultanas P (2004) Nucleic Acids Res 32:2977
91. Lio A, Charych DH, Salmeron M (1997) J Phys Chem 101:3800
92. Hansma HG, Laney DE (1996) Biophys J 70:1933
93. Rivetti C, Codeluppi S (2001) Ultramicroscopy 87:55
94. Shlyakhtenko LS, Gall AA, Weimer JJ, Hawn DD, Lyubchenko YL (1999) Biophys J 77:568
95. Shlyakhtenko LS, Potaman VN, Sinden RR, Lyubchenko YL (1998) J Mol Biol 280:61
96. Ono MY, Spain EM (1999) J Am Chem Soc 121:7330
97. Hansma HG, Laney DE (1996) Biophys J 70:1933
98. Umemura K, Ishikawa M, Kuroda R (2001) Anal Biochem 290:232
99. Zheng HP, Li Z, Wu AG, Zhou HL (2003) Biophys Chem 104:37
100. Manning GS (1996) Physica A 231:236
101. Manning GS (1996) Ber Bunsen Phys Chem 100:909
102. Manning GS (1996) J Polym Sci B 34:393
103. Romanowski G, Lorenz MG, Wackernagel W (1991) Appl Environ Microbiol 57
104. Hansma HG, Pietrasanta LI, Auerbach ID, Sorenson C, Golan R, Holden P (2000) Journal of Biomater Sci Polym Ed 11:7
105. Dunlap DD, Maggi A, Soria MR, Monaco L (1997) Nucleic Acids Res 25:3095
106. Fang Y, Hoh JH (1998) J Am Chem Soc 120:8903
107. Hansma HG, Golan R, Hsieh W, Lollo CP, Mullen-Ley P, Kwoh D (1998) Nucleic Acids Res 26:2481
108. Kichler A (2004) J Gene Med 6:S3
109. Merlin JL, N'Doye A, Bouriez T, Dolivet G (2002) Drug News Perspect 15:445
110. Kleemann E, Dailey LA, Abdelhady HG, Gessler T, Schmehl T, Roberts CJ, Davies MC, Seeger W, Kissel T (2004) J Control Release 100:437
111. Brus C, Kleemann E, Aigner A, Czubayko F, Kissel T (2004) J Control Release 95:119
112. Bennink ML, Nikova DN, van der Werf KO, Greve J (2003) Anal Chim Acta 479:3
113. Saecker RM, Record J, Thomas M (2002) Curr Opin Struct Biol 12:311
114. Rippe K, Guthold M, vonHippel PH, Bustamante C (1997) J Mol Biol 270:125
115. Ebright RH (1998) Cold Spring Harbor Symp Quant Biol 63:11
116. Turner IJ, Scott DJ, Allen S, Roberts CJ, Soultanas P (2004) FEBS Lett 577:460
117. van Noort J, Orsini F, Eker A, Wyman C, de Grooth B, Greve J (1999) Nucleic Acids Res 27:3875
118. Yokota H, Nickerson DA, Trask BJ, van den Engh G, Hirst M, Sadowski I, Aebersold R (1998) Anal Biochem 264:158
119. Umemura K, Komatsu J, Uchihashi T, Choi N, Ikawa S, Nishinaka T, Shibata T, Nakayama Y, Katsura S, Mizuno A, Tokumoto H, Ishikawa M, Kuroda R (2001) Biochemical and Biophys Res Commun 281:390
120. Jiao YK, Cherny DI, Heim G, Jovin TM, Schaffer TE (2001) J Mol Biol 314:233
121. Liu WG, Yao KD, Liu QG (2001) J Appl Polym Sci 82:3391–3395
122. Ros R, Eckel R, Bartels F, Sischka A, Baumgarth B, Wilking SD, Puhler A, Sewald N, Becker A, Anselmetti D (2004) Biotechnol 112:5
123. Krautbauer R, Pope LH, Schrader TE, Allen S, Gaub HE (2002) FEBS Lett 510:154
124. Krautbauer R, Fischerlander S, Allen S, Gaub HE (2002) Single Molecules 3:97
125. Krautbauer R, Allen S, Tessmer I, Skinner G, Pope LH, Schrader TE, Molloy JE, Tendler SJB, Gaub HE (2001) Abstr Papers Am Chem Soc 221:247
126. Krautbauer R, Clausen-Schaumann H, Gaub HE (2000) Angew Chem Int Ed 39:3912

127. Chaires JB (1998) Curr Opin Struct Biol 8:314
128. Berge T, Jenkins NS, Hopkirk RB, Waring MJ, Edwardson JM, Henderson RM (2002) Nucleic Acids Res 30:2980
129. Hegner M, Grange W (2002) J Muscle Res Cell Motil 23:367
130. Eckel R, Ros R, Ros A, Wilking SD, Sewald N, Anselmetti D (2003) Biophys J 85:1968
131. Krautbauer R, Schrader TE, Pope LH, Allen S, Gaub HE (2002) Biophys J 82:692
132. Mitry E, Rougier P (2001) Crit Rev Oncol Hematol 37:47
133. Wang JC (1974) J Mol Biol 89:783
134. Baraldi PG, Tabrizi MA, Preti D, Fruttarolo F, Avitabile B, Bovero A, Pavani G, Carretero MDN, Romagnoli R (2003) Pure Appl Chem 75:187
135. Utsuno K, Tsuboi M, Katsumata S, Iwamoto T (2001) Chem Pharm Bull 49:413
136. Kaji N, Ueda M, Baba Y (2001) Electrophoresis 22:3357
137. Pyle AM (2002) J Biol Inorg Chem 7:679
138. Jaeger L, Leontis NB (2000) Angew Chem Int Ed 39:2521
139. Hansma HG, Oroudjev E, Baudrey S, Jaeger L (2003) J Microsc-Oxford 212:273
140. Bonin M, Zhu R, Klaue Y, Oberstrass J, Oesterschulze E, Nellen W (2002) Nucleic Acids Res 30:e81
141. Burnett JP, Frank BH, Douthart RJ (1975) Nucleic Acids Res 2:759
142. Lyubchenko YL, Jacobs BL, Lindsay SM (1992) Nucleic Acids Res 20:3983
143. Kienberger F, Zhu R, Moser R, Blaas D, Hinterdorfer P (2004) J Virol 78:3203
144. Kasas S, Thomson NH, Smith BL, Hansma HG, Zhu XS, Guthold M, Bustamante C, Kool ET, Kashlev M, Hansma PK (1997) Biochemistry 36:461
145. Daubendiek SL, Ryan K, Kool ET (1995) J Am Chem Soc 117:7818
146. Hansma HG, Golan R, Hsieh W, Daubendiek SL, Kool ET (1999) J Struct Biol 127:240
147. Chworos A, Severcan I, Koyfman AY, Weinkam P, Oroudjev E, Hansma HG, Jaeger L (2004) Science 306:2068
148. Ng JD, Kuznetsov YG, Malkin AJ, Keith G, Giege R, McPherson A (1997) Nucleic Acids Res 25:2582
149. Rief M, Clausen-Schaumann H, Gaub HE (1999) Nat Struct Biol 6:346
150. Kirby AR, Gunning AP, Morris VJ (1996) Biopolymers 38:355
151. Guerrera Cidade GA, Terezinha Costa L, Weissmuller G, Da Silva Neto AJ, Costa Roberty N, Bueno De Moraes M, Piedade Prazeres GM, Marques Hill CE, Machado Ribeiro SJ, Bastos De Souza GG, Da Silva Pinto Teixeira L, Da Costa Moncores M, Mascarello Bisch P (2003) Artif Organs 27:447
152. Abu-Lail NI, Camesano TA (2003) J Microsc 212:217
153. Brant DA (1999) Curr Opin Struct Biol 9:556
154. Morris VJ, Kirby AR, Gunning AP (1999) Scanning 21:287
155. Baker AA, Helbert W, Sugiyama J, Miles MJ (1998) Appl Phys A 66:S559
156. Yan LF, Li W, Yang JL, Zhu QS (2004) Macromol Biosci 4:112
157. Baldwin PM, Adler J, Davies MC, Melia CD (1998) J Cereal Sci 27:255
158. Gallant DJ, Bouchet B, Baldwin PM (1997) Carbohydr Polym 32:177
159. Ridout MJ, Gunning AP, Parker ML, Wilson RH, Morris VJ (2002) Carbohydr Polym 50:123
160. Ridout MJ, Parker ML, Hedley CL, Bogracheva TY, Morris VJ (2003) Carbohydr Res 338:2135
161. Thomson NH, Miles MJ, Ring SG, Shewry PR, Tatham AS (1994) J Vac Sci B 12:1565
162. Ridout MJ, Parker ML, Hedley CL, Bogracheva TY, Morris VJ (2004) Biomacromolecules 5:1519
163. Tasker S, Matthijs G, Davies MC, Roberts CJ, Schacht EH, Tendler SJB (1996) Langmuir 12:6436

164. Frazier RA, Davies MC, Matthijs G, Roberts CJ, Schacht E, Tendler SJB, Williams PM (1997) Langmuir 13:4795
165. Frazier RA, Matthijs G, Davies MC, Roberts CJ, Schacht E, Tendler SJB (2000) Biomaterials 21:957
166. Lee I, Evans BR, Woodward J (2000) Ultramicroscopy 82:213
167. Schneider SW, Larmer J, Henderson RM, Oberleithner H (1998) Pflugers Arch 435:362
168. Jandt KD (2001) Surf Sci 491:303
169. Worcester DL, Miller RG, Bryant PJ (1988) J Microsc-Oxford 152:817
170. Mogi T, Stern LJ, Chao BH, Khorana HG (1989) J Biol Chem 264:14192
171. Muller DJ, Engel A (2002) Atomic force microscopy in cell biology (Methods in cell biology), vol 68. Academic, New York, p 257
172. Muller DJ, Heymann JB, Oesterhelt F, Möller C, Gaub H, Buldt G, Engel A (2000) Biochim Biophys Acta 1460:27
173. Muller DJ, Sass HJ, Muller SA, Buldt G, Engel A (1999) J Mol Biol 285:1903
174. Muller DJ, Amrein M, Engel A (1997) J Struct Biol 119:172
175. Muller DJ, Schoenenberger CA, Buldt G, Engel A (1996) Biophys J 70:1796
176. Muller DJ, Buldt G, Engel A (1995) J Mol Biol 249:239
177. Muller DJ, Schabert FA, Buldt G, Engel A (1995) Biophys J 68:1681
178. Kates M, Kushwaha SC, Sprott GD (1982) Methods Enzymol 88:98
179. Worcester DL, Kim HS, Miller RG, Bryant PJ (1990) J Vac Sci Technol A 8:403
180. Persike N, Pfeiffer M, Guckenberger R, Fritz M (2000) Colloids Surf B 19:325
181. Muller DJ, Engel A, Matthey U, Meier T, Dimroth P, Suda K (2003) J Mol Biol 327:925
182. Muller DJ, Fotiadis D, Engel A (1998) FEBS Lett 430:105
183. Moloney M, McDonnell L, O'Shea H (2002) Ultramicroscopy 91:275
184. Kuznetsov YG, Low A, Fan H, McPherson A (2004) Virology 323:189
185. Moloney M, McDonnell L, O'Shea H (2004) Ultramicroscopy 100:163
186. Scheuring S, Fotiadis D, Möller C, Muller SA, Engel A, Muller DJ (2001) Single Molecules 2:59
187. Furuno T, Sasabe H, Ikegami A (1998) Ultramicroscopy 70:125
188. Schaper A, Starink JPP, Jovin TM (1994) FEBS Lett 355:91
189. Klein DCG, Stroh CM, Jensenius H, van Es M, Kamruzzahan ASM, Stamouli A, Gruber HJ, Oosterkamp TH, Hinterdorfer P (2003) Chemphyschem 4:1367
190. Allen S, Chen X, Davies J, Davies MC, Dawkes AC, Edwards JC, Roberts CJ, Tendler SJB, Williams PM (1998) Appl Phys A 66:s255
191. Chen X, Davies MC, Roberts CJ, Tendler SJB, Williams PM, Davies J, Dawkes AC, Edwards JC (1997) Langmuir 13:4106
192. Su M, Pan Z, Dravid VP (2004) J Microsc-Oxford 216:194
193. Wu AG, Li Z, Yu LH, Wang HD, Wang EK (2002) Ultramicroscopy 92:201
194. Markiewicz P, Goh MC (1997) Ultramicroscopy 68:215
195. Patel N, Davies MC, Hartshorne M, Heaton RJ, Roberts CJ, Tendler SJB, Williams PM (1997) Langmuir 13:6485
196. Williams PM, Davies MC, Jackson DE, Roberts CJ, Tendler SJB (1994) J Vac Sc Technol B 12:1456
197. Raab A, Han WH, Badt D, Smith-Gill SJ, Lindsay SM, Schindler H, Hinterdorfer P (1999) Nat Biotechnol 17:902
198. Snellings G, Vansteenkiste SO, Corneillie SI, Davies MC, Schacht EH (2000) Adv Mater 12:1959
199. Vansteenkiste SO, Corneillie SI, Schacht EH, Chen X, Davies MC, Moens M, Van Vaeck L (2000) Langmuir 16:3330

200. Holland NB, Marchant RE (2000) J Biomed Mater Res 51
201. Cacciafesta P, Hallam KR, Watkinson AC, Allen GC, Miles MJ, Jandt KD (2001) Surf Sci 491:405
202. Marchant RE, Kang I, Sit PS, Zhou Y, Todd BA, Eppell SJ, Lee I (2002) Curr Protein Pept Sci 3:249
203. Huang N, Yang P, Leng YX, Chen JY, Sun H, Wang J, Wang GJ, Ding PD, Xi TF, Leng Y (2003) Biomaterials 24:2177
204. Revzin A, Tompkins RG, Toner M (2003) Langmuir 19:9855
205. Hussain MA, Siedlecki CA (2004) Micron 35:565
206. Desai SM, Singh RP (2004) Long-term properties of polyolefins (Advances in Polymer Science), vol 169. Springer, Berlin Heidelberg New York, p 231
207. Dobson CM (2001) Philos Trans R Soc Lond B 356:133
208. Chiti F, Webster P, Taddei N, Clark A, Stefani M, Ramponi G, Dobson CM (1999) Proc Natl Acad Sci USA 96:3590
209. Blackley HKL, Patel N, Davies MC, Roberts CJ, Tendler SJB, Wilkinson MJ, Williams PM (1999) Exp Neurol 158:437
210. Blackley HKL, Sanders GHW, Davies MC, Roberts CJ, Tendler SJB, Wilkinson MJ (2000) J Mol Biol 298:833
211. Green JD, Kreplak L, Goldsbury C, Blatter XL, Stolz M, Cooper GS, Seelig A, Kistler J, Aebi U (2004) J Mol Biol 342:877
212. Heal JR, Roberts GW, Christie G, Miller AD (2002) Chembiochem 3:86
213. Wang Z, Zhou C, Wang C, Wan L, Fang X, Bai C (2003) Ultramicroscopy 97:73
214. Stolz M, Stoffler D, Aebi U, Goldsbury C (2000) J Struct Biol 131:171
215. Sedman VL, Allen S, Chan WC, Davies MC, Roberts CJ, Tendler SJB, Williams PM (2005) Protein Pept Lett 12:79
216. Mou JX, Yang J, Shao ZF (1995) J Mol Biol 248:507
217. Best RB, Brockwell DJ, Toca-Herrera JL, Blake AW, Smith DA, Radford SE, Clarke J (2003) Anal Chim Acta 479:87
218. Leckband D (2000) Annu Rev Biophys Biomol Struct 29:1
219. Carrion-Vazquez M, Oberhauser AF, Fisher TE, Marszalek PE, Li H, Fernandez JM (2000) Prog Biophys Mol Biol 74:63
220. Allen S, Rigby-Singleton SM, Harris H, Davies MC, O'Shea P (2003) Biochem Soc Trans 31:1052
221. Rief M, Gautel M, Oesterhelt F, Fernandez JM, Gaub HE (1997) Science 276:1109
222. Williams PM, Fowler SB, Best RB, Toca-Herrera JL, Scott KA, Steward A, Clarke J (2003) Nature 422:446
223. Marszalek PE, Lu H, Li HB, Carrion-Vazquez M, Oberhauser AF, Schulten K, Fernandez JM (1999) Nature 402:100
224. Kellermayer MSZ, Smith SB, Granzier HL, Bustamante C (1997) Science 276:1112
225. Willemsen OH, Snel MME, Cambi A, Greve J, De Grooth BG, Figdor CG (2000) Biophys J 79:3267
226. Lee GU, Kidwell DA, Colton RJ (1994) Langmuir 10:354
227. Wong J, Chilkoti A, Moy VT (1999) Biomol Eng 16:45
228. Benoit M, Gabriel D, Gerisch G, Gaub HE (2000) Nat Cell Biol 2:313
229. Kaur J, Singh KV, Schmid AH, Varshney GC, Suri CR, Raje M (2004) Biosens Bioelectron 20:284
230. Abdelhady HG (2004) Investigation Of Polyamidoamine Dendrimers Induced Dna Condensation And Enzymatic Degradation Of These Complexes: An Atomic Force Microscopy Study. Thesis, University of Nottingham, School of Pharmacy

Polymer Genomics

Alexander V. Kabanov[1] (✉) · Elena V. Batrakova[1] · Simon Sherman[2] · Valery Y. Alakhov[3]

[1] Center for Drug Delivery and Nanomedicine and Department of Pharmaceutical Sciences, University of Nebraska Medical Center, 985830 Nebraska Medical Center, Omaha, Nebraska, 8198-5830, USA
akabanov@unmc.edu

[2] Eppley Institute for Research in Cancer and Allied Diseases, University of Nebraska Medical Center, Nebraska Medical Center, Omaha, Nebraska, 9868050, USA

[3] Supratek Pharma Inc., 215 Bvd Bouchard, Suite 1315, H9S1A9 Laval, Quebec, Canada

1	Introduction	174
2	Phenotypic selectivity of polymer-drug formulations for MDR cancers	175
3	Alteration of signal transduction pathways by polymers in cancer cells	179
4	Polymer effects on genomic profiles in drug-selected cancer cells	180
5	Analysis of the effects of polymers on gene expression profiles	183
6	Transcriptional activation of gene expression by synthetic polymers	185
7	Phenotypic correction of immune response by synthetic polyelectrolytes	189
8	Effects of polymer-coated surfaces on signal transduction in adherent cells	191
9	Polymer genomics hypothesis and future studies	192
10	Conclusions	194
	References	195

Abstract Biological activities of biocompatible synthetic polymers used in drug delivery, gene delivery, vaccine development or biomaterial surface modification are discussed. Synthetic polymers display selective phenotypic effects in cells and in the body, affecting signal transduction mechanisms involving inflammation, differentiation, proliferation, and apoptosis. These effects are realized as a result of interactions of water-soluble polymers with plasma cell membranes, delivery of polymers to intracellular organelles, and at the sites of cell contacts with polymer-coated surfaces. The ability of the cells and organisms to respond to the effects of these polymers can be dependent on phenotype or genotype. In selected cases, polymer agents can bypass limitations on biological response imposed by the genotype; for example, achievement of phenotypic correction of immune response by polyelectrolytes. Overall, these effects are relatively weak as they do not result in cytotoxicity or major toxicities in the body. However, when combined with specific biological agents, such as cytotoxic agents, bacterial DNA or antigens, either by mixing or by covalent conjugation, the polymers can drastically alter specific genetically controlled responses to these agents. Collectively these studies propose the need for thorough assessment of pharmacogenomic effects of polymer materials in order to maximize the clinical outcomes and understand the pharmacological and toxicological effects of polymer formulations of biological agents – *polymer genomics*.

Keywords Artificial vaccines · DNA microarray · Drug resistance · Phenotype · Signal transduction

1
Introduction

Pharmacogenomics has emerged as an important field at the interface of pharmaceutics and genetics, which studies how an individual's genetic inheritance affects the body's response to drugs [1]. Pharmacogenomics holds the promise that drugs might one day be tailor-made for individuals and adapted to each person's own genetic make-up. Of equal importance are studies of cellular responses to drugs, particularly in cancer chemotherapy and other areas where chemotherapeutic agents can select genetic mutations that result in acquired resistance to these agents [2]. The subject of this review, "*polymer genomics*", addresses the effects of biocompatible synthetic polymer materials in responses to biological agents. This consideration may be generally applicable to many types of biocompatible materials that come into contact with cells and body tissues. However, the relevance and urgency of this consideration in pharmaceutics has become more obvious because of the tremendous growth of work using polymer-based drug and gene delivery systems. Polymer-based drug and gene delivery systems emerged from the laboratory bench in the 1990s as a promising therapeutic strategy for the treatment of devastating human diseases [3–9]. There is a fundamental reason why polymer materials are useful for solving drug delivery and gene delivery problems. Polymers are relatively large compared to low molecular mass drugs, and when combined with these drugs they can augment the drug's performance and change their bioavailability. Moreover, synthetic polymers are perfectly suited for producing formulations with biopolymers, such as proteins and nucleic acids, since they can self-assemble with these molecules, forming nanosized complexes that are useful in drug delivery and gene therapy applications [10, 11]. A number of polymer-based therapeutics are presently on the market or are undergoing clinical evaluations as treatments for cancer and other diseases [7]. The central paradigm of drug delivery developed in these studies considers the polymers used in these formulations as *biologically inert* excipients that protect biological agents from degradation, prolong their exposure to tissues, and enhance the transport of biological agents into cells. However, such a view is undergoing major revision due to the growing evidence that selected synthetic polymers, when combined with biological agents (low molecular mass drugs, DNA or antigens), can alter genetically controlled cellular responses to these agents [12]. This paper provides an overview of such studies. First, we consider the effects of water-soluble amphiphilic block copolymers in multidrug resistant (MDR) cancers and demonstrate that these effects are selective for MDR phenotype. Second, we discuss the effects involving alterations of apoptotic signal transduction pathways in cancer cells by water-soluble polymers mixed or covalently conjugated with antineoplastic drugs. Third, we present evidence that amphiphilic block copolymers can drastically alter gene expression profiles and prevent the development of drug resistance during the biological selection of cancer cells with anticancer agents. Fourth, we demonstrate that these polymers can change gene expression profiles and improve gene therapy by activating transcription by affecting selected inflammatory signal transduction pathways. Fifth, we analyze the effects of synthetic polyelectrolytes on immune response, including its phenotypic correction, and studies developing fully synthetic vaccines on the basis of polyelectrolytes. Sixth, we demonstrate that signal transduction in adherent cells of polymer-modified surfaces can strongly depend on the surface chemistry. Based on these studies, we discuss the significance and future directions of polymer genomics research.

2
Phenotypic selectivity of polymer-drug formulations for MDR cancers

In contrast to genotype, which is defined as the inherited genetic makeup of the body, the phenotype defines properties acquired as a result of environmental effects. In an evolutionary sense, biological selection acts on the phenotypes, because differential reproduction and survival depend on the phenotype. Thus, alterations in phenotype precede genotype changes, but they still play a critical role in the response and survival of the cells. One example of a phenotypic change of high significance for cancer chemotherapy and treatment of other diseases is the development of MDR in response to chemotherapy. Tumors with MDR phenotype overexpress efflux transporters belonging to a superfamily of ATP binding cassette (ABC) proteins, such as P-glycoprotein (Pgp), multidrug resistance-associated proteins (MRP), and breast cancer resistance protein (BCRP), which pump drugs out of a cell and thus reduce their cytotoxicity [13, 14]. Furthermore, the glutathione (GSH)/glutathione S-transferase (GST) detoxification system is frequently activated in MDR cells, contributing to drug resistance [15, 16]. Another impediment to treatment, which is present in MDR cells, involves the sequestration of drugs within cytoplasmic vesicles, followed by drug extrusion out of the cell [17]. Drug sequestration in MDR cells is achieved through the maintenance of abnormally elevated pH gradients across organelle membranes – by the activity of H^+-ATPase, an ATP-dependent pump [18, 19]. The combination of several mechanisms of drug resistance complicates chemotherapy and reinforces the need to develop novel drug formulations that are effective against drug-resistant cancers.

One approach to overcoming MDR is to use Pluronic block copolymers in formulations used to treat drug-resistant cancers [20–22]. Pluronic block copolymers consist of ethylene oxide (EO) and propylene oxide (PO) segments arranged in a basic A-B-A structure: $EO_a – PO_b – EO_a$ (Fig. 1). This arrangement results in an amphiphilic molecule where altering the numbers of the EO and PO units can vary the size, hydrophilicity and lipophilicity of the molecule. Due to the self-assembly of hydrophobic PO chains, Pluronic molecules can form micelles and gels (Fig. 2). Some early studies using Pluronic focused on the use of polymeric micelles as nanocontainers for targeted delivery of hydrophobic drugs [23–25]. However, it soon became clear that Pluronic molecules themselves display unique properties that have important implications for the delivery of drugs into cells [4]. One notable example is the effect of Pluronic on MDR cancer cells. Pluronic results in the chemosensitization of these cells to antineoplastic agents [20, 22, 26, 27]. Pluronic molecules have been shown to inhibit drug efflux transporters, such as Pgp, expressed in MDR cells [27, 28]. The block copolymers exert a "double punch" effect on Pgp by 1) inducing conformational changes that decrease the affinity of the Pgp for drugs and ATP, and 2) causing intracellular depletion of the ATP necessary for proper Pgp function [27, 29]. The synergy between these two effects results in the inhibition of the Pgp efflux system, as well as some other ATP-dependent drug efflux transporters, such as MRP [30]. Furthermore, Pluronic inhibited other energy-dependent drug resistance mechanisms, such as the GSH/GST detoxification system and cy-

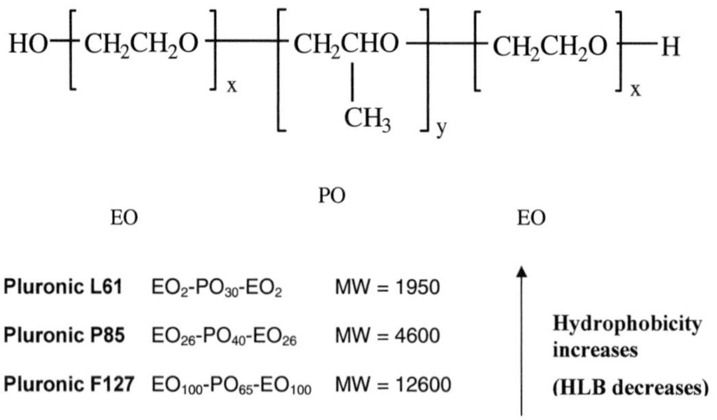

$$HO-[CH_2CH_2O]_x-[CH_2CHO(CH_3)]_y-[CH_2CH_2O]_x-H$$

EO   PO   EO

Pluronic L61 EO_2-PO_{30}-EO_2 MW = 1950

Pluronic P85 EO_{26}-PO_{40}-EO_{26} MW = 4600

Pluronic F127 EO_{100}-PO_{65}-EO_{100} MW = 12600

Hydrophobicity increases

(HLB decreases)

Fig. 1 Structures of selected Pluronic block copolymers available from BASF

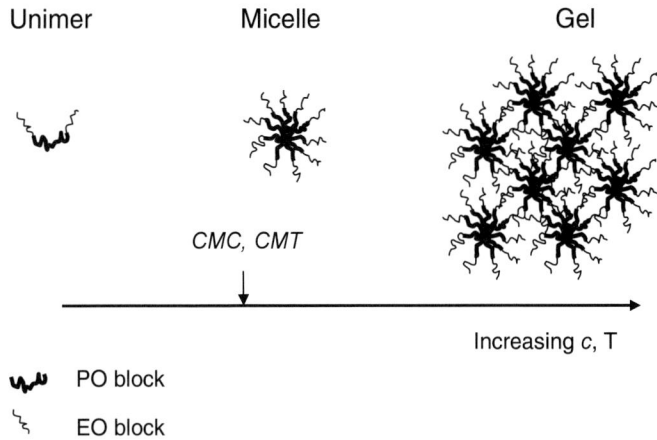

Fig. 2 Self-assembly of Pluronic unimers into micelles occurs as the concentration of the block copolymer reaches the critical micellization concentration (CMC) and the temperature reaches critical micellization temperature (CMT). A further increase in the block copolymer concentration results in the formation of Pluronic gel

toplasmic drug sequestration in MDR cells [22, 30, 31]. This means that highly potent chemosensitization of MDR cancer cells by Pluronic block copolymers has been accomplished.

This combination of activities of Pluronic in MDR cells is due to the amphiphilic surfactant architecture of the block copolymer molecules, which can incorporate into cellular membranes resulting in changes in membrane permeability, membrane potential, and membrane transport of various compounds. The likely targets of Pluronic effects are the cell plasma membrane (where Pgp is localized) and the mitochondria membrane, where Pluronic affects normal functioning of the drug efflux transporters and respiratory chain, respectively [28, 32]. Selected Pluronics, which can translocate inside the cell and thus affect intracellular compartments, in particular mitochondria, are the most efficacious modulators in both Pgp inhibition and ATP depletion [28].

The membrane-active properties of Pluronic are rather nonspecific. Therefore, it is quite remarkable that both the ATP depletion and sensitization of MDR cells correlate with the expression of drug resistance proteins in these cells [33]. Figure 3 represents the correlation between the EC_{50} values, determined as the concentrations of the block copolymer that induced 50% depletion of ATP, and the levels of Pgp expression in various cells. As the expression of Pgp was increased, the ATP-depleting potency of the block copolymer was also elevated. This was related to the abnormally high energy consumption observed in MDR cells, which rendered them very responsive to respiratory inhibition by the block copolymer [27]. Conversely, the nonresistant cells were more likely to adjust to the reduced respiration conditions

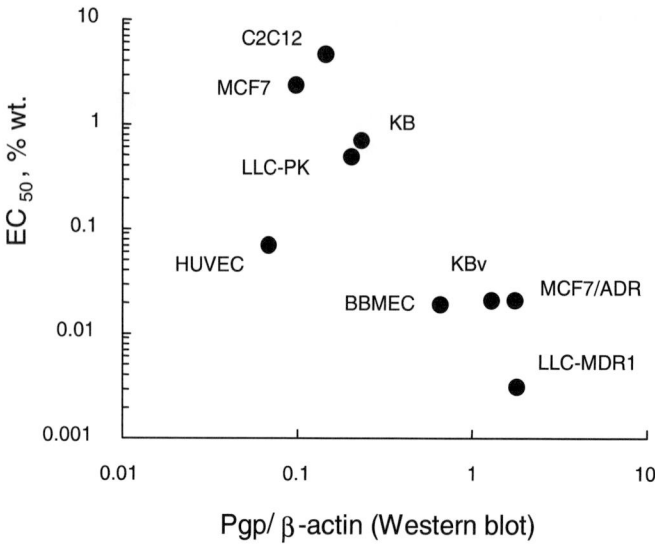

Fig. 3 Relationship between ATP depletion and Pgp expression levels in various cells. The studies included human breast carcinoma cells, MCF-7, and their MDR subline, MCF-7/ADR; human oral epidermoid carcinoma cells, KB, and their MDR subline, KBv; wild-type porcine kidney epithelial cells, LLC-PK1, and human MDR1-transfected cells, LLC-MDR1; human umbilical vein endothelial cells, HUVEC; bovine brain microvessel endothelial cells, BBMEC; and murine myoblast cells, C2C12. EC_{50} values for each of the cell lines were determined from the dose response curves as the concentration of Pluronic P85 inducing a 50% decrease in intracellular ATP following 2 h exposure of the cells to the block copolymer. The drug efflux transport proteins were identified using the immunoblot technique and normalized to constitutively expressed β-actin. Plotted using data reported in [33]

and, thus, were less vulnerable to the block copolymer effects. This selectivity of Pluronic with respect to the cells displaying MDR1 phenotype is unique. It serves as an example that the biochemical activities of polymer-drug formulations in cells can be determined by the cell phenotype. Based on these studies, MDR gene expression could be a valuable pharmacogenomic marker for the prediction of successful outcomes of chemotherapy with drug-Pluronic formulations [33]. Indeed, out of several dozen different cancer lines examined in our laboratory, the cancers overexpressing MDR1 were found to be the most sensitive to the sensitization effect of Pluronic formulated with Doxorubicin (Dox) (paper in preparation). This reinforces the use of a Dox-Pluronic formulation in the ongoing clinical trials involving cancers with a high incidence of MDR [34]. Future studies will extend the pharmacogenomic approach to correlate gene expression profiles and chemotherapeutic responses to Pluronic-based drug formulations in different tumors. This will allow genetic markers that can predict the greatest response to such formulations to be determined in order to maximize the clinical benefits. It is possible that similar studies could also be useful to improve therapeutic outcomes using other polymer-drug formulations, including drugs incorporated in liposomes, polymer micelles, and polymer-drug conjugates.

3
Alteration of signal transduction pathways by polymers in cancer cells

Although polymer-drug formulations are used primarily to improve the *delivery* of the drugs to their molecular targets, evidence has begun to mount that polymer excipients in these formulations may also affect the intracellular signal transduction pathways that underlie specific genetically controlled responses of cells to the drugs. Current examples include anticancer drugs that are chemically bound to the polymer carriers, those incorporated into polymer micelles, or simply coadministered with the polymers, such as Pluronic [35–38]. For example, conjugation of Dox with poly(2-hydroxypropylmethacrylate) (PHPMA) resulted in the induction of additional (compared to the free drug) caspase-dependent apoptosis signaling pathways in the sensitive and resistant human ovarian carcinoma cell lines A2780 and A2780/AD [35]. Specifically, the polymer-drug conjugate induced apoptosis by up-regulating *caspases 3, 6, 7, 8*, and *9*, while the free drug caused induction of only three *caspases 9, 3*, and *7*. Moreover, following treatment of cells with PHPMA-Dox conjugate, the *bcl-2* gene responsible for the cellular defense mechanism was down-regulated, resulting in further stimulation of the proapoptotic route. Conversely, free Dox increased expression of the *bcl-2* gene, which partially mitigated its cytotoxic activity.

Alterations in signal transduction pathways were also demonstrated for another polymer-drug conjugate in the A2780 cancer cell line [36]. It was shown that conjugation of camptothecin with poly(ethyleneglycol) (PEG) considerably increased expression of pro-apoptotic genes encoding caspase activators (*SMAC* and *APAF1*), as well as *caspases 9* and *3*, compared to the effects of the free drug. Polymer-conjugated drug also induced up-regulation of genes related to the cellular defense, *bcl-2* and *bcl-xl*. Nevertheless, induction of apoptosis with the polymer-conjugated drug was developed more rapidly and to a greater extent than with the free drug, which resulted in the enhancement of cell death pathways.

In collaboration with T. Minko (Rutgers University), we have recently demonstrated that Pluronic P85 formulated with Dox also enhanced the net proapoptotic response in MDR human breast carcinoma cell line MCF7/ADR [39]. Specifically, in the presence of the block copolymer, expression of the proapoptotic genes *BAX, P53, APAF1, caspase 9* and *caspase 3* was up-regulated, while the antiapoptotic cellular defense, *BCL2*, was decreased compared to Dox alone. Therefore, in addition to enhanced drug delivery in the cells due to the inhibition of the drug efflux systems, the block copolymer also enhanced the cytotoxic effect of the delivered drug by increasing the cell death pathways in the MDR cells.

A DNA microarray technique has been a useful tool for elucidating complex and interrelated cellular responses to polymeric drug formulations [40, 41]. This technique was applied by the group of Kataoka in order to analyze the responses to cisplatin (CDDP) incorporated into polymeric micelles with poly(ethylene glycol)-poly(glutamic acid) block copolymer in non-small-cell lung cancer PC-14 cells [37]. It was reported that the expression levels of genes related to cell cycle regulation, apoptosis, detoxification, and DNA repair were different in the cells treated with free CDDP and CDDP formulated with the micelles (CDDP/m). For example, incorporation of CDDP into polymer micelles resulted in down-regulation of the gene expression of integrin and matrix metalloprotease (MMP) families, while free CDDP up-regulated these genes. This may reduce tumor invasion, metastasis, and angiogenesis, leading to additional therapeutic benefits of CDDP/m.

Overall, these studies suggest that various polymer components in different drug delivery formulations can substantially modulate the biochemical effects of the anticancer drugs in the cells by altering the extent and direction of signal transduction pathways.

4
Polymer effects on genomic profiles in drug-selected cancer cells

Selection of cells in the presence of chemotherapeutic agents results in drastic changes in the gene expression profiles and development of drug resistance. The principal question in the current consideration of polymer genomics is whether polymer excipients can affect the outcomes of these changes and prevent (or redirect) the development of drug resistance. An affirmative answer to this question has been obtained in a recent study that demonstrated that Pluronic P85 altered gene expression profiles in human breast carcinoma cells (MCF7) selected with Dox [38]. Two cell sublines were established from non-resistant parental MCF7 cells. The first subline, MCF7/Dox, was derived by exposure to increasing concentrations of Dox. The second subline, MCF7/Dox-P85, was selected with Dox in the presence of 0.001 % P85 in the culture medium. As expected, the exposure of MCF7 cells to Dox resulted in up-regulation of expression of the MDR1 gene and overexpression of functionally active Pgp, (the appearance of the MDR1 phenotype). These cells were able to grow in the presence of as much as 10 000 ng/ml of Dox. In contrast, MCF7/Dox-P85 cells could only tolerate 1000-fold lower concentrations of about 10 ng/ml of Dox. These cells did not display alteration in MDR1 gene expression or increases in active Pgp compared to the parental, nonresistant cancer cells.

The global 20 K gene expression profiles in selected cells were analyzed by a cDNA microarray technique. For each cell subline, the relative levels of gene expression were plotted against levels of expression of the same genes in the parental MCF7 cells (Fig. 4a,b). Deviations from the linear dependences indicated by dotted lines suggested up- or down-regulation of specific genes compared to the parental cells. There were significant changes in the expression of several hundreds of genes in each selected cell subline. However, the genes implicated in MDR were differentially expressed in these cells. In particular, GST pi gene (*GSTP1*), a clinical prognostic indicator for resistance to chemotherapy in breast cancer, and MDR3 gene, a member of the MDR/TAP subfamily involved in MDR (*ABCB4*), were up-regulated in MCF7/Dox but not in MCF7/Dox-P85 cells. In addition, transcriptional regulation factor, *NSEP1*, involved in the control of MDR1, was up-regulated in MCF7/Dox, but not in MCF7/Dox-P85 cells. Practically no changes in the gene expression levels of other drug efflux proteins, MRPs, BCRP, and lung resistance protein (LRP) were found in either cell subline. Genes involved in metabolic drug resistance (cytochrome P450, thioredoxin reductase, superoxide dismutase 1), apoptosis (cytochrome C), and transcriptional factors (zinc finger proteins, *ZNF22, ZNF198*) were also up-regulated in MCF7/Dox, but not in MCF7/Dox-P85 cells. Finally, the connective tissue growth factor (*CTGF*), which plays a role in the progression of breast cancer growth, was up-regulated in MCF7/Dox, but not in MCF7/Dox-P85 cells.

Conversely, several genes involved in drug resistance mechanisms, such as members of the metallothionein family (*MT2A, MT1R, MT1g*), heat shock proteins (*HSJ2, HSC71*), and the vacuolar proton pump group, were up-regulated in both cell sublines, indicating that formulation of the drug with Pluronic can prevent development of some drug resistance mechanisms, and not others. The β-tubulin expression levels were also increased in both cell lines indicating the possible appearance of drug resistance to antineoplastic agents targeting microtubule assembly, drug binding, and dynamics. Estrogen-dependent factor *TFF1* was down-regulated in both cell sublines, which indicated the possibility of increased cell proliferation and tumor-invasive character. Importantly, MCF-7 parental cells cultured with Pluronic P85 alone (MCF7/P85) did not show changes in the expression of these genes. As is evident from the scatterplot presented in Fig. 4c, few, if any, genes were altered in MDC7/P85 cells compared to the parental MCF7 cells. Therefore block copolymer excipient alone was genetically benign.

Overall, this work demonstrated that formulation of antineoplastic drug Dox with Pluronic P85 prevents development of MDR in breast cancer cells. This reinforces the use of Pluronic block copolymers to improve the chemotherapy of tumors; if resistance is intrinsic, the block copolymers sensitize the tumor, whereas if resistance is acquired, MDR cells no longer have a selective advantage. This supports the clinical use of Pluronic block copolymers in chemotherapy and suggests that formulations using these block copolymers may be useful in the treatment of resistant tumors and may have an additional benefit in preventing the development of drug resistance.

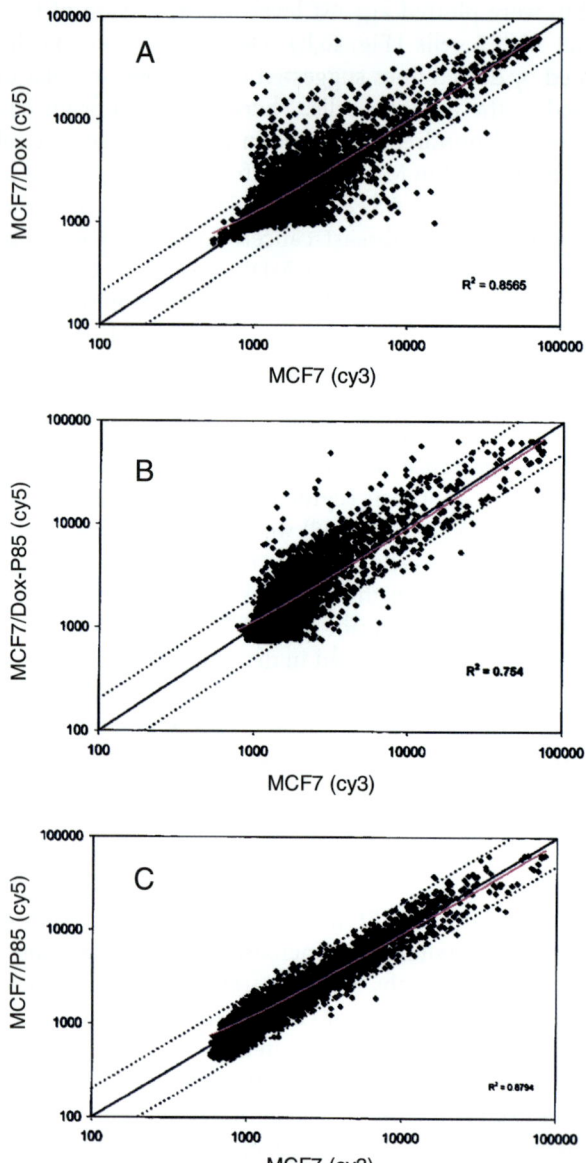

Fig. 4 Expression of 20 K genes in sublines of human breast carcinoma MCF7 cells treated with: **A** Dox alone (MCF7/Dox), **B** Dox formulated with 0.001% P85 (MCF7/Dox-P85) and **C** 0.001% P85 (MCF7/P85), related to the expression of the same genes in parental MCF7 cells. The DNA microarray scatterplots present fluorescence intensities of Cy5 label coupled to the targeted cDNA elements from MCF7 sublines vs. fluorescence intensities of Cy3 label coupled to the cDNA from parental MCF7 cells. The *dotted lines* indicate significant up-regulation (*upper dotted line*) or down-regulation (*lower dotted line*) of the genes. Reproduced with permission from [12]

Another general conclusion is that formulation of a chemotherapeutic drug with a polymer excipient, which is not even covalently bound to this drug and has little if any affect on gene expression on its own, can drastically change the pharmacogenomic responses to the drug. Of considerable interest are genes that are altered by the Pluronic-formulated drug, and not altered by the drug alone. For example, MCF7/Dox-P85 cells displayed elevated levels of mitochondrial creatine kinase (*CKMT2*), nuclear respiratory factor (*NRF1*), succinate dehydrogenase complex II protein (*SDHC*), and cytochrome C oxidase assembly protein (*Cox11*), which were not up-regulated in MCF7/Dox cells. It is worth mentioning that these genes are involved in the functioning of the respiratory chain and that their up-regulation may occur in response to the need to compensate for respiratory inhibition by the block copolymer. Another group of genes up-regulated in MCF7/Dox-P85 cells are involved in signaling and regulation of apoptosis, such as programmed cell death 5 (*PDCD5*) and tumor necrosis factor receptors (*TNFRSF8, TNFSF10A*). These results reinforce the need to consider the pharmacogenomic effects of polymer excipients, including the potential for the development of new resistance mechanisms, which are not found with the low molecular mass drugs. This conclusion may be of general significance for other polymer therapeutics and drug delivery systems, such as polymer-drug conjugates and drugs entrapped in polymer micelles, liposomes and various nanoparticles.

5
Analysis of the effects of polymers on gene expression profiles

Expression of several hundred genes can be up- or down-regulated following selection of cells with biological agents. Thus, robust tools for analyzing the effects of polymers formulated with biological agents on gene expression are needed. For this purpose, the DNA microarray is an appropriate initial high-throughput screening method, which is then followed by other methods to assess the expression of selected genes [40]. Various bioinformatics tools can be applied to the analysis of the DNA microarray data. For example, self-organizing maps (SOM) permit visual examination of gene alterations clustered in different map units [42]. Examples of such maps derived for cells selected with different concentrations of Dox, Dox-P85 or free P85 are presented in Fig. 5. The vertical colored bar chart represents the normalized expression values of the genes in comparison with parental MCF7 cells. Each colored octagon cell contains a specific cluster of genes, which are affected to different extents. Two different patterns can be determined using these maps. One pattern represents cells selected with high concentrations of drug: Dox 200 ng/ml, Dox 1000 ng/ml, and Dox 5000 ng/ml (Fig. 5a–c), and low concentrations of Dox-P85 (Fig. 5d). This pattern displays the overexpressed

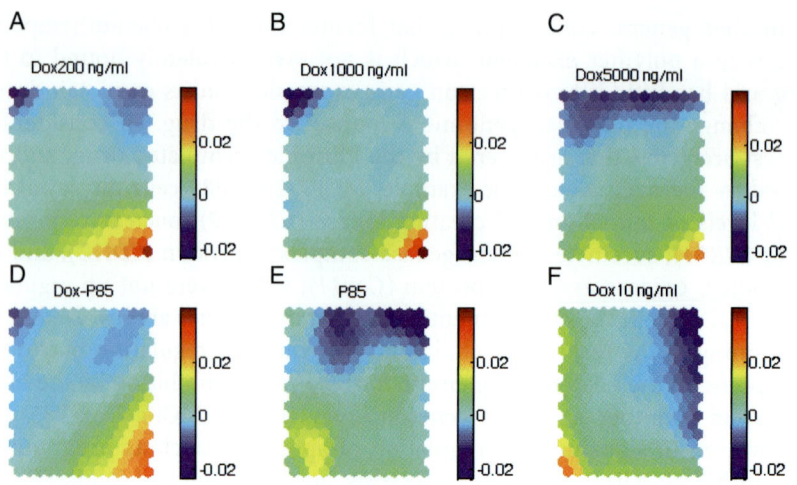

Fig. 5 SOM outputs of genome transcriptional changes from MCF7 cells selected with **A–C** high concentrations of Dox alone, **D** Dox formulated with 0.001% P85, **E** P85 alone, and **F** low concentration of Dox alone. The data analysis was performed by Li Xiao (Eppley Institute for Research in Cancer, UNMC) using MATLAB software (The MathWorks, Inc.) and SOM toolbox for MATLAB (http://www.cis.hut.fi/projects/somtoolbox)

genes (*red color*) in the bottom right corner and the down-regulated genes (*blue*) in the top left corner. Another pattern represents cells selected with Pluronic alone, P85 (Fig. 5e), or with a low concentration of free drug, Dox 10 ng/ml (Fig. 5f). It displays the up-regulated genes in the bottom left corner, and down-regulated genes in the top right corner. Thus, by comparing the SOMs for different selected cell samples, one can clearly see the differences in gene expression and relate the colored areas to the gene clusters affected.

Perhaps even more convenient for the purpose of this analysis is the method employing multivariate scatter plots comparing pairs of selected cell lines. In these plots, the *X*-axis and *Y*-axis present the normalized levels of gene expression for each cell line compared (Fig. 6a). Figure 6b schematically presents the situation where the same genes are up- or down-regulated to the same extent in both cell lines. This situation is most closely exemplified by the comparison of Dox 200 ng/ml and Dox 1000 ng/ml selected cell lines (panel 13). Figure 6c illustrates the situation where only one of the two compared cell lines displays substantial changes in gene expression. This situation, to some extent, is realized for the comparisons of P85 with each of the free drug-selected cells (panels 6, 7, 8, and 9). Finally, Fig. 6d presents the situation where some of the genes are altered in both cell lines, while other genes are affected in only one line. This situation, in particular, is realized for the comparison of Dox-P85 with Dox 1000 ng/ml, which was discussed in greater detail in the previous section.

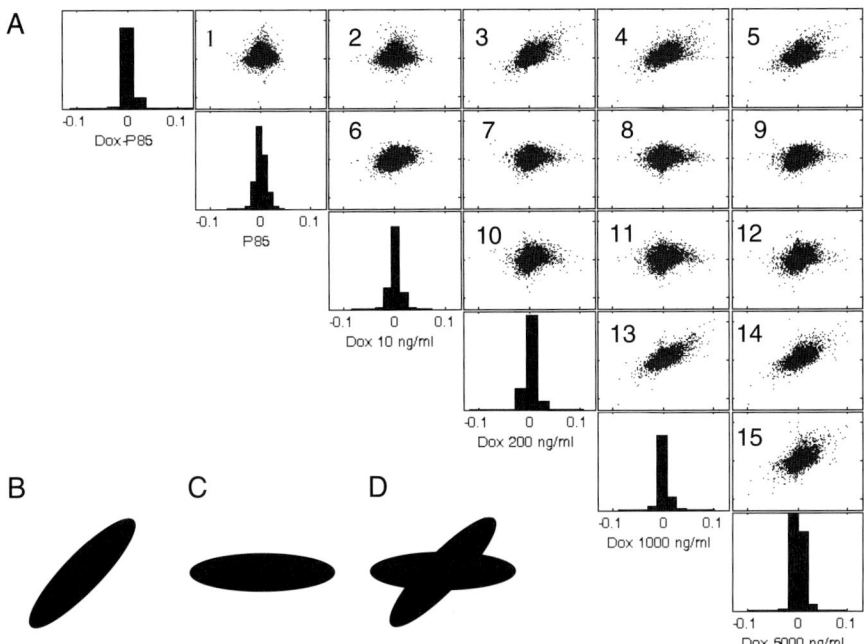

Fig. 6 Multivariate scatter plots of gene expression in selected cell lines (**A**) and schematic representations of different expression patterns (**B–D**). The cases were normalized by array median centering, with the median value of the gene expression in each case centered to zero and the standard deviation to one. The data analysis was performed by Li Xiao (Eppley Institute for Research in Cancer, UNMC) using MATLAB software (The MathWorks, Inc.)

6
Transcriptional activation of gene expression by synthetic polymers

This section discusses the effects of synthetic polymers on the delivery and expression of DNA for gene therapy. Injection of plasmid DNA alone ("naked" DNA) into skeletal muscle results in direct gene transfer into myocytes and sustained gene expression in vivo [43]. Gene therapies involving intramuscular (*i.m.*) injection of naked DNA were evaluated for the treatment of muscular dystrophies (MDs), local or systemic secretion of therapeutic proteins, and elicitation of immune responses against specific antigens of infectious diseases or cancer [44–49]. The use of naked DNA can avoid some inherent problems of recombinant viral vectors, including immune and toxic reactions as well as viral recombination [50]. However, the potential of this technique for gene therapy is limited, due to poor transduction efficiency and high variability of gene expression observed with the naked DNA. Nonviral vectors based on cationic lipids and polycations have been developed to enhance DNA delivery [51, 52]. These agents act by 1) binding with the DNA,

2) condensing the DNA, 3) protecting the DNA from nuclease degradation, and 4) enhancing transport of the DNA into the target cell. These combined effects result in higher levels of transgene expression in the cells. However, the efficiency of gene delivery by these cationic molecules following injection into the muscle is usually quite low, and in many cases, less than that of the naked DNA [4, 53]. Conversely, selected nonionic polymer carriers, such as polyvinyl pyrrolidone (PVP), were shown to improve gene expression of naked DNA in the muscle [54, 55]. Another nonionic polymer, Pluronic, has also shown substantial promise by increasing gene expression of a plasmid DNA in skeletal and cardiac muscle, as well as in solid tumors [56–62]. The levels of gene expression in the muscle observed with Pluronic-formulated DNA were much greater than those reported for PVP-DNA formulations [56] and comparable with the expression achieved by electroporation [59]. Furthermore, the Pluronic-based formulation of a plasmid DNA applied in combination with electrotransfer resulted in a further increase in gene expression in the muscle [62].

The mechanism by which Pluronic enhanced gene expression of plasmid DNA was very different from those associated with the cationic lipids or polycations. Pluronic did not bind with and condense the plasmid DNA, and therefore did not protect DNA from degradation or facilitate transport into the cell. However, compelling evidence was obtained that Pluronic effects involved transcriptional activation of the gene expression in the muscle. This was reinforced by the discovery of the promoter selectivity of the effect of Pluronic on gene expression [63]. For example, a set of plasmids expressing a reporter gene, luciferase, under the control of different transcription factors, the activator protein 1 (AP-1), the cyclic AMP response element (CRE), NF-κB, and the tumor suppressor p53 (AP1-Luc, CRE-Luc, NF-κB-Luc, p53-Luc), was used in recent experiments [64]. The studies demonstrated that Pluronic activated the luciferase gene driven by the CMV promoter and the NF-κB response element, while having much less or no effect on the gene driven by the SV40 promoter or the AP-1 and CRE response elements (Fig. 7). Analysis of the transcription factor binding sites in the CMV and SV40 promoters suggested that while AP-1 and CRE sites are present in both promoters, the NF-κB is unique to the CMV promoter. Thus, the differential effects of Pluronic on these two promoters and the selectivity with respect to the NF-κB response element was consistent with the activation of the NF-κB signaling pathway, which plays a central role in the regulation of the cellular defense and immunological responses [65]. The promoter selectivity study further concluded that Pluronic activated the gene driven by the p53 response element, which was not present in either of the promoters used. Thus, Pluronic induced transcriptional activation of gene expression by activating the p53 and NF-κB signaling pathways. Interestingly, some natural amphiphilic agents, such as green tea polyphenol epigallocatechin-3-gallate were shown to activate both p53 and NF-κB in the muscle cells [66].

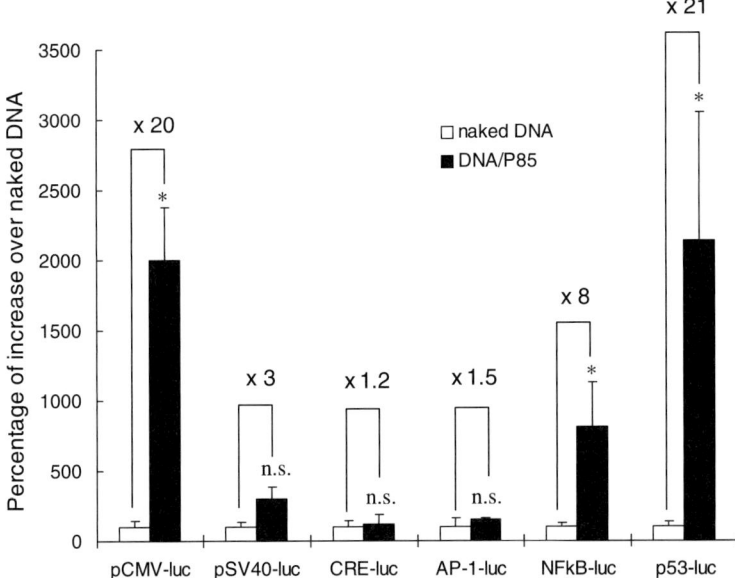

Fig. 7 Promoter selective effect of Pluronic P85 on luciferase reported gene expression in skeletal muscle. Groups of five Balb/c mice were injected *i.m.* with 0.3% P85-formulated (*black bars*) or nonformulated (*white bars*) luciferase encoding plasmids driven by CMV, SV40, CRE, AP-1, NF-κB or p53. Luciferase activity was determined 24 h after injections. Data are the percentage of increase of P85-formulated DNA over the naked DNA, mean $\pm$ SEM ($n = 10$). *$p < 0.05$, n.s. not significant ($p = 0.06$)

The mouse strain studies showed that the effect of Pluronic on *i.m.* transgene expression was exerted in both normal mouse strains Balb/c and C57Bl/7, but was lost in the athymic nude mouse strain (Fig. 8). The athymic nude mouse has a defective immune system lacking many immune and inflammation related components. Thus, the fact that the nude mouse did not respond to Pluronic was generally consistent with the inflammatory promoter selectivity of the block copolymer. On the other hand, NF-κB and frequently p53 transcription factors are involved in inflammatory events. Combined, these results suggest that the transcriptional activity of Pluronic may be directly related to the inflammation signaling pathways. Importantly, AP-1, a proapoptotic factor that is also frequently regulated by the NF-κB system, was not responding, which indicated that the Pluronic-mediated effect on transcription was selective and not a result of a general nonspecific activation of the immune defense system such as with a NO-mediated burst.

Pluronic block copolymers have long been known as adjuvants, increasing immune responses and inflammation in vivo [67–72]. The inflammatory responses caused by Pluronic are related to histamine release in the mediator containing cells, which in turn correlates with the ionophore properties of the block copolymers selective for monovalent cations [73, 74]. There is

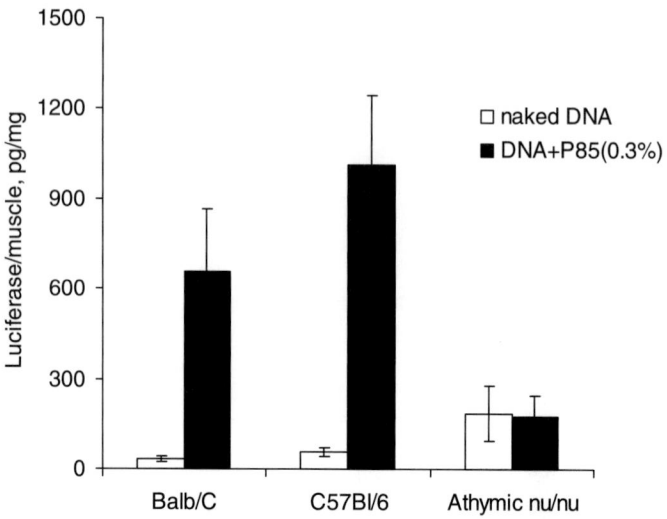

Fig. 8 Effect of P85 formulation on the expression of luciferase after *i.m.* administration of the plasmid DNA in mice. Bilateral tibialis anterior muscles of Balb/c, C57Bl/6 or athymic nu/nu mice were injected with 5 μg of gWiz Luc formulated with 0.3% w/v P85 or the plasmid DNA alone

a considerable evidence that the amphiphilicity of Pluronic molecules resulting in the membrane activity is responsible for a variety of immunologic and inflammatory activities of these compounds.

Though the formulations of plasmid DNA with Pluronic did not enhance DNA uptake by cells in vitro, the recent studies suggested that Pluronic P85 and other block copolymers enhanced expression of reporter genes under the control of CMV promoter and the NF-κB response element in stably transfected mouse fibroblasts and myoblasts [75]. This study further reinforced the conclusion that Pluronic block copolymers acted as biological response modifying agents by up-regulating the transcription of genes through the activation of selected signaling pathways.

The fact that Pluronic can affect the expression of genes that are already delivered into cells raises the question of whether other polymer-based formulations for gene delivery can also display similar effects. Recently, the impact on global gene expression of another DNA delivery system comprised of the cationic polymer formulations Lipofectin and Oligofectamine was studied in human epithelial cells A431 [76]. In this study, microarray expression profiling of 200 genes revealed marked changes in the expression of several genes for both formulations in the treated cells. Lipofectin and Oligofectamine treatment resulted in the (more than two-fold) altered expression of 10 and 27 genes, respectively. Particularly, of the genes represented in the array, Lipofectin initiated up-regulation of the replication protein a1 (*rpa1*), whereas Oligofectamine induced the overexpression of genes involved in apoptosis, such as the bcl2-related protein a1 (*bcl2a1*), caspase 8 isoform c (*casp8*), heat shock protein 70 (*hsp70*), and heat shock 60KDa protein 1 chaperonin (*hspd1*). Overall, the downstream functional consequences of the cationic polymer-induced gene expression alterations led to an increased tendency of cells for early apoptosis. Further, Lipofectin triggered suppression of endothelin receptor type b isoform 2 (*ednrb*), ribosomal protein 16 (*rp16*), and endothelin receptor type b, isoform 1 (ednrb). At the same time, Oligofectamine inhibited the expression of the genes annexin a2 (*anxa2*), retinoid x receptor alpha (*rxra*), and s100 calcium-binding protein a8 (*s100a8*). Nevertheless, the observed gene expression changes were not sufficient to induce any significant DNA damage, as assessed by single cell gel electrophoresis assay. Interestingly, these data suggested that gene changes – and thus the genotoxicity elicited by the cationic polymer formulations – were lipid-dependent, as no extensive overlap was observed in the cells treated with Lipofectin or Oligofectamine. It is emphasized that cationic polymer DNA delivery systems may also affect the outcome of gene therapy experiments inducing, attenuating, or even masking the desired effects of the nucleic acid [76].

7
Phenotypic correction of immune response by synthetic polyelectrolytes

The studies presented in this section were initiated by a group of polymer chemists in collaboration with immunologists in the Soviet Union over thirty years ago [77–80]. Linear synthetic polyelectrolytes of diverse chemical structure, when introduced into organisms, were shown to markedly intensify bone marrow formation and subsequent migration and settling of stem cells, which are precursors of all immune cells, in particular macrophages and T- and B-lymphocytes. The examples included negatively charged polyacrylic acid (PAA) and positively charged poly-4-vinylpyridine (PVP), as well as other polyelectrolytes. These synthetic polyelectrolytes had no structural relationship to biopolymers, and so they were not antigenic. However, when introduced in mixtures with typical antigens, such as proteins, natural microbial polysaccharides, or their synthetic analogs, they acted as immunostimulants, enhancing the immune response and increasing antibody production several-fold.

Thereafter, it was shown that polyelectrolytes trigger immune system response by a mechanism not encountered in nature. In particular, very small amounts of these polyelectrolytes added to a suspension of isolated B-lymphocytes activated DNA synthesis, induced cell fission, and, in the presence of antigens, triggered antigen-dependent differentiation of cells. Thus, the polyelectrolytes actually initiated immune response in B-lymphocyte cell suspension, notably, without the assistance of T-lymphocytes, antigen presenting cells (APC), or other cells of the immune system. Furthermore, the ability of polyelectrolytes to trigger activation of B-lymphocytes was demonstrated in vivo using mice devoid of T-cells. These mice did not respond to the administration of ordinary antigens. However, when the same antigens were introduced into the mixture with the polyelectrolytes, B-lymphocytes were activated and then antibodies were produced in the spleen that exhibited a strong immune response that was almost the same as seen in immunocompetent animals. Thus, synthetic polyelectrolytes can activate the immune response through an alternative pathway bypassing the control programmed by the immune response genes (Ir-genes) of the main complex of histocompatibility. As a result, either well or poorly genetically protected species reacted practically equally as strongly, which provides a unique example of the *phenotypic correction* of the Ir-genic control of immune response [81]. The alternative triggering is based on the general ability of linear polyelectrolytes to bind with cellular membranes and aggregate membrane protein molecules. It was shown that polyelectrolytes induce potassium vs. sodium and calcium ion exchange, which activated (K^+, Na^+)- and Ca^{2+}-ATPases, and served as a signal for the activation of the division and differentiation of the B-lymphocytes [82, 83].

This group developed a nontoxic biodegradable polycation, "polyoxidonium", a water-soluble ternary copolymer of 1,4-ethylenepiperazine, 1,4-ethylenepiperazine-N-oxide, and (N-carboxymethylene)-1,4-ethylenepiperazinium bromide, permitted for administration as an immunostimulant to humans in Russia [80]. The potency of polyoxidonium (and other polyelectrolytes) as an immunostimulator was greatly increased when it was covalently conjugated to antigens. It was found that such conjugates enhanced specific immune responses by orders of magnitude. Furthermore, a copolymer of acrylic acid (AA) with N-vinylpyrrolidone (VP) was shown to significantly increase immune response to the conjugated low-molecular mass trinitrophenyl (TNP) group, which is not even immunogenic [84]. The repeated administration of the conjugate into mice was accompanied by a strong production of TNP-specific antibody-forming cells in the animal's spleen. A similar effect was observed when bovine serum albumin was attached to the synthetic polyelectrolyte [85]. Discovery of the biological effect of conjugation of polyelectrolytes with antigens has opened the door to a novel generation of synthetic vaccines. In particular, individual bacterial or viral antigens, which are not sufficiently active on their own, induced specific immune responses when chemically conjugated to polyoxidonium [80]. The immunization of animals with such conjugates elicited protection against mortal doses of the corresponding bacteria or viruses [86, 87]. These studies led to the development of polymer-subunit human vaccines. One such vaccine, "Grippol", was obtained by conjugation of polyoxidonium with protein subunits of influenza viruses, hemagglutinin, and neuraminidase. About 50 million recipients have been vaccinated with Grippol over the last seven years, and extensive statistical data indicates the high efficiency and innocuous character of the vaccine obtained [80, 88].

8
Effects of polymer-coated surfaces on signal transduction in adherent cells

The studies discussed above refer to the use of water-soluble or amphiphilic membrane-active polymers, which were administered to cells or in the body in aqueous dispersions. There is, however, an increasing body of evidence that polymers immobilized at surfaces can also have significant phenotypic effects at signal transduction and gene expression levels. It has long been known that surface chemistry can modulate many critical functions of adherent cells, such as adhesion, fusion, spreading, phagocytosis, and secretion. These effects can reflect changes in expression levels of the corresponding genes. Several studies have described such alterations in cells growing on surfaces coated with different polymers [89, 90]. In particular, Chinese hamster ovary fibroblasts, CHO, were cultured on standard polystyrene dishes covered with poly-2-hydroxyethyl methacrylate (PHEMA) or untreated dishes used as a control [89]. The cells growing on PHEMA dishes acquired spheroidal morphology and were of reduced size compared to the control cells. Furthermore, these cells had round and reduced nuclear areas and displayed a high degree of chromatin condensation compared to the control cells, which had large ellipsoidal nuclei. Interestingly, the thicker the layer of the polymer coating, the more pronounced the changes in the cells. These changes in the cell morphology were accompanied by different levels of gene expression, as determined by reverse transcriptase polymerase chain reaction (RT-PCR) and RNA dot blot. In particular, transcriptional induction of oncogenes (*c-myc*, *c-fos*, and *c-jun*) and a constitutive gene, collagen $\alpha 1$ (VI) chain, were observed in cells grown on PHEMA surfaces. A remodeling of the structure and organization of chromatin were suggested as a possible mechanism for signal transduction from the cell surface interfacing the polymer layer to the cell gene expression machinery. Unfortunately, this study did not examine whether any polymer molecules can cross the cellular membranes and enter the cells. In this case, PHEMA can also target intracellular organelles, for example, mitochondria or nuclei, and exercise various approaches to gene alteration.

The effects of various polymer coatings on signal transduction were also studied in adherent monocyte/macrophages [91–93]. In these studies, polyethylene terephthalate base surfaces were coated with poly(benzyl-*N*,*N*-diethyldithio-carbamate-co-styrene) (BDEDTC) and then modified using photograft-copolymerization with polyacrylamide, PAA sodium salt, or methiodide of poly(dimethylaminopropyl-acrylamide), to yield hydrophilic, hydrophobic, anionic, and cationic surfaces. Expression levels of genes responsible for anti-inflammatory response, specifically, interleukins 8 and 10 (IL-8, IL-10), were evaluated by the RT-PCR method [92]. Expression of IL-10 was significantly increased in the cells growing on hydrophilic and anionic surfaces, but decreased in cells grown at the cationic surfaces. Conversely, expression of IL-8 was significantly decreased in the cells adherent to the hydrophilic and anionic surfaces. Further analysis revealed that these surfaces inhibited monocyte adhesion and IL-4 mediated macrophage fusion into foreign body giant cells. Overall, the hydrophilic and anionic surfaces promoted an anti-inflammatory response by "imposing" selective cytokine production in adherent monocytes and macrophages. It was also shown that the chemical composition of the polymers coating can affect apoptosis in the adherent cells [91, 93]. Thus, hydrophilic and anionic surfaces induced apoptosis of human adherent macrophages to a greater extent than hydrophobic or cationic surfaces. The inverse correlation between induction of apoptosis and the ability of the surface to promote adhesion and IL-4-mediated fusion of adherent macrophages was also demonstrated [91]. Finally, the effects of the surface chemistries of model implants on leukocyte cytokine mRNA expression were demonstrated in vivo [94]. This study demonstrated that hydrophilic surface chemistries had significant effects on leukocyte cytokine responses by decreasing the expression of inflammatory and wound healing cytokines by inflammatory cells adherent to the biomaterial, as well as present in the surrounding exudates. Overall, these findings can contribute to the design of bioengineered implant surfaces that are less susceptible to failure due to the host-foreign-body response.

9
Polymer genomics hypothesis and future studies

The studies described in this paper suggest that synthetic polymers can have rather selective phenotypic effects in cells and in the body, affecting signal transduction mechanisms involving inflammation, differentiation, proliferation and apoptosis. These effects can be realized as a result of interactions of water-soluble polymers with plasma cell membranes, delivery of polymers to intracellular organelles and at the sites of cell contacts with polymer-coated surfaces. The ability of the cells and organisms to respond to the effects of these polymers can be dependent on phenotype (such as ATP depletion in MDR cells) or genotype (as with the effects of Pluronic on gene expression in muscle in immunocompetent and nude mouse). In selected cases, polymer agents can bypass limitations to biological response imposed by the genotype (as in phenotypic correction of immune response by polyelectrolytes). Overall, these effects are relatively "weak" or restricted, as they do not result in cytotoxicity or major toxicities in the body. Indeed the polymers described in these studies are generally safe, and some of them, such as Pluronic, PHPMA or polyoxydonium, are used in humans without major side effects. Furthermore, selection of cells with these polymers, such as Pluronic, does not induce major genomic alterations, suggesting that these polymers are genetically benign – the cells can withstand the phenotypic effects of these polymers without drastic changes in cell function and life cycle. However, when combined with specific biological agents, such as cytotoxic agents, bacterial DNA, or antigens, either by mixing or covalent conjugation, these polymers can drastically alter specific genetically controlled responses to these agents. Thus, the central hypothesis of polymer genomics is that *weak phenotypic effects of polymer materials can be greatly amplified during the response of a cell or organism to biologically active agents by changing the magnitude, and in selected cases, the direction of these responses.* Several examples have shown that combinations of synthetic polymers and biological agents can be beneficial to health, for example preventing the development of drug resistance during chemotherapy or enhancing the protective immune response after vaccination. In other cases, these effects may result in unexpected or unwanted consequences, and these should be analyzed from the standpoint of toxicology, immune reactions, and development of new resistance mechanisms.

Future studies should determine the specific mechanisms by which polymer materials modify biological responses. It should be expected, based on available work (such as the effects of Pluronic block copolymers in MDR cancer cells), that these mechanisms could be quite complex and include multiple and interrelated components. Therefore, mechanistic studies should focus on identifying structural determinants of the polymer materials that render them biologically active in various specific applications. One unifying property of polymer materials is the ability to engage in cooperative interactions with biomacromolecules and subcellular components. Due to their polymeric structure, these materials contain numerous chemical groups, each of which can engage in relatively weak and nonspecific van der Waals, electrostatic or hydrophobic interactions with other macromolecular objects. Combining all of these groups in polymeric species can result in profound effects on biomacromolecules and subcellular components, such as changes in the conformations and functions of membrane proteins, alterations in the permeability of biological membranes, and other biophysical properties. From this standpoint, even polymeric species assembled from many small molecules (such as lipid vesicles or nanosuspensions of hydrophobic drugs) can also engage in substantial cooperative surface interactions within the cells and modulate the activities of biological agents in the body. This consideration becomes particularly significant because of the increasing use of polymer therapeutics in humans and the development of a novel generation of nanomaterials for drug and gene delivery. In addition to being *sufficiently large*, from the standpoint of being capable of cooperative interactions with other macromolecular objects, these nanomaterials are also *sufficiently small*, and can be transported into the body and into cells to the sites where they can reach different organelles and molecular targets. Thus the nanoscale range of sizes is particularly important from the point of view of the abilities of the polymer materials to modify biological responses and their pharmacogenomic effects.

Studies of mechanisms should also focus on identifying the general signal transduction pathways affected by different types of polymer materials. In this respect bioinformatics approaches can be of great help in determining and analyzing multiple alterations in cells in combination with proteomics and functional genomics approaches. It is possible that this analysis will allow several general pathways affected by different polymeric materials to be identified and related to specific chemical compositions and physicochemical properties of the materials. In this respect, the studies relating physicochemical properties of various polymer-modified surfaces and inflammatory cytokine responses to biomaterials described in the previous section are exemplary. Obtaining information about structure-functional properties of polymer materials will further facilitate the engineering of materials tailored for different types of biological applications. It may also allow "responsive" materials to be designed that can be delivered to specific sites in the body and within the cell to locally modulate signaling pathways using external forces, such as applied magnetic field, heat or ultrasound. Even though such a goal may currently appear futuristic, it should be achievable with modern developments in site-specific drug delivery and the expanding possibilities in nanomaterial design, particularly in the area of magnetic and temperature-responsive nanoparticles.

Once substantial clinical information regarding various polymer-based drugs and drug delivery systems becomes available, it will also be possible to relate the toxicological and immunological properties of such materials, as well as their efficacy in specific applications, to their potential biological response-modifying and pharmacogenomic effects. This will underscore the clinical significance of the polymer genomics approach to human health, and allow us to determine its limitations.

10
Conclusions

Studies described in this paper were performed using different polymer materials and diverse biological models. These studies, however, present compelling evidence that polymer materials can have significant effects on signal transduction mechanisms, gene expression, and the effects of biological agents. We have only recently begun to explore the field of polymer genomics [12]. However, as often happens, some of the ideas underlying this new field have been discussed before. This is illustrated by the following citation: "... *exciting prospects are linked to the studies of synthetic polyelectrolytes as novel types of physiologically active agents. Ability for cooperative interaction with charged cell membranes and other components of living systems most likely allows for extremely strong effects of ionizing macromolecules of different chemical compositions on the basic functions of the living organisms, effects that are currently hard to predict, and which can be directed towards the benefit of the organism as well as to its harm.*" (translated from [95]). It became clear that the range of materials may be much broader than just the initially proposed ionizing macromolecules and also includes nonionic water-soluble and amphiphilic polymers, as well as various polymer-coated surfaces. Importantly, these materials include safe polymer excipients, such as the Pluronic block copolymers currently listed in pharmacopeias, PHPMA and others. The pharmaceutical industry and health regulatory authorities refer to some polymer excipients as "GRAS" compounds ("*generally regarded as safe*"). We now suggest that this approach should be used with a degree of caution, since some polymer excipients, particularly in combination with biologically active agents, cannot be considered to be inert. Therefore, the combination of a biological agent and an excipient in a drug or gene delivery system may need to be thought of as a "new" entity rather than a reformulation of an existing one. In selected cases, the activities of polymer excipients may have beneficial effects, such as overcoming drug resistance in chemotherapy or enhancement of protective immune response during vaccination. In any case, the pharmacogenomic effects of polymers require further characterization, since they may play important roles in pharmaceutics and perhaps other areas, such as tissue engineering. Collectively these studies propose the need for thorough assessment of the pharmacogenomic effects of polymer materials in order to maximize the clinical outcomes and to understand the pharmacological and toxicological effects of polymer formulations of biological agents – in other words *polymer genomics.*

Acknowledgements We are grateful to Dr. David Kelly and Dr. Li Xiao (Eppley Institute for Research in Cancer, UNMC) for their assistance during the analysis of the DNA microarray data. The studies described in this paper were in part supported by the United States National Cancer Institute (CA89225).

References

1. Roses AD (2004) Nat Rev Genet 5:645
2. Gottesman MM, Fojo T, Bates SE (2002) Nat Rev Cancer 2:48
3. Langer R (1998) Nature 392:5
4. Kabanov A, Alakhov V (2002) Crit Rev Ther Drug Carrier Syst 19:1
5. Parker AL, Newman C, Briggs S, Seymour L, Sheridan PJ (2003) Expert Rev Mol Med 5:1
6. Nishiyama N, Kataoka K (2003) Adv Exp Med Biol 519:155

7. Duncan R (2003) Nat Rev Drug Discov 2:347
8. Savic R, Luo L, Eisenberg A, Maysinger D (2003) Science 300:615
9. Hubbell JA (2003) Science 300:595
10. Russell-Jones GJ (1998) Crit Rev Ther Drug Carrier Syst 15:557
11. Kabanov AV, Felgner PL, Seymour LW (1998) Self-assembling complexes for gene delivery: From laboratory to clinical trial. Wiley, Chichester, p 442
12. Kabanov AV, Batrakova EV, Sriadibhatla S, Yang Z, Kelly DL, Alakov VY (2005) J Control Release 101:259
13. Krishan A, Fitz CM, Andritsch I (1997) Cytometry 29:297
14. van Veen HW, Konings WN (1997) Semin Cancer Biol 8:183
15. Zaman GJ, Lankelma J, van Tellingen O, Beijnen J, Dekker H, Paulusma C, Oude Elferink RP, Baas F, Borst P (1995) Proc Natl Acad Sci USA 92:7690
16. Hayes JD, Pulford DJ (1995) Crit Rev Biochem Mol Biol 30:445
17. Altan N, Chen Y, Schindler M, Simon SM (1998) J Exp Med 187:1583
18. Benderra Z, Morjani H, Trussardi A, Manfait M (1998) Int J Oncol 12:711
19. Benderra Z, Morjani H, Trussardi A, Manfait M (1999) Adv Exp Med Biol 457:151
20. Alakhov V, Moskaleva E, Batrakova E, Kabanov A (1996) Bioconjug Chem 7:209
21. Batrakova EV, Dorodnych TY, Klinskii EY, Kliushnenkova EN, Shemchukova OB, Goncharova ON, Arjakov SA, Alakhov VY, Kabanov AV (1996) Br J Cancer 74:1545
22. Venne A, Li S, Mandeville R, Kabanov A, Alakhov V (1996) Cancer Res 56:3626
23. Kabanov AV, Chekhonin VP, Alakhov VY, Batrakova EV, Lebedev AS, Melik-Nubarov NS, Arzhakov SA, Levashov AV, Morozov GV, Severin ES, Kabanov VA (1989) FEBS Lett 258:343
24. Kabanov A, Batrakova E, Melik-Nubarov N, Fedoseev N, Dorodnich T, Alakhov V, Chekhonin V, Nazarova I, Kabanov V (1992) J Control Release 22:141
25. Kabanov AV, Slepnev VI, Kuznetsova LE, Batrakova EV, Alakhov VY, Melik-Nubarov NS, Sveshnikov PG, Kabanov VA (1992) Biochem Int 26:1035
26. Batrakova E, Lee S, Li S, Venne A, Alakhov VAK (1999) Pharm Res 16:1373
27. Batrakova EV, Li S, Elmquist WF, Miller DW, Alakhov VY, Kabanov AV (2001) Br J Cancer 85:1987
28. Batrakova E, Li S, Alakhov V, Miller D, Kabanov A (2003) J Pharmacol Exp Ther 304:845
29. Batrakova E, Li S, Vinogradov S, Alakhov V, Miller D, Kabanov A (2001) J Pharmacol Exp Ther 299:483
30. Batrakova EV, Li S, Alakhov VY, Elmquist WF, Miller DW, Kabanov AV (2003) Pharm Res 20:1581
31. Kabanov A, Batrakova E, Alakhov V (2002) Adv Drug Deliv Rev 54:759
32. Rapoport N, Marin A, Timoshin A (2000) Arch Biochem Biophys 384:1000
33. Kabanov AV, Batrakova EV, Alakhov VY (2003) J Control Release 91:75
34. Valle JW, Lawrance J, Brewer J, Clayton A, Corrie P, Alakhov V, Ranson M (2004) J Clin Oncol ASCO Annual Meeting Proc (Post-Meeting Edition) 22:4195
35. Minko T, Kopeckova P, Kopecek J (2001) J Control Release 71:227
36. Dharap SS, Qiu B, Williams GC, Sinko P, Stein S, Minko T (2003) J Control Release 91:61
37. Nishiyama N, Koizumi F, Okazaki S, Matsumura Y, Nishio K, Kataoka K (2003) Bioconjug Chem 14:449
38. Kabanov AV, Batrakova EV, Sriadibhatla S, Yang Z, Kelly DL, Alakov VYu (2004) Polym Prepr 45(2):442
39. Minko T, Batrakova EV, Li S, Li Y, Pakunlu RI, Alakhov VYu, Kabanov AV (2005) J Control Release 105(3):269

40. Chin KV, Kong AN (2002) Pharm Res 19:1773
41. Watts GS, Futscher BW, Isett R, Gleason-Guzman M, Kunkel MW, Salmon SE (2001) J Pharmacol Exp Ther 299:434
42. Xiao L, Wang K, Teng Y, Zhang J (2003) FEBS Lett 538:117
43. Wolff JA, Malone RW, Williams P, Chong W, Acsadi G, Jani A, Felgner PL (1990) Science 247:1465
44. Ragot T, Vincent N, Chafey P, Vigne E, Gilgenkrantz H, Couton D, Cartaud J, Briand P, Kaplan JC, Perricaudet M et al. (1993) Nature 361:647
45. Alila H, Coleman M, Nitta H, French M, Anwer K, Liu Q, Meyer T, Wang J, Mumper R, Oubari D, Long S, Nordstrom J, Rolland A (1997) Hum Gene Ther 8:1785
46. Tokui M, Takei I, Tashiro F, Shimada A, Kasuga A, Ishii M, Ishii T, Takatsu K, Saruta T, Miyazaki J (1997) Biochem Biophys Res Commun 233:527
47. Tripathy SK, Svensson EC, Black HB, Goldwasser E, Margalith M, Hobart PM, Leiden JM (1996) Proc Natl Acad Sci USA 93:10876
48. Manthorpe M, Cornefert-Jensen F, Hartikka J, Felgner J, Rundell A, Margalith M, Dwarki V (1993) Hum Gene Ther 4:419
49. Blezinger P, Wang J, Gondo M, Quezada A, Mehrens D, French M, Singhal A, Sullivan S, Rolland A, Ralston R, Min W (1999) Nat Biotechnol 17:343
50. Cristiano R (1998) Anticancer Res 18:3241
51. Boussif O, Lezoualc'h F, Zanta MA, Mergny MD, Scherman D, Demeneix B, Behr JP (1995) Proc Natl Acad Sci USA 92:7297
52. Kabanov AV, Kabanov VA (1995) Bioconjug Chem 6:7
53. Schwartz B, Benoist C, Abdallah B, Rangara R, Hassan A, Scherman D, Demeneix BA (1996) Gene Ther 3:405
54. Mumper RJ, Duguid JG, Anwer K, Barron MK, Nitta H, Rolland AP (1996) Pharm Res 13:701
55. Rolland AP, Mumper RJ (1998) Adv Drug Deliv Rev 30:151
56. Lemieux P, Guerin N, Paradis G, Proulx R, Chistyakova L, Kabanov A, Alakhov V (2000) Gene Ther 7:986
57. Liaw J, Chang SF, Hsiao FC (2001) Gene Ther 8:999
58. Hartikka J, Sukhu L, Buchner C, Hazard D, Bozoukova V, Margalith M, Nishioka WK, Wheeler CJ, Manthorp M, Sawdey M (2001) Mol Ther 4:407
59. Pitard B, Pollard H, Agbulut O, Lambert O, Vilquin JT, Cherel Y, Abadie J, Samuel JL, Rigaud JL, Menoret S, Anegon I, Escande D (2002) Hum Gene Ther 13:1767
60. Gebhart CL, Alakhov VY, Kabanov AV (2003) Proc Controlled Release Soc 30th Annual Meeting, 19–23 July 2003, Glasgow, UK
61. Lu QL, Bou-Gharios G, Partridge TA (2003) Gene Ther 10:131
62. Riera M, Chillon M, Aran JM, Cruzado JM, Torras J, Grinyo JM, Fillat C (2004) J Gene Med 6:111
63. Alakhov V, Klinski E, Lemieux P, Pietrzynski G, Kabanov A (2001) Expert Opin Biol Ther 1:583
64. Yang Z, Zhu J, Sriadibhatla S, Gebhart C, Alakhov V, Kabanov A (2005) J Control Release (published online: 5 September)
65. Gilmore TD (1999) Oncogene 18:6842
66. Hofmann CS, Sonenshein GE (2003) FASEB J 17:702
67. Hunter R, Olsen M, Buynitzky S (1991) Vaccine 9:250
68. Hunter R, Strickland F, Kezdy F (1981) J Immunol 127:1244
69. Hunter RL, Bennett B (1984) J Immunol 133:3167
70. Hunter RL, Bennett B (1986) Scand J Immunol 23:287
71. Hunter RL, McNicholl J, Lal AA (1994) AIDS Res Hum Retroviruses 10:S95

72. Howerton DA, Hunter RL, Ziegler HK, Check IJ (1990) J Immunol 144:1578
73. Atkinson T, Smith T, Hunter R (1988) J Immunol 141:1307
74. Atkinson TP, Smith TF, Hunter RL (1988) J Immunol 141:1302
75. Kabanov A, Sriadibhatla S, Gebhart C, Yang Z, Alakhov V (2004) Polym Prepr 45:392
76. Omidi Y, Hollins AJ, Benboubetra M, Drayton R, Benter IF, Akhtar S (2003) J Drug Target 11:311
77. Petrov RV, Khaitov RM, Mikhailova AA, Manko VM, Kabanov VA (1984) Cell interactions and vaccines of tomorrow. Mir, Moscow
78. Kabanov V, Petrov R, Khaitov R (1984) Sov Sci Rev D Physicochem Biol 5
79. Petrov RV, Khaitov RM (1986) Am J Reprod Immunol Microbiol 10:105
80. Kabanov V (2004) Pure Appl Chem 76:1659
81. Petrov RV, Khaitov RM, Norimov AS, Nekrasov AV, Koryakin SA (1986) Immunol Lett 12:237
82. Ataullakhanov R, Petrov R, Khaitov R, Ataullakhanov F, Abdullaev D (1984) Dokl Akad Nauk SSSR 274:479
83. Kabanov VA (1986) Macromol Chem Macromol Symp 1:101
84. Petrov RV, Evdakov VP, Khaitov RM, Filatova ED, Alesksevea N (1977) Dokl Akad Nauk SSSR 236:1260
85. Petrov RV, Khaitov RM, Norimov A, Kabanov VA, Mustafaev MI (1979) Dokl Akad Nauk SSSR 249:249
86. Petrov RV, Kabanov VA, Khaitov RM, Nekrasov AV, Alekseeva N (1983) Dokl Akad Nauk SSSR 270:1257
87. Petrov RV, Khaitov RM, Zhdanov VM, Sinyakov MS, Norimov A, Nekrasov AV, Podchernyaeva R, Kharitonenkov IG, Shchipanova MV (1985) Vaccine 3:392
88. Khaitov RM, Nekrasov AV, Puchkova NG, Ivanova AS (2003) Zh Mikrobiol Epidemiol Immunobiol May–Jun(3):83
89. Vergani L, Grattarola M, Nicolini C (2004) Int J Biochem Cell Biol 36:1447
90. Gong YK, Luo L, Petit A, Zukor DJ, Huk OL, Antoniou J, Winnik FM, Mwale F (2004) J Biomed Mater Res
91. Brodbeck WG, Shive MS, Colton E, Nakayama Y, Matsuda T, Anderson JM (2001) J Biomed Mater Res 55:661
92. Brodbeck WG, Nakayama Y, Matsuda T, Colton E, Ziats NP, Anderson JM (2002) Cytokine 18:311
93. Brodbeck WG, Patel J, Voskerician G, Christenson E, Shive MS, Nakayama Y, Matsuda T, Ziats NP, Anderson JM (2002) Proc Natl Acad Sci USA 99:10287
94. Brodbeck WG, Voskerician G, Ziats NP, Nakayama Y, Matsuda T, Anderson JM (2003) J Biomed Mater Res 64A:320
95. Kabanov V, Topchiev D (1975) Polymerization of ionizing monomers. Nauka, Moscow

Author Index Volumes 101–193

Author Index Volumes 1–100 see Volume 100

de Abajo, J. and *de la Campa, J. G.*: Processable Aromatic Polyimides. Vol. 140, pp. 23–60.
Abe, A., Furuya, H., Zhou, Z., Hiejima, T. and *Kobayashi, Y.*: Stepwise Phase Transitions of Chain Molecules: Crystallization/Melting via a Nematic Liquid-Crystalline Phase. Vol. 181, pp. 121–152.
Abetz, V. and *Simon, P. F. W.*: Phase Behaviour and Morphologies of Block Copolymers. Vol. 189, pp. 125–212.
Abetz, V. see Förster, S.: Vol. 166, pp. 173–210.
Adolf, D. B. see Ediger, M. D.: Vol. 116, pp. 73–110.
Aharoni, S. M. and *Edwards, S. F.*: Rigid Polymer Networks. Vol. 118, pp. 1–231.
Alakhov, V. Y. see Kabanov, A. V.: Vol. 193, pp. 173–198.
Albertsson, A.-C. and *Varma, I. K.*: Aliphatic Polyesters: Synthesis, Properties and Applications. Vol. 157, pp. 99–138.
Albertsson, A.-C. see Edlund, U.: Vol. 157, pp. 53–98.
Albertsson, A.-C. see Söderqvist Lindblad, M.: Vol. 157, pp. 139–161.
Albertsson, A.-C. see Stridsberg, K. M.: Vol. 157, pp. 27–51.
Albertsson, A.-C. see Al-Malaika, S.: Vol. 169, pp. 177–199.
Allegra, G. and *Meille, S. V.*: Pre-Crystalline, High-Entropy Aggregates: A Role in Polymer Crystallization? Vol. 191, pp. 87–135.
Allen, S. see Ellis, J. S.: Vol. 193, pp. 123–172.
Al-Malaika, S.: Perspectives in Stabilisation of Polyolefins. Vol. 169, pp. 121–150.
Altstädt, V.: The Influence of Molecular Variables on Fatigue Resistance in Stress Cracking Environments. Vol. 188, pp. 105–152.
Améduri, B., Boutevin, B. and *Gramain, P.*: Synthesis of Block Copolymers by Radical Polymerization and Telomerization. Vol. 127, pp. 87–142.
Améduri, B. and *Boutevin, B.*: Synthesis and Properties of Fluorinated Telechelic Monodispersed Compounds. Vol. 102, pp. 133–170.
Ameduri, B. see Taguet, A.: Vol. 184, pp. 127–211.
Amir, R. J. and *Shabat, D.*: Domino Dendrimers. Vol. 192, pp. 59–94.
Amselem, S. see Domb, A. J.: Vol. 107, pp. 93–142.
Anantawaraskul, S., Soares, J. B. P. and *Wood-Adams, P. M.*: Fractionation of Semicrystalline Polymers by Crystallization Analysis Fractionation and Temperature Rising Elution Fractionation. Vol. 182, pp. 1–54.
Andrady, A. L.: Wavelenght Sensitivity in Polymer Photodegradation. Vol. 128, pp. 47–94.
Andreis, M. and *Koenig, J. L.*: Application of Nitrogen–15 NMR to Polymers. Vol. 124, pp. 191–238.
Angiolini, L. see Carlini, C.: Vol. 123, pp. 127–214.
Anjum, N. see Gupta, B.: Vol. 162, pp. 37–63.
Anseth, K. S., Newman, S. M. and *Bowman, C. N.*: Polymeric Dental Composites: Properties and Reaction Behavior of Multimethacrylate Dental Restorations. Vol. 122, pp. 177–218.

Antonietti, M. see Cölfen, H.: Vol. 150, pp. 67–187.
Aoki, H. see Ito, S.: Vol. 182, pp. 131–170.
Armitage, B. A. see O'Brien, D. F.: Vol. 126, pp. 53–58.
Arnal, M. L. see Müller, A. J.: Vol. 190, pp. 1–63.
Arndt, M. see Kaminski, W.: Vol. 127, pp. 143–187.
Arnold, A. and *Holm, C.*: Efficient Methods to Compute Long-Range Interactions for Soft Matter Systems. Vol. 185, pp. 59–109.
Arnold Jr., F. E. and *Arnold, F. E.*: Rigid-Rod Polymers and Molecular Composites. Vol. 117, pp. 257–296.
Arora, M. see Kumar, M. N. V. R.: Vol. 160, pp. 45–118.
Arshady, R.: Polymer Synthesis via Activated Esters: A New Dimension of Creativity in Macromolecular Chemistry. Vol. 111, pp. 1–42.
Auer, S. and *Frenkel, D.*: Numerical Simulation of Crystal Nucleation in Colloids. Vol. 173, pp. 149–208.
Auriemma, F., de Rosa, C. and *Corradini, P.*: Solid Mesophases in Semicrystalline Polymers: Structural Analysis by Diffraction Techniques. Vol. 181, pp. 1–74.

Bahar, I., Erman, B. and *Monnerie, L.*: Effect of Molecular Structure on Local Chain Dynamics: Analytical Approaches and Computational Methods. Vol. 116, pp. 145–206.
Baietto-Dubourg, M. C. see Chateauminois, A.: Vol. 188, pp. 153–193.
Ballauff, M. see Dingenouts, N.: Vol. 144, pp. 1–48.
Ballauff, M. see Holm, C.: Vol. 166, pp. 1–27.
Ballauff, M. see Rühe, J.: Vol. 165, pp. 79–150.
Balsamo, V. see Müller, A. J.: Vol. 190, pp. 1–63.
Baltá-Calleja, F. J., González Arche, A., Ezquerra, T. A., Santa Cruz, C., Batallón, F., Frick, B. and *López Cabarcos, E.*: Structure and Properties of Ferroelectric Copolymers of Poly(vinylidene) Fluoride. Vol. 108, pp. 1–48.
Baltussen, J. J. M. see Northolt, M. G.: Vol. 178, pp. 1–108.
Barnes, M. D. see Otaigbe, J. U.: Vol. 154, pp. 1–86.
Barnes, C. M. see Satchi-Fainaro, R.: Vol. 193, pp. 1–65.
Barsett, H. see Paulsen, S. B.: Vol. 186, pp. 69–101.
Barshtein, G. R. and *Sabsai, O. Y.*: Compositions with Mineralorganic Fillers. Vol. 101, pp. 1–28.
Barton, J. see Hunkeler, D.: Vol. 112, pp. 115–134.
Baschnagel, J., Binder, K., Doruker, P., Gusev, A. A., Hahn, O., Kremer, K., Mattice, W. L., Müller-Plathe, F., Murat, M., Paul, W., Santos, S., Sutter, U. W. and *Tries, V.*: Bridging the Gap Between Atomistic and Coarse-Grained Models of Polymers: Status and Perspectives. Vol. 152, pp. 41–156.
Bassett, D. C.: On the Role of the Hexagonal Phase in the Crystallization of Polyethylene. Vol. 180, pp. 1–16.
Batallán, F. see Baltá-Calleja, F. J.: Vol. 108, pp. 1–48.
Batog, A. E., Pet'ko, I. P. and *Penczek, P.*: Aliphatic-Cycloaliphatic Epoxy Compounds and Polymers. Vol. 144, pp. 49–114.
Batrakova, E. V. see Kabanov, A. V.: Vol. 193, pp. 173–198.
Baughman, T. W. and *Wagener, K. B.*: Recent Advances in ADMET Polymerization. Vol. 176, pp. 1–42.
Becker, O. and *Simon, G. P.*: Epoxy Layered Silicate Nanocomposites. Vol. 179, pp. 29–82.
Bell, C. L. and *Peppas, N. A.*: Biomedical Membranes from Hydrogels and Interpolymer Complexes. Vol. 122, pp. 125–176.
Bellon-Maurel, A. see Calmon-Decriaud, A.: Vol. 135, pp. 207–226.

Bennett, D. E. see O'Brien, D. F.: Vol. 126, pp. 53–84.
Berry, G. C.: Static and Dynamic Light Scattering on Moderately Concentraded Solutions: Isotropic Solutions of Flexible and Rodlike Chains and Nematic Solutions of Rodlike Chains. Vol. 114, pp. 233–290.
Bershtein, V. A. and *Ryzhov, V. A.*: Far Infrared Spectroscopy of Polymers. Vol. 114, pp. 43–122.
Bhargava, R., Wang, S.-Q. and *Koenig, J. L*: FTIR Microspectroscopy of Polymeric Systems. Vol. 163, pp. 137–191.
Biesalski, M. see Rühe, J.: Vol. 165, pp. 79–150.
Bigg, D. M.: Thermal Conductivity of Heterophase Polymer Compositions. Vol. 119, pp. 1–30.
Binder, K.: Phase Transitions in Polymer Blends and Block Copolymer Melts: Some Recent Developments. Vol. 112, pp. 115–134.
Binder, K.: Phase Transitions of Polymer Blends and Block Copolymer Melts in Thin Films. Vol. 138, pp. 1–90.
Binder, K. see Baschnagel, J.: Vol. 152, pp. 41–156.
Binder, K., Müller, M., Virnau, P. and *González MacDowell, L.*: Polymer+Solvent Systems: Phase Diagrams, Interface Free Energies, and Nucleation. Vol. 173, pp. 1–104.
Bird, R. B. see Curtiss, C. F.: Vol. 125, pp. 1–102.
Biswas, M. and *Mukherjee, A.*: Synthesis and Evaluation of Metal-Containing Polymers. Vol. 115, pp. 89–124.
Biswas, M. and *Sinha Ray, S.*: Recent Progress in Synthesis and Evaluation of Polymer-Montmorillonite Nanocomposites. Vol. 155, pp. 167–221.
Blankenburg, L. see Klemm, E.: Vol. 177, pp. 53–90.
Blumen, A. see Gurtovenko, A. A.: Vol. 182, pp. 171–282.
Bogdal, D., Penczek, P., Pielichowski, J. and *Prociak, A.*: Microwave Assisted Synthesis, Crosslinking, and Processing of Polymeric Materials. Vol. 163, pp. 193–263.
Bohrisch, J., Eisenbach, C. D., Jaeger, W., Mori, H., Müller, A. H. E., Rehahn, M., Schaller, C., Traser, S. and *Wittmeyer, P.*: New Polyelectrolyte Architectures. Vol. 165, pp. 1–41.
Bolze, J. see Dingenouts, N.: Vol. 144, pp. 1–48.
Bosshard, C.: see Gubler, U.: Vol. 158, pp. 123–190.
Boutevin, B. and *Robin, J. J.*: Synthesis and Properties of Fluorinated Diols. Vol. 102, pp. 105–132.
Boutevin, B. see Améduri, B.: Vol. 102, pp. 133–170.
Boutevin, B. see Améduri, B.: Vol. 127, pp. 87–142.
Boutevin, B. see Guida-Pietrasanta, F.: Vol. 179, pp. 1–27.
Boutevin, B. see Taguet, A.: Vol. 184, pp. 127–211.
Bowman, C. N. see Anseth, K. S.: Vol. 122, pp. 177–218.
Boyd, R. H.: Prediction of Polymer Crystal Structures and Properties. Vol. 116, pp. 1–26.
Bracco, S. see Sozzani, P.: Vol. 181, pp. 153–177.
Briber, R. M. see Hedrick, J. L.: Vol. 141, pp. 1–44.
Bronnikov, S. V., Vettegren, V. I. and *Frenkel, S. Y.*: Kinetics of Deformation and Relaxation in Highly Oriented Polymers. Vol. 125, pp. 103–146.
Brown, H. R. see Creton, C.: Vol. 156, pp. 53–135.
Bruza, K. J. see Kirchhoff, R. A.: Vol. 117, pp. 1–66.
Buchmeiser, M. R.: Regioselective Polymerization of 1-Alkynes and Stereoselective Cyclopolymerization of a, w-Heptadiynes. Vol. 176, pp. 89–119.
Budkowski, A.: Interfacial Phenomena in Thin Polymer Films: Phase Coexistence and Segregation. Vol. 148, pp. 1–112.
Bunz, U. H. F.: Synthesis and Structure of PAEs. Vol. 177, pp. 1–52.

Burban, J. H. see Cussler, E. L.: Vol. 110, pp. 67–80.
Burchard, W.: Solution Properties of Branched Macromolecules. Vol. 143, pp. 113–194.
Butté, A. see Schork, F. J.: Vol. 175, pp. 129–255.

Calmon-Decriaud, A., Bellon-Maurel, V., Silvestre, F.: Standard Methods for Testing the Aerobic Biodegradation of Polymeric Materials. Vol. 135, pp. 207–226.
Cameron, N. R. and *Sherrington, D. C.*: High Internal Phase Emulsions (HIPEs)-Structure, Properties and Use in Polymer Preparation. Vol. 126, pp. 163–214.
de la Campa, J. G. see de Abajo, J.: Vol. 140, pp. 23–60.
Candau, F. see Hunkeler, D.: Vol. 112, pp. 115–134.
Canelas, D. A. and *DeSimone, J. M.*: Polymerizations in Liquid and Supercritical Carbon Dioxide. Vol. 133, pp. 103–140.
Canva, M. and *Stegeman, G. I.*: Quadratic Parametric Interactions in Organic Waveguides. Vol. 158, pp. 87–121.
Capek, I.: Kinetics of the Free-Radical Emulsion Polymerization of Vinyl Chloride. Vol. 120, pp. 135–206.
Capek, I.: Radical Polymerization of Polyoxyethylene Macromonomers in Disperse Systems. Vol. 145, pp. 1–56.
Capek, I. and *Chern, C.-S.*: Radical Polymerization in Direct Mini-Emulsion Systems. Vol. 155, pp. 101–166.
Cappella, B. see Munz, M.: Vol. 164, pp. 87–210.
Carlesso, G. see Prokop, A.: Vol. 160, pp. 119–174.
Carlini, C. and *Angiolini, L.*: Polymers as Free Radical Photoinitiators. Vol. 123, pp. 127–214.
Carter, K. R. see Hedrick, J. L.: Vol. 141, pp. 1–44.
Casas-Vazquez, J. see Jou, D.: Vol. 120, pp. 207–266.
Chan, C.-M. and *Li, L.*: Direct Observation of the Growth of Lamellae and Spherulites by AFM. Vol. 188, pp. 1–41.
Chandrasekhar, V.: Polymer Solid Electrolytes: Synthesis and Structure. Vol. 135, pp. 139–206.
Chang, J. Y. see Han, M. J.: Vol. 153, pp. 1–36.
Chang, T.: Recent Advances in Liquid Chromatography Analysis of Synthetic Polymers. Vol. 163, pp. 1–60.
Charleux, B. and *Faust, R.*: Synthesis of Branched Polymers by Cationic Polymerization. Vol. 142, pp. 1–70.
Chateauminois, A. and *Baietto-Dubourg, M. C.*: Fracture of Glassy Polymers Within Sliding Contacts. Vol. 188, pp. 153–193.
Chen, P. see Jaffe, M.: Vol. 117, pp. 297–328.
Chern, C.-S. see Capek, I.: Vol. 155, pp. 101–166.
Chevolot, Y. see Mathieu, H. J.: Vol. 162, pp. 1–35.
Chim, Y. T. A. see Ellis, J. S.: Vol. 193, pp. 123–172.
Choe, E.-W. see Jaffe, M.: Vol. 117, pp. 297–328.
Chow, P. Y. and *Gan, L. M.*: Microemulsion Polymerizations and Reactions. Vol. 175, pp. 257–298.
Chow, T. S.: Glassy State Relaxation and Deformation in Polymers. Vol. 103, pp. 149–190.
Chujo, Y. see Uemura, T.: Vol. 167, pp. 81–106.
Chung, S.-J. see Lin, T.-C.: Vol. 161, pp. 157–193.
Chung, T.-S. see Jaffe, M.: Vol. 117, pp. 297–328.
Clarke, N.: Effect of Shear Flow on Polymer Blends. Vol. 183, pp. 127–173.
Coenjarts, C. see Li, M.: Vol. 190, pp. 183–226.

Cölfen, H. and *Antonietti, M.*: Field-Flow Fractionation Techniques for Polymer and Colloid Analysis. Vol. 150, pp. 67–187.
Colmenero, J. see Richter, D.: Vol. 174, pp. 1–221.
Comanita, B. see Roovers, J.: Vol. 142, pp. 179–228.
Comotti, A. see Sozzani, P.: Vol. 181, pp. 153–177.
Connell, J. W. see Hergenrother, P. M.: Vol. 117, pp. 67–110.
Corradini, P. see Auriemma, F.: Vol. 181, pp. 1–74.
Creton, C., Kramer, E. J., Brown, H. R. and *Hui, C.-Y.*: Adhesion and Fracture of Interfaces Between Immiscible Polymers: From the Molecular to the Continuum Scale. Vol. 156, pp. 53–135.
Criado-Sancho, M. see Jou, D.: Vol. 120, pp. 207–266.
Curro, J. G. see Schweizer, K. S.: Vol. 116, pp. 319–378.
Curtiss, C. F. and *Bird, R. B.*: Statistical Mechanics of Transport Phenomena: Polymeric Liquid Mixtures. Vol. 125, pp. 1–102.
Cussler, E. L., Wang, K. L. and *Burban, J. H.*: Hydrogels as Separation Agents. Vol. 110, pp. 67–80.
Czub, P. see Penczek, P.: Vol. 184, pp. 1–95.

Dalton, L.: Nonlinear Optical Polymeric Materials: From Chromophore Design to Commercial Applications. Vol. 158, pp. 1–86.
Dautzenberg, H. see Holm, C.: Vol. 166, pp. 113–171.
Davidson, J. M. see Prokop, A.: Vol. 160, pp. 119–174.
Davies, M. C. see Ellis, J. S.: Vol. 193, pp. 123–172.
Den Decker, M. G. see Northolt, M. G.: Vol. 178, pp. 1–108.
Desai, S. M. and *Singh, R. P.*: Surface Modification of Polyethylene. Vol. 169, pp. 231–293.
DeSimone, J. M. see Canelas, D. A.: Vol. 133, pp. 103–140.
DeSimone, J. M. see Kennedy, K. A.: Vol. 175, pp. 329–346.
Dhal, P. K., Holmes-Farley, S. R., Huval, C. C. and *Jozefiak, T. H.*: Polymers as Drugs. Vol. 192, pp. 9–58.
DiMari, S. see Prokop, A.: Vol. 136, pp. 1–52.
Dimonie, M. V. see Hunkeler, D.: Vol. 112, pp. 115–134.
Dingenouts, N., Bolze, J., Pötschke, D. and *Ballauf, M.*: Analysis of Polymer Latexes by Small-Angle X-Ray Scattering. Vol. 144, pp. 1–48.
Dodd, L. R. and *Theodorou, D. N.*: Atomistic Monte Carlo Simulation and Continuum Mean Field Theory of the Structure and Equation of State Properties of Alkane and Polymer Melts. Vol. 116, pp. 249–282.
Doelker, E.: Cellulose Derivatives. Vol. 107, pp. 199–266.
Dolden, J. G.: Calculation of a Mesogenic Index with Emphasis Upon LC-Polyimides. Vol. 141, pp. 189–245.
Domb, A. J., Amselem, S., Shah, J. and *Maniar, M.*: Polyanhydrides: Synthesis and Characterization. Vol. 107, pp. 93–142.
Domb, A. J. see Kumar, M. N. V. R.: Vol. 160, pp. 45–118.
Doruker, P. see Baschnagel, J.: Vol. 152, pp. 41–156.
Dubois, P. see Mecerreyes, D.: Vol. 147, pp. 1–60.
Dubrovskii, S. A. see Kazanskii, K. S.: Vol. 104, pp. 97–134.
Dudowicz, J. see Freed, K. F.: Vol. 183, pp. 63–126.
Duncan, R., Ringsdorf, H. and *Satchi-Fainaro, R.*: Polymer Therapeutics: Polymers as Drugs, Drug and Protein Conjugates and Gene Delivery Systems: Past, Present and Future Opportunities. Vol. 192, pp. 1–8.
Duncan, R. see Satchi-Fainaro, R.: Vol. 193, pp. 1–65.

Dunkin, I. R. see *Steinke, J.*: Vol. 123, pp. 81–126.
Dunson, D. L. see *McGrath, J. E.*: Vol. 140, pp. 61–106.
Dziezok, P. see *Rühe, J.*: Vol. 165, pp. 79–150.

Eastmond, G. C.: Poly(e-caprolactone) Blends. Vol. 149, pp. 59–223.
Ebringerová, A., Hromádková, Z. and *Heinze, T.*: Hemicellulose. Vol. 186, pp. 1–67.
Economy, J. and *Goranov, K.*: Thermotropic Liquid Crystalline Polymers for High Performance Applications. Vol. 117, pp. 221–256.
Ediger, M. D. and *Adolf, D. B.*: Brownian Dynamics Simulations of Local Polymer Dynamics. Vol. 116, pp. 73–110.
Edlund, U. and *Albertsson, A.-C.*: Degradable Polymer Microspheres for Controlled Drug Delivery. Vol. 157, pp. 53–98.
Edwards, S. F. see *Aharoni, S. M.*: Vol. 118, pp. 1–231.
Eisenbach, C. D. see *Bohrisch, J.*: Vol. 165, pp. 1–41.
Ellis, J. S., Allen, S., Chim, Y. T. A., Roberts, C. J., Tendler, S. J. B. and *Davies, M. C.*: Molecular-Scale Studies on Biopolymers Using Atomic Force Microscopy. Vol. 193, pp. 123–172.
Endo, T. see *Yagci, Y.*: Vol. 127, pp. 59–86.
Engelhardt, H. and *Grosche, O.*: Capillary Electrophoresis in Polymer Analysis. Vol. 150, pp. 189–217.
Engelhardt, H. and *Martin, H.*: Characterization of Synthetic Polyelectrolytes by Capillary Electrophoretic Methods. Vol. 165, pp. 211–247.
Eriksson, P. see *Jacobson, K.*: Vol. 169, pp. 151–176.
Erman, B. see *Bahar, I.*: Vol. 116, pp. 145–206.
Eschner, M. see *Spange, S.*: Vol. 165, pp. 43–78.
Estel, K. see *Spange, S.*: Vol. 165, pp. 43–78.
Estevez, R. and *Van der Giessen, E.*: Modeling and Computational Analysis of Fracture of Glassy Polymers. Vol. 188, pp. 195–234.
Ewen, B. and *Richter, D.*: Neutron Spin Echo Investigations on the Segmental Dynamics of Polymers in Melts, Networks and Solutions. Vol. 134, pp. 1–130.
Ezquerra, T. A. see *Baltá-Calleja, F. J.*: Vol. 108, pp. 1–48.

Fatkullin, N. see *Kimmich, R.*: Vol. 170, pp. 1–113.
Faust, R. see *Charleux, B.*: Vol. 142, pp. 1–70.
Faust, R. see *Kwon, Y.*: Vol. 167, pp. 107–135.
Fekete, E. see *Pukánszky, B.*: Vol. 139, pp. 109–154.
Fendler, J. H.: Membrane-Mimetic Approach to Advanced Materials. Vol. 113, pp. 1–209.
Fetters, L. J. see *Xu, Z.*: Vol. 120, pp. 1–50.
Fontenot, K. see *Schork, F. J.*: Vol. 175, pp. 129–255.
Förster, S., Abetz, V. and *Müller, A. H. E.*: Polyelectrolyte Block Copolymer Micelles. Vol. 166, pp. 173–210.
Förster, S. and *Schmidt, M.*: Polyelectrolytes in Solution. Vol. 120, pp. 51–134.
Freed, K. F. and *Dudowicz, J.*: Influence of Monomer Molecular Structure on the Miscibility of Polymer Blends. Vol. 183, pp. 63–126.
Freire, J. J.: Conformational Properties of Branched Polymers: Theory and Simulations. Vol. 143, pp. 35–112.
Frenkel, D. see *Hu, W.*: Vol. 191, pp. 1–35.
Frenkel, S. Y. see *Bronnikov, S. V.*: Vol. 125, pp. 103–146.
Frick, B. see *Baltá-Calleja, F. J.*: Vol. 108, pp. 1–48.
Fridman, M. L.: see *Terent'eva, J. P.*: Vol. 101, pp. 29–64.
Fuchs, G. see *Trimmel, G.*: Vol. 176, pp. 43–87.

Fukui, K. see Otaigbe, J. U.: Vol. 154, pp. 1–86.
Funke, W.: Microgels-Intramolecularly Crosslinked Macromolecules with a Globular Structure. Vol. 136, pp. 137–232.
Furusho, Y. see Takata, T.: Vol. 171, pp. 1–75.
Furuya, H. see Abe, A.: Vol. 181, pp. 121–152.

Galina, H.: Mean-Field Kinetic Modeling of Polymerization: The Smoluchowski Coagulation Equation. Vol. 137, pp. 135–172.
Gan, L. M. see Chow, P. Y.: Vol. 175, pp. 257–298.
Ganesh, K. see Kishore, K.: Vol. 121, pp. 81–122.
Gaw, K. O. and *Kakimoto, M.*: Polyimide-Epoxy Composites. Vol. 140, pp. 107–136.
Geckeler, K. E. see Rivas, B.: Vol. 102, pp. 171–188.
Geckeler, K. E.: Soluble Polymer Supports for Liquid-Phase Synthesis. Vol. 121, pp. 31–80.
Gedde, U. W. and *Mattozzi, A.*: Polyethylene Morphology. Vol. 169, pp. 29–73.
Gehrke, S. H.: Synthesis, Equilibrium Swelling, Kinetics Permeability and Applications of Environmentally Responsive Gels. Vol. 110, pp. 81–144.
Geil, P. H., Yang, J., Williams, R. A., Petersen, K. L., Long, T.-C. and *Xu, P.*: Effect of Molecular Weight and Melt Time and Temperature on the Morphology of Poly(tetrafluorethylene). Vol. 180, pp. 89–159.
de Gennes, P.-G.: Flexible Polymers in Nanopores. Vol. 138, pp. 91–106.
Georgiou, S.: Laser Cleaning Methodologies of Polymer Substrates. Vol. 168, pp. 1–49.
Geuss, M. see Munz, M.: Vol. 164, pp. 87–210.
Giannelis, E. P., Krishnamoorti, R. and *Manias, E.*: Polymer-Silicate Nanocomposites: Model Systems for Confined Polymers and Polymer Brushes. Vol. 138, pp. 107–148.
Van der Giessen, E. see Estevez, R.: Vol. 188, pp. 195–234.
Godovsky, D. Y.: Device Applications of Polymer-Nanocomposites. Vol. 153, pp. 163–205.
Godovsky, D. Y.: Electron Behavior and Magnetic Properties Polymer-Nanocomposites. Vol. 119, pp. 79–122.
Gohy, J.-F.: Block Copolymer Micelles. Vol. 190, pp. 65–136.
González Arche, A. see Baltá-Calleja, F. J.: Vol. 108, pp. 1–48.
Goranov, K. see Economy, J.: Vol. 117, pp. 221–256.
Gramain, P. see Améduri, B.: Vol. 127, pp. 87–142.
Grein, C.: Toughness of Neat, Rubber Modified and Filled β-Nucleated Polypropylene: From Fundamentals to Applications. Vol. 188, pp. 43–104.
Greish, K. see Maeda, H.: Vol. 193, pp. 103–121.
Grest, G. S.: Normal and Shear Forces Between Polymer Brushes. Vol. 138, pp. 149–184.
Grigorescu, G. and *Kulicke, W.-M.*: Prediction of Viscoelastic Properties and Shear Stability of Polymers in Solution. Vol. 152, p. 1–40.
Gröhn, F. see Rühe, J.: Vol. 165, pp. 79–150.
Grosberg, A. and *Nechaev, S.*: Polymer Topology. Vol. 106, pp. 1–30.
Grosche, O. see Engelhardt, H.: Vol. 150, pp. 189–217.
Grubbs, R., Risse, W. and *Novac, B.*: The Development of Well-defined Catalysts for Ring-Opening Olefin Metathesis. Vol. 102, pp. 47–72.
Gubler, U. and *Bosshard, C.*: Molecular Design for Third-Order Nonlinear Optics. Vol. 158, pp. 123–190.
Guida-Pietrasanta, F. and *Boutevin, B.*: Polysilalkylene or Silarylene Siloxanes Said Hybrid Silicones. Vol. 179, pp. 1–27.
van Gunsteren, W. F. see Gusev, A. A.: Vol. 116, pp. 207–248.
Gupta, B. and *Anjum, N.*: Plasma and Radiation-Induced Graft Modification of Polymers for Biomedical Applications. Vol. 162, pp. 37–63.

Gurtovenko, A. A. and *Blumen, A.*: Generalized Gaussian Structures: Models for Polymer Systems with Complex Topologies. Vol. 182, pp. 171–282.
Gusev, A. A., Müller-Plathe, F., van Gunsteren, W. F. and *Suter, U. W.*: Dynamics of Small Molecules in Bulk Polymers. Vol. 116, pp. 207–248.
Gusev, A. A. see Baschnagel, J.: Vol. 152, pp. 41–156.
Guillot, J. see Hunkeler, D.: Vol. 112, pp. 115–134.
Guyot, A. and *Tauer, K.*: Reactive Surfactants in Emulsion Polymerization. Vol. 111, pp. 43–66.

Hadjichristidis, N., Pispas, S., Pitsikalis, M., Iatrou, H. and *Vlahos, C.*: Asymmetric Star Polymers Synthesis and Properties. Vol. 142, pp. 71–128.
Hadjichristidis, N., Pitsikalis, M. and *Iatrou, H.*: Synthesis of Block Copolymers. Vol. 189, pp. 1–124.
Hadjichristidis, N. see Xu, Z.: Vol. 120, pp. 1–50.
Hadjichristidis, N. see Pitsikalis, M.: Vol. 135, pp. 1–138.
Hahn, O. see Baschnagel, J.: Vol. 152, pp. 41–156.
Hakkarainen, M.: Aliphatic Polyesters: Abiotic and Biotic Degradation and Degradation Products. Vol. 157, pp. 1–26.
Hakkarainen, M. and *Albertsson, A.-C.*: Environmental Degradation of Polyethylene. Vol. 169, pp. 177–199.
Halary, J. L. see Monnerie, L.: Vol. 187, pp. 35–213.
Halary, J. L. see Monnerie, L.: Vol. 187, pp. 215–364.
Hall, H. K. see Penelle, J.: Vol. 102, pp. 73–104.
Hamley, I. W.: Crystallization in Block Copolymers. Vol. 148, pp. 113–138.
Hammouda, B.: SANS from Homogeneous Polymer Mixtures: A Unified Overview. Vol. 106, pp. 87–134.
Han, M. J. and *Chang, J. Y.*: Polynucleotide Analogues. Vol. 153, pp. 1–36.
Harada, A.: Design and Construction of Supramolecular Architectures Consisting of Cyclodextrins and Polymers. Vol. 133, pp. 141–192.
Haralson, M. A. see Prokop, A.: Vol. 136, pp. 1–52.
Harding, S. E.: Analysis of Polysaccharides by Ultracentrifugation. Size, Conformation and Interactions in Solution. Vol. 186, pp. 211–254.
Hasegawa, N. see Usuki, A.: Vol. 179, pp. 135–195.
Hassan, C. M. and *Peppas, N. A.*: Structure and Applications of Poly(vinyl alcohol) Hydrogels Produced by Conventional Crosslinking or by Freezing/Thawing Methods. Vol. 153, pp. 37–65.
Hawker, C. J.: Dentritic and Hyperbranched Macromolecules Precisely Controlled Macromolecular Architectures. Vol. 147, pp. 113–160.
Hawker, C. J. see Hedrick, J. L.: Vol. 141, pp. 1–44.
He, G. S. see Lin, T.-C.: Vol. 161, pp. 157–193.
Hedrick, J. L., Carter, K. R., Labadie, J. W., Miller, R. D., Volksen, W., Hawker, C. J., Yoon, D. Y., Russell, T. P., McGrath, J. E. and *Briber, R. M.*: Nanoporous Polyimides. Vol. 141, pp. 1–44.
Hedrick, J. L., Labadie, J. W., Volksen, W. and *Hilborn, J. G.*: Nanoscopically Engineered Polyimides. Vol. 147, pp. 61–112.
Hedrick, J. L. see Hergenrother, P. M.: Vol. 117, pp. 67–110.
Hedrick, J. L. see Kiefer, J.: Vol. 147, pp. 161–247.
Hedrick, J. L. see McGrath, J. E.: Vol. 140, pp. 61–106.
Heine, D. R., Grest, G. S. and *Curro, J. G.*: Structure of Polymer Melts and Blends: Comparison of Integral Equation theory and Computer Sumulation. Vol. 173, pp. 209–249.
Heinrich, G. and *Klüppel, M.*: Recent Advances in the Theory of Filler Networking in Elastomers. Vol. 160, pp. 1–44.

Heinze, T. see Ebringerová, A.: Vol. 186, pp. 1–67.
Heinze, T. see El Seoud, O. A.: Vol. 186, pp. 103–149.
Heller, J.: Poly (Ortho Esters). Vol. 107, pp. 41–92.
Helm, C. A. see Möhwald, H.: Vol. 165, pp. 151–175.
Hemielec, A. A. see Hunkeler, D.: Vol. 112, pp. 115–134.
Hergenrother, P. M., Connell, J. W., Labadie, J. W. and *Hedrick, J. L.*: Poly(arylene ether)s Containing Heterocyclic Units. Vol. 117, pp. 67–110.
Hernández-Barajas, J. see Wandrey, C.: Vol. 145, pp. 123–182.
Hervet, H. see Léger, L.: Vol. 138, pp. 185–226.
Hiejima, T. see Abe, A.: Vol. 181, pp. 121–152.
Hikosaka, M., Watanabe, K., Okada, K. and *Yamazaki, S.*: Topological Mechanism of Polymer Nucleation and Growth – The Role of Chain Sliding Diffusion and Entanglement. Vol. 191, pp. 137–186.
Hilborn, J. G. see Hedrick, J. L.: Vol. 147, pp. 61–112.
Hilborn, J. G. see Kiefer, J.: Vol. 147, pp. 161–247.
Hillborg, H. see Vancso, G. J.: Vol. 182, pp. 55–129.
Hillmyer, M. A.: Nanoporous Materials from Block Copolymer Precursors. Vol. 190, pp. 137–181.
Hiramatsu, N. see Matsushige, M.: Vol. 125, pp. 147–186.
Hirasa, O. see Suzuki, M.: Vol. 110, pp. 241–262.
Hirotsu, S.: Coexistence of Phases and the Nature of First-Order Transition in Poly-N-isopropylacrylamide Gels. Vol. 110, pp. 1–26.
Höcker, H. see Klee, D.: Vol. 149, pp. 1–57.
Holm, C. see Arnold, A.: Vol. 185, pp. 59–109.
Holm, C., Hofmann, T., Joanny, J. F., Kremer, K., Netz, R. R., Reineker, P., Seidel, C., Vilgis, T. A. and *Winkler, R. G.*: Polyelectrolyte Theory. Vol. 166, pp. 67–111.
Holm, C., Rehahn, M., Oppermann, W. and *Ballauff, M.*: Stiff-Chain Polyelectrolytes. Vol. 166, pp. 1–27.
Holmes-Farley, S. R. see Dhal, P. K.: Vol. 192, pp. 9–58.
Hornsby, P.: Rheology, Compounding and Processing of Filled Thermoplastics. Vol. 139, pp. 155–216.
Houbenov, N. see Rühe, J.: Vol. 165, pp. 79–150.
Hromádková, Z. see Ebringerová, A.: Vol. 186, pp. 1–67.
Hu, W. and *Frenkel, D.*: Polymer Crystallization Driven by Anisotropic Interactions. Vol. 191, pp. 1–35.
Huber, K. see Volk, N.: Vol. 166, pp. 29–65.
Hugenberg, N. see Rühe, J.: Vol. 165, pp. 79–150.
Hui, C.-Y. see Creton, C.: Vol. 156, pp. 53–135.
Hult, A., Johansson, M. and *Malmström, E.*: Hyperbranched Polymers. Vol. 143, pp. 1–34.
Hünenberger, P. H.: Thermostat Algorithms for Molecular-Dynamics Simulations. Vol. 173, pp. 105–147.
Hunkeler, D., Candau, F., Pichot, C., Hemielec, A. E., Xie, T. Y., Barton, J., Vaskova, V., Guillot, J., Dimonie, M. V. and *Reichert, K. H.*: Heterophase Polymerization: A Physical and Kinetic Comparision and Categorization. Vol. 112, pp. 115–134.
Hunkeler, D. see Macko, T.: Vol. 163, pp. 61–136.
Hunkeler, D. see Prokop, A.: Vol. 136, pp. 1–52; 53–74.
Hunkeler, D. see Wandrey, C.: Vol. 145, pp. 123–182.
Huval, C. C. see Dhal, P. K.: Vol. 192, pp. 9–58.

Iatrou, H. see *Hadjichristidis, N.*: Vol. 142, pp. 71–128.
Iatrou, H. see *Hadjichristidis, N.*: Vol. 189, pp. 1–124.
Ichikawa, T. see *Yoshida, H.*: Vol. 105, pp. 3–36.
Ihara, E. see *Yasuda, H.*: Vol. 133, pp. 53–102.
Ikada, Y. see *Uyama, Y.*: Vol. 137, pp. 1–40.
Ikehara, T. see *Jinnuai, H.*: Vol. 170, pp. 115–167.
Ilavsky, M.: Effect on Phase Transition on Swelling and Mechanical Behavior of Synthetic Hydrogels. Vol. 109, pp. 173–206.
Imai, M. see *Kaji, K.*: Vol. 191, pp. 187–240.
Imai, Y.: Rapid Synthesis of Polyimides from Nylon-Salt Monomers. Vol. 140, pp. 1–23.
Inomata, H. see *Saito, S.*: Vol. 106, pp. 207–232.
Inoue, S. see *Sugimoto, H.*: Vol. 146, pp. 39–120.
Irie, M.: Stimuli-Responsive Poly(N-isopropylacrylamide), Photo- and Chemical-Induced Phase Transitions. Vol. 110, pp. 49–66.
Ise, N. see *Matsuoka, H.*: Vol. 114, pp. 187–232.
Ishikawa, T.: Advances in Inorganic Fibers. Vol. 178, pp. 109–144.
Ito, H.: Chemical Amplification Resists for Microlithography. Vol. 172, pp. 37–245.
Ito, K. and *Kawaguchi, S.*: Poly(macronomers), Homo- and Copolymerization. Vol. 142, pp. 129–178.
Ito, K. see *Kawaguchi, S.*: Vol. 175, pp. 299–328.
Ito, S. and *Aoki, H.*: Nano-Imaging of Polymers by Optical Microscopy. Vol. 182, pp. 131–170.
Ito, Y. see *Suginome, M.*: Vol. 171, pp. 77–136.
Ivanov, A. E. see *Zubov, V. P.*: Vol. 104, pp. 135–176.

Jacob, S. and *Kennedy, J.*: Synthesis, Characterization and Properties of OCTA-ARM Polyisobutylene-Based Star Polymers. Vol. 146, pp. 1–38.
Jacobson, K., Eriksson, P., Reitberger, T. and *Stenberg, B.*: Chemiluminescence as a Tool for Polyolefin. Vol. 169, pp. 151–176.
Jaeger, W. see *Bohrisch, J.*: Vol. 165, pp. 1–41.
Jaffe, M., Chen, P., Choe, E.-W., Chung, T.-S. and *Makhija, S.*: High Performance Polymer Blends. Vol. 117, pp. 297–328.
Jancar, J.: Structure-Property Relationships in Thermoplastic Matrices. Vol. 139, pp. 1–66.
Jen, A. K.-Y. see *Kajzar, F.*: Vol. 161, pp. 1–85.
Jerome, R. see *Mecerreyes, D.*: Vol. 147, pp. 1–60.
de Jeu, W. H. see *Li, L.*: Vol. 181, pp. 75–120.
Jiang, M., Li, M., Xiang, M. and *Zhou, H.*: Interpolymer Complexation and Miscibility and Enhancement by Hydrogen Bonding. Vol. 146, pp. 121–194.
Jin, J. see *Shim, H.-K.*: Vol. 158, pp. 191–241.
Jinnai, H., Nishikawa, Y., Ikehara, T. and *Nishi, T.*: Emerging Technologies for the 3D Analysis of Polymer Structures. Vol. 170, pp. 115–167.
Jo, W. H. and *Yang, J. S.*: Molecular Simulation Approaches for Multiphase Polymer Systems. Vol. 156, pp. 1–52.
Joanny, J.-F. see *Holm, C.*: Vol. 166, pp. 67–111.
Joanny, J.-F. see *Thünemann, A. F.*: Vol. 166, pp. 113–171.
Johannsmann, D. see *Rühe, J.*: Vol. 165, pp. 79–150.
Johansson, M. see *Hult, A.*: Vol. 143, pp. 1–34.
Joos-Müller, B. see *Funke, W.*: Vol. 136, pp. 137–232.
Jou, D., Casas-Vazquez, J. and *Criado-Sancho, M.*: Thermodynamics of Polymer Solutions under Flow: Phase Separation and Polymer Degradation. Vol. 120, pp. 207–266.
Jozefiak, T. H. see *Dhal, P. K.*: Vol. 192, pp. 9–58.

Kabanov, A. V., Batrakova, E. V., Sherman, S. and *Alakhov, V. Y.*: Polymer Genomics. Vol. 193, pp. 173–198.
Kaetsu, I.: Radiation Synthesis of Polymeric Materials for Biomedical and Biochemical Applications. Vol. 105, pp. 81–98.
Kaji, K., Nishida, K., Kanaya, T., Matsuba, G., Konishi, T. and *Imai, M.*: Spinodal Crystallization of Polymers: Crystallization from the Unstable Melt. Vol. 191, pp. 187–240.
Kaji, K. see Kanaya, T.: Vol. 154, pp. 87–141.
Kajzar, F., Lee, K.-S. and *Jen, A. K.-Y.*: Polymeric Materials and their Orientation Techniques for Second-Order Nonlinear Optics. Vol. 161, pp. 1–85.
Kakimoto, M. see Gaw, K. O.: Vol. 140, pp. 107–136.
Kaminski, W. and *Arndt, M.*: Metallocenes for Polymer Catalysis. Vol. 127, pp. 143–187.
Kammer, H. W., Kressler, H. and *Kummerloewe, C.*: Phase Behavior of Polymer Blends – Effects of Thermodynamics and Rheology. Vol. 106, pp. 31–86.
Kanaya, T. and *Kaji, K.*: Dynamcis in the Glassy State and Near the Glass Transition of Amorphous Polymers as Studied by Neutron Scattering. Vol. 154, pp. 87–141.
Kanaya, T. see Kaji, K.: Vol. 191, pp. 187–240.
Kandyrin, L. B. and *Kuleznev, V. N.*: The Dependence of Viscosity on the Composition of Concentrated Dispersions and the Free Volume Concept of Disperse Systems. Vol. 103, pp. 103–148.
Kaneko, M. see Ramaraj, R.: Vol. 123, pp. 215–242.
Kang, E. T., Neoh, K. G. and *Tan, K. L.*: X-Ray Photoelectron Spectroscopic Studies of Electroactive Polymers. Vol. 106, pp. 135–190.
Karlsson, S. see Söderqvist Lindblad, M.: Vol. 157, pp. 139–161.
Karlsson, S.: Recycled Polyolefins. Material Properties and Means for Quality Determination. Vol. 169, pp. 201–229.
Kataoka, K. see Nishiyama, N.: Vol. 193, pp. 67–101.
Kato, K. see Uyama, Y.: Vol. 137, pp. 1–40.
Kato, M. see Usuki, A.: Vol. 179, pp. 135–195.
Kausch, H.-H. and *Michler, G. H.*: The Effect of Time on Crazing and Fracture. Vol. 187, pp. 1–33.
Kausch, H.-H. see Monnerie, L. Vol. 187, pp. 215–364.
Kautek, W. see Krüger, J.: Vol. 168, pp. 247–290.
Kawaguchi, S. see Ito, K.: Vol. 142, pp. 129–178.
Kawaguchi, S. and *Ito, K.*: Dispersion Polymerization. Vol. 175, pp. 299–328.
Kawata, S. see Sun, H.-B.: Vol. 170, pp. 169–273.
Kazanskii, K. S. and *Dubrovskii, S. A.*: Chemistry and Physics of Agricultural Hydrogels. Vol. 104, pp. 97–134.
Kennedy, J. P. see Jacob, S.: Vol. 146, pp. 1–38.
Kennedy, J. P. see Majoros, I.: Vol. 112, pp. 1–113.
Kennedy, K. A., Roberts, G. W. and *DeSimone, J. M.*: Heterogeneous Polymerization of Fluoroolefins in Supercritical Carbon Dioxide. Vol. 175, pp. 329–346.
Khokhlov, A., Starodybtzev, S. and *Vasilevskaya, V.*: Conformational Transitions of Polymer Gels: Theory and Experiment. Vol. 109, pp. 121–172.
Kiefer, J., Hedrick, J. L. and *Hiborn, J. G.*: Macroporous Thermosets by Chemically Induced Phase Separation. Vol. 147, pp. 161–247.
Kihara, N. see Takata, T.: Vol. 171, pp. 1–75.
Kilian, H. G. and *Pieper, T.*: Packing of Chain Segments. A Method for Describing X-Ray Patterns of Crystalline, Liquid Crystalline and Non-Crystalline Polymers. Vol. 108, pp. 49–90.
Kim, J. see Quirk, R. P.: Vol. 153, pp. 67–162.

Kim, K.-S. see *Lin, T.-C.*: Vol. 161, pp. 157–193.
Kimmich, R. and *Fatkullin, N.*: Polymer Chain Dynamics and NMR. Vol. 170, pp. 1–113.
Kippelen, B. and *Peyghambarian, N.*: Photorefractive Polymers and their Applications. Vol. 161, pp. 87–156.
Kirchhoff, R. A. and *Bruza, K. J.*: Polymers from Benzocyclobutenes. Vol. 117, pp. 1–66.
Kishore, K. and *Ganesh, K.*: Polymers Containing Disulfide, Tetrasulfide, Diselenide and Ditelluride Linkages in the Main Chain. Vol. 121, pp. 81–122.
Kitamaru, R.: Phase Structure of Polyethylene and Other Crystalline Polymers by Solid-State 13C/MNR. Vol. 137, pp. 41–102.
Klapper, M. see *Rusanov, A. L.*: Vol. 179, pp. 83–134.
Klee, D. and *Höcker, H.*: Polymers for Biomedical Applications: Improvement of the Interface Compatibility. Vol. 149, pp. 1–57.
Klemm, E., Pautzsch, T. and *Blankenburg, L.*: Organometallic PAEs. Vol. 177, pp. 53–90.
Klier, J. see *Scranton, A. B.*: Vol. 122, pp. 1–54.
v. Klitzing, R. and *Tieke, B.*: Polyelectrolyte Membranes. Vol. 165, pp. 177–210.
Kloeckner, J. see *Wagner, E.*: Vol. 192, pp. 135–173.
Klüppel, M.: The Role of Disorder in Filler Reinforcement of Elastomers on Various Length Scales. Vol. 164, pp. 1–86.
Klüppel, M. see *Heinrich, G.*: Vol. 160, pp. 1–44.
Knuuttila, H., Lehtinen, A. and *Nummila-Pakarinen, A.*: Advanced Polyethylene Technologies – Controlled Material Properties. Vol. 169, pp. 13–27.
Kobayashi, S., Shoda, S. and *Uyama, H.*: Enzymatic Polymerization and Oligomerization. Vol. 121, pp. 1–30.
Kobayashi, T. see *Abe, A.*: Vol. 181, pp. 121–152.
Köhler, W. and *Schäfer, R.*: Polymer Analysis by Thermal-Diffusion Forced Rayleigh Scattering. Vol. 151, pp. 1–59.
Koenig, J. L. see *Bhargava, R.*: Vol. 163, pp. 137–191.
Koenig, J. L. see *Andreis, M.*: Vol. 124, pp. 191–238.
Koike, T.: Viscoelastic Behavior of Epoxy Resins Before Crosslinking. Vol. 148, pp. 139–188.
Kokko, E. see *Löfgren, B.*: Vol. 169, pp. 1–12.
Kokufuta, E.: Novel Applications for Stimulus-Sensitive Polymer Gels in the Preparation of Functional Immobilized Biocatalysts. Vol. 110, pp. 157–178.
Konishi, T. see *Kaji, K.*: Vol. 191, pp. 187–240.
Konno, M. see *Saito, S.*: Vol. 109, pp. 207–232.
Konradi, R. see *Rühe, J.*: Vol. 165, pp. 79–150.
Kopecek, J. see *Putnam, D.*: Vol. 122, pp. 55–124.
Koßmehl, G. see *Schopf, G.*: Vol. 129, pp. 1–145.
Kostoglodov, P. V. see *Rusanov, A. L.*: Vol. 179, pp. 83–134.
Kozlov, E. see *Prokop, A.*: Vol. 160, pp. 119–174.
Kramer, E. J. see *Creton, C.*: Vol. 156, pp. 53–135.
Kremer, K. see *Baschnagel, J.*: Vol. 152, pp. 41–156.
Kremer, K. see *Holm, C.*: Vol. 166, pp. 67–111.
Kressler, J. see *Kammer, H. W.*: Vol. 106, pp. 31–86.
Kricheldorf, H. R.: Liquid-Cristalline Polyimides. Vol. 141, pp. 83–188.
Krishnamoorti, R. see *Giannelis, E. P.*: Vol. 138, pp. 107–148.
Krüger, J. and *Kautek, W.*: Ultrashort Pulse Laser Interaction with Dielectrics and Polymers, Vol. 168, pp. 247–290.
Kuchanov, S. I.: Modern Aspects of Quantitative Theory of Free-Radical Copolymerization. Vol. 103, pp. 1–102.

Kuchanov, S. I.: Principles of Quantitive Description of Chemical Structure of Synthetic Polymers. Vol. 152, pp. 157–202.
Kudaibergennow, S. E.: Recent Advances in Studying of Synthetic Polyampholytes in Solutions. Vol. 144, pp. 115–198.
Kuleznev, V. N. see Kandyrin, L. B.: Vol. 103, pp. 103–148.
Kulichkhin, S. G. see Malkin, A. Y.: Vol. 101, pp. 217–258.
Kulicke, W.-M. see Grigorescu, G.: Vol. 152, pp. 1–40.
Kumar, M. N. V. R., Kumar, N., Domb, A. J. and *Arora, M.*: Pharmaceutical Polymeric Controlled Drug Delivery Systems. Vol. 160, pp. 45–118.
Kumar, N. see Kumar, M. N. V. R.: Vol. 160, pp. 45–118.
Kummerloewe, C. see Kammer, H. W.: Vol. 106, pp. 31–86.
Kuznetsova, N. P. see Samsonov, G. V.: Vol. 104, pp. 1–50.
Kwon, Y. and *Faust, R.*: Synthesis of Polyisobutylene-Based Block Copolymers with Precisely Controlled Architecture by Living Cationic Polymerization. Vol. 167, pp. 107–135.

Labadie, J. W. see Hergenrother, P. M.: Vol. 117, pp. 67–110.
Labadie, J. W. see Hedrick, J. L.: Vol. 141, pp. 1–44.
Labadie, J. W. see Hedrick, J. L.: Vol. 147, pp. 61–112.
Lamparski, H. G. see O'Brien, D. F.: Vol. 126, pp. 53–84.
Laschewsky, A.: Molecular Concepts, Self-Organisation and Properties of Polysoaps. Vol. 124, pp. 1–86.
Laso, M. see Leontidis, E.: Vol. 116, pp. 283–318.
Laupêtre, F. see Monnerie, L.: Vol. 187, pp. 35–213.
Lazár, M. and *Rychl, R.*: Oxidation of Hydrocarbon Polymers. Vol. 102, pp. 189–222.
Lechowicz, J. see Galina, H.: Vol. 137, pp. 135–172.
Léger, L., Raphaël, E. and *Hervet, H.*: Surface-Anchored Polymer Chains: Their Role in Adhesion and Friction. Vol. 138, pp. 185–226.
Lenz, R. W.: Biodegradable Polymers. Vol. 107, pp. 1–40.
Leontidis, E., de Pablo, J. J., Laso, M. and *Suter, U. W.*: A Critical Evaluation of Novel Algorithms for the Off-Lattice Monte Carlo Simulation of Condensed Polymer Phases. Vol. 116, pp. 283–318.
Lee, B. see Quirk, R. P.: Vol. 153, pp. 67–162.
Lee, K.-S. see Kajzar, F.: Vol. 161, pp. 1–85.
Lee, Y. see Quirk, R. P.: Vol. 153, pp. 67–162.
Lehtinen, A. see Knuuttila, H.: Vol. 169, pp. 13–27.
Leónard, D. see Mathieu, H. J.: Vol. 162, pp. 1–35.
Lesec, J. see Viovy, J.-L.: Vol. 114, pp. 1–42.
Levesque, D. see Weis, J.-J.: Vol. 185, pp. 163–225.
Li, L. and *de Jeu, W. H.*: Flow-induced mesophases in crystallizable polymers. Vol. 181, pp. 75–120.
Li, L. see Chan, C.-M.: Vol. 188, pp. 1–41.
Li, M., Coenjarts, C. and *Ober, C. K.*: Patternable Block Copolymers. Vol. 190, pp. 183–226.
Li, M. see Jiang, M.: Vol. 146, pp. 121–194.
Liang, G. L. see Sumpter, B. G.: Vol. 116, pp. 27–72.
Lienert, K.-W.: Poly(ester-imide)s for Industrial Use. Vol. 141, pp. 45–82.
Likhatchev, D. see Rusanov, A. L.: Vol. 179, pp. 83–134.
Lin, J. and *Sherrington, D. C.*: Recent Developments in the Synthesis, Thermostability and Liquid Crystal Properties of Aromatic Polyamides. Vol. 111, pp. 177–220.

Lin, T.-C., Chung, S.-J., Kim, K.-S., Wang, X., He, G. S., Swiatkiewicz, J., Pudavar, H. E. and *Prasad, P. N.*: Organics and Polymers with High Two-Photon Activities and their Applications. Vol. 161, pp. 157–193.
Linse, P.: Simulation of Charged Colloids in Solution. Vol. 185, pp. 111–162.
Lippert, T.: Laser Application of Polymers. Vol. 168, pp. 51–246.
Liu, Y. see Söderqvist Lindblad, M.: Vol. 157, pp. 139–161.
Long, T.-C. see Geil, P. H.: Vol. 180, pp. 89–159.
López Cabarcos, E. see Baltá-Calleja, F. J.: Vol. 108, pp. 1–48.
Lotz, B.: Analysis and Observation of Polymer Crystal Structures at the Individual Stem Level. Vol. 180, pp. 17–44.
Löfgren, B., Kokko, E. and *Seppälä, J.*: Specific Structures Enabled by Metallocene Catalysis in Polyethenes. Vol. 169, pp. 1–12.
Löwen, H. see Thünemann, A. F.: Vol. 166, pp. 113–171.
Luo, Y. see Schork, F. J.: Vol. 175, pp. 129–255.

Macko, T. and *Hunkeler, D.*: Liquid Chromatography under Critical and Limiting Conditions: A Survey of Experimental Systems for Synthetic Polymers. Vol. 163, pp. 61–136.
Maeda, H., Greish, K. and *Fang, J.*: The EPR Effect and Polymeric Drugs: A Paradigm Shift for Cancer Chemotherapy in the 21st Century. Vol. 193, pp. 103–121.
Majoros, I., Nagy, A. and *Kennedy, J. P.*: Conventional and Living Carbocationic Polymerizations United. I. A Comprehensive Model and New Diagnostic Method to Probe the Mechanism of Homopolymerizations. Vol. 112, pp. 1–113.
Makhija, S. see Jaffe, M.: Vol. 117, pp. 297–328.
Malmström, E. see Hult, A.: Vol. 143, pp. 1–34.
Malkin, A. Y. and *Kulichkhin, S. G.*: Rheokinetics of Curing. Vol. 101, pp. 217–258.
Maniar, M. see Domb, A. J.: Vol. 107, pp. 93–142.
Manias, E. see Giannelis, E. P.: Vol. 138, pp. 107–148.
Martin, H. see Engelhardt, H.: Vol. 165, pp. 211–247.
Marty, J. D. and *Mauzac, M.*: Molecular Imprinting: State of the Art and Perspectives. Vol. 172, pp. 1–35.
Mashima, K., Nakayama, Y. and *Nakamura, A.*: Recent Trends in Polymerization of a-Olefins Catalyzed by Organometallic Complexes of Early Transition Metals. Vol. 133, pp. 1–52.
Mathew, D. see Reghunadhan Nair, C. P.: Vol. 155, pp. 1–99.
Mathieu, H. J., Chevolot, Y., Ruiz-Taylor, L. and *Léonard, D.*: Engineering and Characterization of Polymer Surfaces for Biomedical Applications. Vol. 162, pp. 1–35.
Matsuba, G. see Kaji, K.: Vol. 191, pp. 187–240.
Matsumoto, A.: Free-Radical Crosslinking Polymerization and Copolymerization of Multivinyl Compounds. Vol. 123, pp. 41–80.
Matsumoto, A. see Otsu, T.: Vol. 136, pp. 75–138.
Matsuoka, H. and *Ise, N.*: Small-Angle and Ultra-Small Angle Scattering Study of the Ordered Structure in Polyelectrolyte Solutions and Colloidal Dispersions. Vol. 114, pp. 187–232.
Matsushige, K., Hiramatsu, N. and *Okabe, H.*: Ultrasonic Spectroscopy for Polymeric Materials. Vol. 125, pp. 147–186.
Mattice, W. L. see Rehahn, M.: Vol. 131/132, pp. 1–475.
Mattice, W. L. see Baschnagel, J.: Vol. 152, pp. 41–156.
Mattozzi, A. see Gedde, U. W.: Vol. 169, pp. 29–73.
Mauzac, M. see Marty, J. D.: Vol. 172, pp. 1–35.
Mays, W. see Xu, Z.: Vol. 120, pp. 1–50.
Mays, J. W. see Pitsikalis, M.: Vol. 135, pp. 1–138.
McGrath, J. E. see Hedrick, J. L.: Vol. 141, pp. 1–44.

McGrath, J. E., Dunson, D. L. and *Hedrick, J. L.*: Synthesis and Characterization of Segmented Polyimide-Polyorganosiloxane Copolymers. Vol. 140, pp. 61–106.
McLeish, T. C. B. and *Milner, S. T.*: Entangled Dynamics and Melt Flow of Branched Polymers. Vol. 143, pp. 195–256.
Mecerreyes, D., Dubois, P. and *Jerome, R.*: Novel Macromolecular Architectures Based on Aliphatic Polyesters: Relevance of the Coordination-Insertion Ring-Opening Polymerization. Vol. 147, pp. 1–60.
Mecham, S. J. see McGrath, J. E.: Vol. 140, pp. 61–106.
Meille, S. V. see Allegra, G.: Vol. 191, pp. 87–135.
Menzel, H. see Möhwald, H.: Vol. 165, pp. 151–175.
Meyer, T. see Spange, S.: Vol. 165, pp. 43–78.
Michler, G. H. see Kausch, H.-H.: Vol. 187, pp. 1–33.
Mikos, A. G. see Thomson, R. C.: Vol. 122, pp. 245–274.
Milner, S. T. see McLeish, T. C. B.: Vol. 143, pp. 195–256.
Mison, P. and *Sillion, B.*: Thermosetting Oligomers Containing Maleimides and Nadiimides End-Groups. Vol. 140, pp. 137–180.
Miyasaka, K.: PVA-Iodine Complexes: Formation, Structure and Properties. Vol. 108, pp. 91–130.
Miller, R. D. see Hedrick, J. L.: Vol. 141, pp. 1–44.
Minko, S. see Rühe, J.: Vol. 165, pp. 79–150.
Möhwald, H., Menzel, H., Helm, C. A. and *Stamm, M.*: Lipid and Polyampholyte Monolayers to Study Polyelectrolyte Interactions and Structure at Interfaces. Vol. 165, pp. 151–175.
Monkenbusch, M. see Richter, D.: Vol. 174, pp. 1–221.
Monnerie, L., Halary, J. L. and *Kausch, H.-H.*: Deformation, Yield and Fracture of Amorphous Polymers: Relation to the Secondary Transitions. Vol. 187, pp. 215–364.
Monnerie, L., Laupêtre, F. and *Halary, J. L.*: Investigation of Solid-State Transitions in Linear and Crosslinked Amorphous Polymers. Vol. 187, pp. 35–213.
Monnerie, L. see Bahar, I.: Vol. 116, pp. 145–206.
Moore, J. S. see Ray, C. R.: Vol. 177, pp. 99–149.
Mori, H. see Bohrisch, J.: Vol. 165, pp. 1–41.
Morishima, Y.: Photoinduced Electron Transfer in Amphiphilic Polyelectrolyte Systems. Vol. 104, pp. 51–96.
Morton, M. see Quirk, R. P.: Vol. 153, pp. 67–162.
Motornov, M. see Rühe, J.: Vol. 165, pp. 79–150.
Mours, M. see Winter, H. H.: Vol. 134, pp. 165–234.
Müllen, K. see Scherf, U.: Vol. 123, pp. 1–40.
Müller, A. H. E. see Bohrisch, J.: Vol. 165, pp. 1–41.
Müller, A. H. E. see Förster, S.: Vol. 166, pp. 173–210.
Müller, A. J., Balsamo, V. and *Arnal, M. L.*: Nucleation and Crystallization in Diblock and Triblock Copolymers. Vol. 190, pp. 1–63.
Müller, M. and *Schmid, F.*: Incorporating Fluctuations and Dynamics in Self-Consistent Field Theories for Polymer Blends. Vol. 185, pp. 1–58.
Müller, M. see Thünemann, A. F.: Vol. 166, pp. 113–171.
Müller-Plathe, F. see Gusev, A. A.: Vol. 116, pp. 207–248.
Müller-Plathe, F. see Baschnagel, J.: Vol. 152, p. 41–156.
Mukerherjee, A. see Biswas, M.: Vol. 115, pp. 89–124.
Munz, M., Cappella, B., Sturm, H., Geuss, M. and *Schulz, E.*: Materials Contrasts and Nanolithography Techniques in Scanning Force Microscopy (SFM) and their Application to Polymers and Polymer Composites. Vol. 164, pp. 87–210.

Murat, M. see Baschnagel, J.: Vol. 152, p. 41–156.
Muthukumar, M.: Modeling Polymer Crystallization. Vol. 191, pp. 241–274.
Muzzarelli, C. see Muzzarelli, R. A. A.: Vol. 186, pp. 151–209.
Muzzarelli, R. A. A. and *Muzzarelli, C.*: Chitosan Chemistry: Relevance to the Biomedical Sciences. Vol. 186, pp. 151–209.
Mylnikov, V.: Photoconducting Polymers. Vol. 115, pp. 1–88.

Nagy, A. see Majoros, I.: Vol. 112, pp. 1–11.
Naka, K. see Uemura, T.: Vol. 167, pp. 81–106.
Nakamura, A. see Mashima, K.: Vol. 133, pp. 1–52.
Nakayama, Y. see Mashima, K.: Vol. 133, pp. 1–52.
Narasinham, B. and *Peppas, N. A.*: The Physics of Polymer Dissolution: Modeling Approaches and Experimental Behavior. Vol. 128, pp. 157–208.
Nechaev, S. see Grosberg, A.: Vol. 106, pp. 1–30.
Neoh, K. G. see Kang, E. T.: Vol. 106, pp. 135–190.
Netz, R. R. see Holm, C.: Vol. 166, pp. 67–111.
Netz, R. R. see Rühe, J.: Vol. 165, pp. 79–150.
Newman, S. M. see Anseth, K. S.: Vol. 122, pp. 177–218.
Nijenhuis, K. te: Thermoreversible Networks. Vol. 130, pp. 1–252.
Ninan, K. N. see Reghunadhan Nair, C. P.: Vol. 155, pp. 1–99.
Nishi, T. see Jinnai, H.: Vol. 170, pp. 115–167.
Nishida, K. see Kaji, K.: Vol. 191, pp. 187–240.
Nishikawa, Y. see Jinnai, H.: Vol. 170, pp. 115–167.
Nishiyama, N. and *Kataoka, K.*: Nanostructured Devices Based on Block Copolymer Assemblies for Drug Delivery: Designing Structures for Enhanced Drug Function. Vol. 193, pp. 67–101.
Noid, D. W. see Otaigbe, J. U.: Vol. 154, pp. 1–86.
Noid, D. W. see Sumpter, B. G.: Vol. 116, pp. 27–72.
Nomura, M., Tobita, H. and *Suzuki, K.*: Emulsion Polymerization: Kinetic and Mechanistic Aspects. Vol. 175, pp. 1–128.
Northolt, M. G., Picken, S. J., Den Decker, M. G., Baltussen, J. J. M. and *Schlatmann, R.*: The Tensile Strength of Polymer Fibres. Vol. 178, pp. 1–108.
Novac, B. see Grubbs, R.: Vol. 102, pp. 47–72.
Novikov, V. V. see Privalko, V. P.: Vol. 119, pp. 31–78.
Nummila-Pakarinen, A. see Knuuttila, H.: Vol. 169, pp. 13–27.

Ober, C. K. see Li, M.: Vol. 190, pp. 183–226.
O'Brien, D. F., Armitage, B. A., Bennett, D. E. and *Lamparski, H. G.*: Polymerization and Domain Formation in Lipid Assemblies. Vol. 126, pp. 53–84.
Ogasawara, M.: Application of Pulse Radiolysis to the Study of Polymers and Polymerizations. Vol.105, pp. 37–80.
Okabe, H. see Matsushige, K.: Vol. 125, pp. 147–186.
Okada, M.: Ring-Opening Polymerization of Bicyclic and Spiro Compounds. Reactivities and Polymerization Mechanisms. Vol. 102, pp. 1–46.
Okada, K. see Hikosaka, M.: Vol. 191, pp. 137–186.
Okano, T.: Molecular Design of Temperature-Responsive Polymers as Intelligent Materials. Vol. 110, pp. 179–198.
Okay, O. see Funke, W.: Vol. 136, pp. 137–232.
Onuki, A.: Theory of Phase Transition in Polymer Gels. Vol. 109, pp. 63–120.
Oppermann, W. see Holm, C.: Vol. 166, pp. 1–27.

Oppermann, W. see *Volk, N.*: Vol. 166, pp. 29–65.
Osad'ko, I. S.: Selective Spectroscopy of Chromophore Doped Polymers and Glasses. Vol. 114, pp. 123–186.
Osakada, K. and *Takeuchi, D.*: Coordination Polymerization of Dienes, Allenes, and Methylenecycloalkanes. Vol. 171, pp. 137–194.
Otaigbe, J. U., Barnes, M. D., Fukui, K., Sumpter, B. G. and *Noid, D. W.*: Generation, Characterization, and Modeling of Polymer Micro- and Nano-Particles. Vol. 154, pp. 1–86.
Otsu, T. and *Matsumoto, A.*: Controlled Synthesis of Polymers Using the Iniferter Technique: Developments in Living Radical Polymerization. Vol. 136, pp. 75–138.

de Pablo, J. J. see *Leontidis, E.*: Vol. 116, pp. 283–318.
Padias, A. B. see *Penelle, J.*: Vol. 102, pp. 73–104.
Pascault, J.-P. see *Williams, R. J. J.*: Vol. 128, pp. 95–156.
Pasch, H.: Analysis of Complex Polymers by Interaction Chromatography. Vol. 128, pp. 1–46.
Pasch, H.: Hyphenated Techniques in Liquid Chromatography of Polymers. Vol. 150, pp. 1–66.
Pasut, G. and *Veronese, F. M.*: PEGylation of Proteins as Tailored Chemistry for Optimized Bioconjugates. Vol. 192, pp. 95–134.
Paul, W. see *Baschnagel, J.*: Vol. 152, pp. 41–156.
Paulsen, S. B. and *Barsett, H.*: Bioactive Pectic Polysaccharides. Vol. 186, pp. 69–101.
Pautzsch, T. see *Klemm, E.*: Vol. 177, pp. 53–90.
Penczek, P., Czub, P. and *Pielichowski, J.*: Unsaturated Polyester Resins: Chemistry and Technology. Vol. 184, pp. 1–95.
Penczek, P. see *Batog, A. E.*: Vol. 144, pp. 49–114.
Penczek, P. see *Bogdal, D.*: Vol. 163, pp. 193–263.
Penelle, J., Hall, H. K., Padias, A. B. and *Tanaka, H.*: Captodative Olefins in Polymer Chemistry. Vol. 102, pp. 73–104.
Peppas, N. A. see *Bell, C. L.*: Vol. 122, pp. 125–176.
Peppas, N. A. see *Hassan, C. M.*: Vol. 153, pp. 37–65.
Peppas, N. A. see *Narasimhan, B.*: Vol. 128, pp. 157–208.
Petersen, K. L. see *Geil, P. H.*: Vol. 180, pp. 89–159.
Pet'ko, I. P. see *Batog, A. E.*: Vol. 144, pp. 49–114.
Pheyghambarian, N. see *Kippelen, B.*: Vol. 161, pp. 87–156.
Pichot, C. see *Hunkeler, D.*: Vol. 112, pp. 115–134.
Picken, S. J. see *Northolt, M. G.*: Vol. 178, pp. 1–108.
Pielichowski, J. see *Bogdal, D.*: Vol. 163, pp. 193–263.
Pielichowski, J. see *Penczek, P.*: Vol. 184, pp. 1–95.
Pieper, T. see *Kilian, H. G.*: Vol. 108, pp. 49–90.
Pispas, S. see *Pitsikalis, M.*: Vol. 135, pp. 1–138.
Pispas, S. see *Hadjichristidis, N.*: Vol. 142, pp. 71–128.
Pitsikalis, M., Pispas, S., Mays, J. W. and *Hadjichristidis, N.*: Nonlinear Block Copolymer Architectures. Vol. 135, pp. 1–138.
Pitsikalis, M. see *Hadjichristidis, N.*: Vol. 142, pp. 71–128.
Pitsikalis, M. see *Hadjichristidis, N.*: Vol. 189, pp. 1–124.
Pleul, D. see *Spange, S.*: Vol. 165, pp. 43–78.
Plummer, C. J. G.: Microdeformation and Fracture in Bulk Polyolefins. Vol. 169, pp. 75–119.
Pötschke, D. see *Dingenouts, N.*: Vol. 144, pp. 1–48.
Pokrovskii, V. N.: The Mesoscopic Theory of the Slow Relaxation of Linear Macromolecules. Vol. 154, pp. 143–219.

Pospíšil, J.: Functionalized Oligomers and Polymers as Stabilizers for Conventional Polymers. Vol. 101, pp. 65–168.
Pospíšil, J.: Aromatic and Heterocyclic Amines in Polymer Stabilization. Vol. 124, pp. 87–190.
Powers, A. C. see Prokop, A.: Vol. 136, pp. 53–74.
Prasad, P. N. see Lin, T.-C.: Vol. 161, pp. 157–193.
Priddy, D. B.: Recent Advances in Styrene Polymerization. Vol. 111, pp. 67–114.
Priddy, D. B.: Thermal Discoloration Chemistry of Styrene-co-Acrylonitrile. Vol. 121, pp. 123–154.
Privalko, V. P. and *Novikov, V. V.*: Model Treatments of the Heat Conductivity of Heterogeneous Polymers. Vol. 119, pp. 31–78.
Prociak, A. see Bogdal, D.: Vol. 163, pp. 193–263.
Prokop, A., Hunkeler, D., DiMari, S., Haralson, M. A. and *Wang, T. G.*: Water Soluble Polymers for Immunoisolation I: Complex Coacervation and Cytotoxicity. Vol. 136, pp. 1–52.
Prokop, A., Hunkeler, D., Powers, A. C., Whitesell, R. R. and *Wang, T. G.*: Water Soluble Polymers for Immunoisolation II: Evaluation of Multicomponent Microencapsulation Systems. Vol. 136, pp. 53–74.
Prokop, A., Kozlov, E., Carlesso, G. and *Davidsen, J. M.*: Hydrogel-Based Colloidal Polymeric System for Protein and Drug Delivery: Physical and Chemical Characterization, Permeability Control and Applications. Vol. 160, pp. 119–174.
Pruitt, L. A.: The Effects of Radiation on the Structural and Mechanical Properties of Medical Polymers. Vol. 162, pp. 65–95.
Pudavar, H. E. see Lin, T.-C.: Vol. 161, pp. 157–193.
Pukánszky, B. and *Fekete, E.*: Adhesion and Surface Modification. Vol. 139, pp. 109–154.
Putnam, D. and *Kopecek, J.*: Polymer Conjugates with Anticancer Acitivity. Vol. 122, pp. 55–124.
Putra, E. G. R. see Ungar, G.: Vol. 180, pp. 45–87.

Quirk, R. P., Yoo, T., Lee, Y., M., Kim, J. and *Lee, B.*: Applications of 1,1-Diphenylethylene Chemistry in Anionic Synthesis of Polymers with Controlled Structures. Vol. 153, pp. 67–162.

Ramaraj, R. and *Kaneko, M.*: Metal Complex in Polymer Membrane as a Model for Photosynthetic Oxygen Evolving Center. Vol. 123, pp. 215–242.
Rangarajan, B. see Scranton, A. B.: Vol. 122, pp. 1–54.
Ranucci, E. see Söderqvist Lindblad, M.: Vol. 157, pp. 139–161.
Raphaël, E. see Léger, L.: Vol. 138, pp. 185–226.
Rastogi, S. and *Terry, A. E.*: Morphological implications of the interphase bridging crystalline and amorphous regions in semi-crystalline polymers. Vol. 180, pp. 161–194.
Ray, C. R. and *Moore, J. S.*: Supramolecular Organization of Foldable Phenylene Ethynylene Oligomers. Vol. 177, pp. 99–149.
Reddinger, J. L. and *Reynolds, J. R.*: Molecular Engineering of p-Conjugated Polymers. Vol. 145, pp. 57–122.
Reghunadhan Nair, C. P., Mathew, D. and *Ninan, K. N.*: Cyanate Ester Resins, Recent Developments. Vol. 155, pp. 1–99.
Reichert, K. H. see Hunkeler, D.: Vol. 112, pp. 115–134.
Rehahn, M., Mattice, W. L. and *Suter, U. W.*: Rotational Isomeric State Models in Macromolecular Systems. Vol. 131/132, pp. 1–475.
Rehahn, M. see Bohrisch, J.: Vol. 165, pp. 1–41.
Rehahn, M. see Holm, C.: Vol. 166, pp. 1–27.
Reineker, P. see Holm, C.: Vol. 166, pp. 67–111.

Reitberger, T. see Jacobson, K.: Vol. 169, pp. 151–176.
Reynolds, J. R. see Reddinger, J. L.: Vol. 145, pp. 57–122.
Richter, D. see Ewen, B.: Vol. 134, pp. 1–130.
Richter, D., Monkenbusch, M. and *Colmenero, J.*: Neutron Spin Echo in Polymer Systems. Vol. 174, pp. 1–221.
Riegler, S. see Trimmel, G.: Vol. 176, pp. 43–87.
Ringsdorf, H. see Duncan, R.: Vol. 192, pp. 1–8.
Risse, W. see Grubbs, R.: Vol. 102, pp. 47–72.
Rivas, B. L. and *Geckeler, K. E.*: Synthesis and Metal Complexation of Poly(ethyleneimine) and Derivatives. Vol. 102, pp. 171–188.
Roberts, C. J. see Ellis, J. S.: Vol. 193, pp. 123–172.
Roberts, G. W. see Kennedy, K. A.: Vol. 175, pp. 329–346.
Robin, J. J.: The Use of Ozone in the Synthesis of New Polymers and the Modification of Polymers. Vol. 167, pp. 35–79.
Robin, J. J. see Boutevin, B.: Vol. 102, pp. 105–132.
Rodríguez-Pérez, M. A.: Crosslinked Polyolefin Foams: Production, Structure, Properties, and Applications. Vol. 184, pp. 97–126.
Roe, R.-J.: MD Simulation Study of Glass Transition and Short Time Dynamics in Polymer Liquids. Vol. 116, pp. 111–114.
Roovers, J. and *Comanita, B.*: Dendrimers and Dendrimer-Polymer Hybrids. Vol. 142, pp. 179–228.
Rothon, R. N.: Mineral Fillers in Thermoplastics: Filler Manufacture and Characterisation. Vol. 139, pp. 67–108.
de Rosa, C. see Auriemma, F.: Vol. 181, pp. 1–74.
Rozenberg, B. A. see Williams, R. J. J.: Vol. 128, pp. 95–156.
Rühe, J., Ballauff, M., Biesalski, M., Dziezok, P., Gröhn, F., Johannsmann, D., Houbenov, N., Hugenberg, N., Konradi, R., Minko, S., Motornov, M., Netz, R. R., Schmidt, M., Seidel, C., Stamm, M., Stephan, T., Usov, D. and *Zhang, H.*: Polyelectrolyte Brushes. Vol. 165, pp. 79–150.
Ruckenstein, E.: Concentrated Emulsion Polymerization. Vol. 127, pp. 1–58.
Ruiz-Taylor, L. see Mathieu, H. J.: Vol. 162, pp. 1–35.
Rusanov, A. L.: Novel Bis (Naphtalic Anhydrides) and Their Polyheteroarylenes with Improved Processability. Vol. 111, pp. 115–176.
Rusanov, A. L., Likhatchev, D., Kostoglodov, P. V., Müllen, K. and *Klapper, M.*: Proton-Exchanging Electrolyte Membranes Based on Aromatic Condensation Polymers. Vol. 179, pp. 83–134.
Russel, T. P. see Hedrick, J. L.: Vol. 141, pp. 1–44.
Russum, J. P. see Schork, F. J.: Vol. 175, pp. 129–255.
Rychly, J. see Lazár, M.: Vol. 102, pp. 189–222.
Ryner, M. see Stridsberg, K. M.: Vol. 157, pp. 27–51.
Ryzhov, V. A. see Bershtein, V. A.: Vol. 114, pp. 43–122.

Sabsai, O. Y. see Barshtein, G. R.: Vol. 101, pp. 1–28.
Saburov, V. V. see Zubov, V. P.: Vol. 104, pp. 135–176.
Saito, S., Konno, M. and *Inomata, H.*: Volume Phase Transition of N-Alkylacrylamide Gels. Vol. 109, pp. 207–232.
Samsonov, G. V. and *Kuznetsova, N. P.*: Crosslinked Polyelectrolytes in Biology. Vol. 104, pp. 1–50.
Santa Cruz, C. see Baltá-Calleja, F. J.: Vol. 108, pp. 1–48.
Santos, S. see Baschnagel, J.: Vol. 152, p. 41–156.

Satchi-Fainaro, R., Duncan, R. and *Barnes, C. M.*: Polymer Therapeutics for Cancer: Current Status and Future Challenges. Vol. 193, pp. 1–65.
Satchi-Fainaro, R. see Duncan, R.: Vol. 192, pp. 1–8.
Sato, T. and *Teramoto, A.*: Concentrated Solutions of Liquid-Christalline Polymers. Vol. 126, pp. 85–162.
Schaller, C. see Bohrisch, J.: Vol. 165, pp. 1–41.
Schäfer, R. see Köhler, W.: Vol. 151, pp. 1–59.
Scherf, U. and *Müllen, K.*: The Synthesis of Ladder Polymers. Vol. 123, pp. 1–40.
Sherman, S. see Kabanov, A. V.: Vol. 193, pp. 173–198.
Schlatmann, R. see Northolt, M. G.: Vol. 178, pp. 1–108.
Schmid, F. see Müller, M.: Vol. 185, pp. 1–58.
Schmidt, M. see Förster, S.: Vol. 120, pp. 51–134.
Schmidt, M. see Rühe, J.: Vol. 165, pp. 79–150.
Schmidt, M. see Volk, N.: Vol. 166, pp. 29–65.
Scholz, M.: Effects of Ion Radiation on Cells and Tissues. Vol. 162, pp. 97–158.
Schönherr, H. see Vancso, G. J.: Vol. 182, pp. 55–129.
Schopf, G. and *Koßmehl, G.*: Polythiophenes – Electrically Conductive Polymers. Vol. 129, pp. 1–145.
Schork, F. J., Luo, Y., Smulders, W., Russum, J. P., Butté, A. and *Fontenot, K.*: Miniemulsion Polymerization. Vol. 175, pp. 127–255.
Schulz, E. see Munz, M.: Vol. 164, pp. 97–210.
Schwahn, D.: Critical to Mean Field Crossover in Polymer Blends. Vol. 183, pp. 1–61.
Seppälä, J. see Löfgren, B.: Vol. 169, pp. 1–12.
Sturm, H. see Munz, M.: Vol. 164, pp. 87–210.
Schweizer, K. S.: Prism Theory of the Structure, Thermodynamics, and Phase Transitions of Polymer Liquids and Alloys. Vol. 116, pp. 319–378.
Scranton, A. B., Rangarajan, B. and *Klier, J.*: Biomedical Applications of Polyelectrolytes. Vol. 122, pp. 1–54.
Sefton, M. V. and *Stevenson, W. T. K.*: Microencapsulation of Live Animal Cells Using Polycrylates. Vol. 107, pp. 143–198.
Seidel, C. see Holm, C.: Vol. 166, pp. 67–111.
Seidel, C. see Rühe, J.: Vol. 165, pp. 79–150.
El Seoud, O. A. and *Heinze, T.*: Organic Esters of Cellulose: New Perspectives for Old Polymers. Vol. 186, pp. 103–149.
Shabat, D. see Amir, R. J.: Vol. 192, pp. 59–94.
Shamanin, V. V.: Bases of the Axiomatic Theory of Addition Polymerization. Vol. 112, pp. 135–180.
Shcherbina, M. A. see Ungar, G.: Vol. 180, pp. 45–87.
Sheiko, S. S.: Imaging of Polymers Using Scanning Force Microscopy: From Superstructures to Individual Molecules. Vol. 151, pp. 61–174.
Sherrington, D. C. see Cameron, N. R.: Vol. 126, pp. 163–214.
Sherrington, D. C. see Lin, J.: Vol. 111, pp. 177–220.
Sherrington, D. C. see Steinke, J.: Vol. 123, pp. 81–126.
Shibayama, M. see Tanaka, T.: Vol. 109, pp. 1–62.
Shiga, T.: Deformation and Viscoelastic Behavior of Polymer Gels in Electric Fields. Vol. 134, pp. 131–164.
Shim, H.-K. and *Jin, J.*: Light-Emitting Characteristics of Conjugated Polymers. Vol. 158, pp. 191–241.
Shoda, S. see Kobayashi, S.: Vol. 121, pp. 1–30.

Siegel, R. A.: Hydrophobic Weak Polyelectrolyte Gels: Studies of Swelling Equilibria and Kinetics. Vol. 109, pp. 233–268.
de Silva, D. S. M. see Ungar, G.: Vol. 180, pp. 45–87.
Silvestre, F. see Calmon-Decriaud, A.: Vol. 207, pp. 207–226.
Sillion, B. see Mison, P.: Vol. 140, pp. 137–180.
Simon, F. see Spange, S.: Vol. 165, pp. 43–78.
Simon, G. P. see Becker, O.: Vol. 179, pp. 29–82.
Simon, P. F. W. see Abetz, V.: Vol. 189, pp. 125–212.
Simonutti, R. see Sozzani, P.: Vol. 181, pp. 153–177.
Singh, R. P. see Sivaram, S.: Vol. 101, pp. 169–216.
Singh, R. P. see Desai, S. M.: Vol. 169, pp. 231–293.
Sinha Ray, S. see Biswas, M.: Vol. 155, pp. 167–221.
Sivaram, S. and *Singh, R. P.*: Degradation and Stabilization of Ethylene-Propylene Copolymers and Their Blends: A Critical Review. Vol. 101, pp. 169–216.
Slugovc, C. see Trimmel, G.: Vol. 176, pp. 43–87.
Smulders, W. see Schork, F. J.: Vol. 175, pp. 129–255.
Soares, J. B. P. see Anantawaraskul, S.: Vol. 182, pp. 1–54.
Sozzani, P., Bracco, S., Comotti, A. and *Simonutti, R.*: Motional Phase Disorder of Polymer Chains as Crystallized to Hexagonal Lattices. Vol. 181, pp. 153–177.
Söderqvist Lindblad, M., Liu, Y., Albertsson, A.-C., Ranucci, E. and *Karlsson, S.*: Polymer from Renewable Resources. Vol. 157, pp. 139–161.
Spange, S., Meyer, T., Voigt, I., Eschner, M., Estel, K., Pleul, D. and *Simon, F.*: Poly(Vinylformamide-co-Vinylamine)/Inorganic Oxid Hybrid Materials. Vol. 165, pp. 43–78.
Stamm, M. see Möhwald, H.: Vol. 165, pp. 151–175.
Stamm, M. see Rühe, J.: Vol. 165, pp. 79–150.
Starodybtzev, S. see Khokhlov, A.: Vol. 109, pp. 121–172.
Stegeman, G. I. see Canva, M.: Vol. 158, pp. 87–121.
Steinke, J., Sherrington, D. C. and *Dunkin, I. R.*: Imprinting of Synthetic Polymers Using Molecular Templates. Vol. 123, pp. 81–126.
Stelzer, F. see Trimmel, G.: Vol. 176, pp. 43–87.
Stenberg, B. see Jacobson, K.: Vol. 169, pp. 151–176.
Stenzenberger, H. D.: Addition Polyimides. Vol. 117, pp. 165–220.
Stephan, T. see Rühe, J.: Vol. 165, pp. 79–150.
Stevenson, W. T. K. see Sefton, M. V.: Vol. 107, pp. 143–198.
Stridsberg, K. M., Ryner, M. and *Albertsson, A.-C.*: Controlled Ring-Opening Polymerization: Polymers with Designed Macromolecular Architecture. Vol. 157, pp. 27–51.
Sturm, H. see Munz, M.: Vol. 164, pp. 87–210.
Suematsu, K.: Recent Progress of Gel Theory: Ring, Excluded Volume, and Dimension. Vol. 156, pp. 136–214.
Sugimoto, H. and *Inoue, S.*: Polymerization by Metalloporphyrin and Related Complexes. Vol. 146, pp. 39–120.
Suginome, M. and *Ito, Y.*: Transition Metal-Mediated Polymerization of Isocyanides. Vol. 171, pp. 77–136.
Sumpter, B. G., Noid, D. W., Liang, G. L. and *Wunderlich, B.*: Atomistic Dynamics of Macromolecular Crystals. Vol. 116, pp. 27–72.
Sumpter, B. G. see Otaigbe, J. U.: Vol. 154, pp. 1–86.
Sun, H.-B. and *Kawata, S.*: Two-Photon Photopolymerization and 3D Lithographic Microfabrication. Vol. 170, pp. 169–273.
Suter, U. W. see Gusev, A. A.: Vol. 116, pp. 207–248.
Suter, U. W. see Leontidis, E.: Vol. 116, pp. 283–318.

Suter, U. W. see Rehahn, M.: Vol. 131/132, pp. 1–475.
Suter, U. W. see Baschnagel, J.: Vol. 152, pp. 41–156.
Suzuki, A.: Phase Transition in Gels of Sub-Millimeter Size Induced by Interaction with Stimuli. Vol. 110, pp. 199–240.
Suzuki, A. and *Hirasa, O.*: An Approach to Artifical Muscle by Polymer Gels due to Micro-Phase Separation. Vol. 110, pp. 241–262.
Suzuki, K. see Nomura, M.: Vol. 175, pp. 1–128.
Swiatkiewicz, J. see Lin, T.-C.: Vol. 161, pp. 157–193.

Tagawa, S.: Radiation Effects on Ion Beams on Polymers. Vol. 105, pp. 99–116.
Taguet, A., Ameduri, B. and *Boutevin, B.*: Crosslinking of Vinylidene Fluoride-Containing Fluoropolymers. Vol. 184, pp. 127–211.
Takata, T., Kihara, N. and *Furusho, Y.*: Polyrotaxanes and Polycatenanes: Recent Advances in Syntheses and Applications of Polymers Comprising of Interlocked Structures. Vol. 171, pp. 1–75.
Takeuchi, D. see Osakada, K.: Vol. 171, pp. 137–194.
Tan, K. L. see Kang, E. T.: Vol. 106, pp. 135–190.
Tanaka, H. and *Shibayama, M.*: Phase Transition and Related Phenomena of Polymer Gels. Vol. 109, pp. 1–62.
Tanaka, T. see Penelle, J.: Vol. 102, pp. 73–104.
Tauer, K. see Guyot, A.: Vol. 111, pp. 43–66.
Tendler, S. J. B. see Ellis, J. S.: Vol. 193, pp. 123–172.
Teramoto, A. see Sato, T.: Vol. 126, pp. 85–162.
Terent'eva, J. P. and *Fridman, M. L.*: Compositions Based on Aminoresins. Vol. 101, pp. 29–64.
Terry, A. E. see Rastogi, S.: Vol. 180, pp. 161–194.
Theodorou, D. N. see Dodd, L. R.: Vol. 116, pp. 249–282.
Thomson, R. C., Wake, M. C., Yaszemski, M. J. and *Mikos, A. G.*: Biodegradable Polymer Scaffolds to Regenerate Organs. Vol. 122, pp. 245–274.
Thünemann, A. F., Müller, M., Dautzenberg, H., Joanny, J.-F. and *Löwen, H.*: Polyelectrolyte complexes. Vol. 166, pp. 113–171.
Tieke, B. see v. Klitzing, R.: Vol. 165, pp. 177–210.
Tobita, H. see Nomura, M.: Vol. 175, pp. 1–128.
Tokita, M.: Friction Between Polymer Networks of Gels and Solvent. Vol. 110, pp. 27–48.
Traser, S. see Bohrisch, J.: Vol. 165, pp. 1–41.
Tries, V. see Baschnagel, J.: Vol. 152, p. 41–156.
Trimmel, G., Riegler, S., Fuchs, G., Slugovc, C. and *Stelzer, F.*: Liquid Crystalline Polymers by Metathesis Polymerization. Vol. 176, pp. 43–87.
Tsuruta, T.: Contemporary Topics in Polymeric Materials for Biomedical Applications. Vol. 126, pp. 1–52.

Uemura, T., Naka, K. and *Chujo, Y.*: Functional Macromolecules with Electron-Donating Dithiafulvene Unit. Vol. 167, pp. 81–106.
Ungar, G., Putra, E. G. R., de Silva, D. S. M., Shcherbina, M. A. and *Waddon, A. J.*: The Effect of Self-Poisoning on Crystal Morphology and Growth Rates. Vol. 180, pp. 45–87.
Usov, D. see Rühe, J.: Vol. 165, pp. 79–150.
Usuki, A., Hasegawa, N. and *Kato, M.*: Polymer-Clay Nanocomposites. Vol. 179, pp. 135–195.
Uyama, H. see Kobayashi, S.: Vol. 121, pp. 1–30.
Uyama, Y.: Surface Modification of Polymers by Grafting. Vol. 137, pp. 1–40.

Vancso, G. J., Hillborg, H. and *Schönherr, H.*: Chemical Composition of Polymer Surfaces Imaged by Atomic Force Microscopy and Complementary Approaches. Vol. 182, pp. 55–129.
Varma, I. K. see Albertsson, A.-C.: Vol. 157, pp. 99–138.
Vasilevskaya, V. see Khokhlov, A.: Vol. 109, pp. 121–172.
Vaskova, V. see Hunkeler, D.: Vol. 112, pp. 115–134.
Verdugo, P.: Polymer Gel Phase Transition in Condensation-Decondensation of Secretory Products. Vol. 110, pp. 145–156.
Veronese, F. M. see Pasut, G.: Vol. 192, pp. 95–134.
Vettegren, V. I. see Bronnikov, S. V.: Vol. 125, pp. 103–146.
Vilgis, T. A. see Holm, C.: Vol. 166, pp. 67–111.
Viovy, J.-L. and *Lesec, J.*: Separation of Macromolecules in Gels: Permeation Chromatography and Electrophoresis. Vol. 114, pp. 1–42.
Vlahos, C. see Hadjichristidis, N.: Vol. 142, pp. 71–128.
Voigt, I. see Spange, S.: Vol. 165, pp. 43–78.
Volk, N., Vollmer, D., Schmidt, M., Oppermann, W. and *Huber, K.*: Conformation and Phase Diagrams of Flexible Polyelectrolytes. Vol. 166, pp. 29–65.
Volksen, W.: Condensation Polyimides: Synthesis, Solution Behavior, and Imidization Characteristics. Vol. 117, pp. 111–164.
Volksen, W. see Hedrick, J. L.: Vol. 141, pp. 1–44.
Volksen, W. see Hedrick, J. L.: Vol. 147, pp. 61–112.
Vollmer, D. see Volk, N.: Vol. 166, pp. 29–65.
Voskerician, G. and *Weder, C.*: Electronic Properties of PAEs. Vol. 177, pp. 209–248.

Waddon, A. J. see Ungar, G.: Vol. 180, pp. 45–87.
Wagener, K. B. see Baughman, T. W.: Vol. 176, pp. 1–42.
Wagner, E. and *Kloeckner, J.*: Gene Delivery Using Polymer Therapeutics. Vol. 192, pp. 135–173.
Wake, M. C. see Thomson, R. C.: Vol. 122, pp. 245–274.
Wandrey, C., Hernández-Barajas, J. and *Hunkeler, D.*: Diallyldimethylammonium Chloride and its Polymers. Vol. 145, pp. 123–182.
Wang, K. L. see Cussler, E. L.: Vol. 110, pp. 67–80.
Wang, S.-Q.: Molecular Transitions and Dynamics at Polymer/Wall Interfaces: Origins of Flow Instabilities and Wall Slip. Vol. 138, pp. 227–276.
Wang, S.-Q. see Bhargava, R.: Vol. 163, pp. 137–191.
Wang, T. G. see Prokop, A.: Vol. 136, pp. 1–52; 53–74.
Wang, X. see Lin, T.-C.: Vol. 161, pp. 157–193.
Watanabe, K. see Hikosaka, M.: Vol. 191, pp. 137–186.
Webster, O. W.: Group Transfer Polymerization: Mechanism and Comparison with Other Methods of Controlled Polymerization of Acrylic Monomers. Vol. 167, pp. 1–34.
Weder, C. see Voskerician, G.: Vol. 177, pp. 209–248.
Weis, J.-J. and *Levesque, D.*: Simple Dipolar Fluids as Generic Models for Soft Matter. Vol. 185, pp. 163–225.
Whitesell, R. R. see Prokop, A.: Vol. 136, pp. 53–74.
Williams, R. A. see Geil, P. H.: Vol. 180, pp. 89–159.
Williams, R. J. J., Rozenberg, B. A. and *Pascault, J.-P.*: Reaction Induced Phase Separation in Modified Thermosetting Polymers. Vol. 128, pp. 95–156.
Winkler, R. G. see Holm, C.: Vol. 166, pp. 67–111.
Winter, H. H. and *Mours, M.*: Rheology of Polymers Near Liquid-Solid Transitions. Vol. 134, pp. 165–234.

Wittmeyer, P. see Bohrisch, J.: Vol. 165, pp. 1–41.
Wood-Adams, P. M. see Anantawaraskul, S.: Vol. 182, pp. 1–54.
Wu, C.: Laser Light Scattering Characterization of Special Intractable Macromolecules in Solution. Vol. 137, pp. 103–134.
Wunderlich, B. see Sumpter, B. G.: Vol. 116, pp. 27–72.

Xiang, M. see Jiang, M.: Vol. 146, pp. 121–194.
Xie, T. Y. see Hunkeler, D.: Vol. 112, pp. 115–134.
Xu, P. see Geil, P. H.: Vol. 180, pp. 89–159.
Xu, Z., Hadjichristidis, N., Fetters, L. J. and *Mays, J. W.*: Structure/Chain-Flexibility Relationships of Polymers. Vol. 120, pp. 1–50.

Yagci, Y. and *Endo, T.*: N-Benzyl and N-Alkoxy Pyridium Salts as Thermal and Photochemical Initiators for Cationic Polymerization. Vol. 127, pp. 59–86.
Yamaguchi, I. see Yamamoto, T.: Vol. 177, pp. 181–208.
Yamamoto, T.: Molecular Dynamics Modeling of the Crystal-Melt Interfaces and the Growth of Chain Folded Lamellae. Vol. 191, pp. 37–85.
Yamamoto, T., Yamaguchi, I. and *Yasuda, T.*: PAEs with Heteroaromatic Rings. Vol. 177, pp. 181–208.
Yamaoka, H.: Polymer Materials for Fusion Reactors. Vol. 105, pp. 117–144.
Yamazaki, S. see Hikosaka, M.: Vol. 191, pp. 137–186.
Yannas, I. V.: Tissue Regeneration Templates Based on Collagen-Glycosaminoglycan Copolymers. Vol. 122, pp. 219–244.
Yang, J. see Geil, P. H.: Vol. 180, pp. 89–159.
Yang, J. S. see Jo, W. H.: Vol. 156, pp. 1–52.
Yasuda, H. and *Ihara, E.*: Rare Earth Metal-Initiated Living Polymerizations of Polar and Nonpolar Monomers. Vol. 133, pp. 53–102.
Yasuda, T. see Yamamoto, T.: Vol. 177, pp. 181–208.
Yaszemski, M. J. see Thomson, R. C.: Vol. 122, pp. 245–274.
Yoo, T. see Quirk, R. P.: Vol. 153, pp. 67–162.
Yoon, D. Y. see Hedrick, J. L.: Vol. 141, pp. 1–44.
Yoshida, H. and *Ichikawa, T.*: Electron Spin Studies of Free Radicals in Irradiated Polymers. Vol. 105, pp. 3–36.

Zhang, H. see Rühe, J.: Vol. 165, pp. 79–150.
Zhang, Y.: Synchrotron Radiation Direct Photo Etching of Polymers. Vol. 168, pp. 291–340.
Zheng, J. and *Swager, T. M.*: Poly(arylene ethynylene)s in Chemosensing and Biosensing. Vol. 177, pp. 151–177.
Zhou, H. see Jiang, M.: Vol. 146, pp. 121–194.
Zhou, Z. see Abe, A.: Vol. 181, pp. 121–152.
Zubov, V. P., Ivanov, A. E. and *Saburov, V. V.*: Polymer-Coated Adsorbents for the Separation of Biopolymers and Particles. Vol. 104, pp. 135–176.

Subject Index

Active targeting *II* 7–9
Adamantane *I* 145
Adaptor units, De Groot *I* 69
– McGrath *I* 70
– Shabat *I* 68
Alcohol dehydrogenase *I* 99
2-(4-Aminobenzylidene propane)-1,3-diol *I* 69
Aminomethyl-pyrene *I* 72
Aminopropyltriethoxysilane *II* 135
Amyloid formation *II* 160–162
Angiogenesis *II* 9–11
– drug targeting *II* 48–49
– markers *II* 43–45
Angiogenesis inhibitors *II* 9–11, 22
Angiotensin-induced hypertension *II* 108–109
Anthrax toxin, sequestration *I* 36
Anti-anthrax agents *I* 36
Antibodies, PEGylation *I* 118, 124
Antibody-G5 PAMAM *I* 63
Anti-HIV agents, anionic polymers *I* 40
Anti-infective polymeric drugs *I* 30
Antimicrobial agents, polyvalent *I* 40
Anti-obesity drugs *I* 47
Antiviral agents, polyvalent ligands *I* 37
Arginine deaminase, PEGylation *I* 121
Arginine-grafted dendrimer *I* 149
Artificial vaccines *II* 173–198
Asialoglycoprotein *I* 140
Atomic force microscopy *II* 123–172
– amyloid formation *II* 160–162
– contact mode *II* 128–129
– cryogenic *II* 130
– DNA *II* 134–146
– force-displacement measurements *II* 132–134
– imaging proteins on substrates *II* 157–160
– instrumentation *II* 126–128
– liquid imaging *II* 130–131
– membrane proteins *II* 154–155
– nontopographical applications *II* 131–132
– phase imaging *II* 129–130
– polysaccharides *II* 150–153
– RNA *II* 146–150
– single molecule biopolymers *II* 134–146
– tapping mode *II* 129
– tip geometry and carbon nanotubes *II* 131
– viral proteins *II* 156–157
ATP binding cassette proteins *II* 175

Bacillus anthracis *I* 36
Bacterial toxins *I* 32
Bile acid sequestrants *I* 25, 28
Bioavailability, oral *I* 12
Bioconjugates, PEG *I* 95
Biopolymers *II* 134–146
– DNA condensation *II* 137–138
– DNA immobilisation *II* 134–137
– DNA-drug interactions *II* 142–144
– DNA-protein interactions *II* 138–142
Bioresponsive polymers *I* 164
2,4-Bis(hydroxymethyl)phenol *I* 70
2,6-Bishydroxymethyl-*p*-cresol *I* 68
Block copolymers *II* 69–70
– synthesis *II* 72–76
Bradykinin *II* 109–110
Brain tumour implants *II* 38–39
Breast cancer resistance protein *II* 175
Bystander effect *II* 25

Camptothecin *I* 79
Camptothecin conjugates *II* 35–36, 116
Cancer chemotherapy *II* 103–121

Carboxylic groups, PEGylation   *I* 127
Carboxymethyl dextran-CPT II analogues   39–40
Catalase   *II* 116
Catecholates   *I* 23
CDDP-incorporated micelles   *II* 3–4
Cellulose   *II* 150
Chitosan   *I* 143
Cholesterol, LDLc   *I* 25
Cholesterol-lowering drugs   *I* 25
Cholestyramine   *I* 26, 33
Clostridium difficile toxin, sequestration   *I* 32
Colestipol   *I* 26, 33
Coronary heart disease   *I* 25
Critical micelle concentration   *II* 76
Cyclodextrin dendrimer   *I* 66
Cyclodextrins   *I* 144, 148

DAB-64   *I* 64
DE-310   *II* 39
De Groot adaptor unit   *I* 69
Dendrimers   *I* 59
– cascade-release   *I* 72
– self-immolative   *I* 72
Dendrons, domino   *I* 67
– multi-triggered   *I* 89
Desferrioxamine   *I* 22
Dextran   *II* 152–153
Dextran-doxorubicin   *II* 37
Diaminobutane poly(propylene imine)   *I* 64
Diarrhea   *I* 32
Divalent cations   *II* 135–136
Divinylether-maleic anhydride   *I* 3
DNA   *II* 134–138
DNA condensation   *II* 137–138
DNA crosslinkers   *II* 142–143
DNA delivery   *I* 135
DNA-drug interactions   *II* 142–144
– direct imaging   *II* 145–146
– DNA crosslinkers   *II* 142–143
– intercalators   *II* 143–144
– minor groove binders   *II* 144
DNA immobilisation   *II* 134–137
– divalent cations   *II* 135–136
– monovalent ions   *II* 136–137
– silanes   *II* 135
DNA microarrays   *II* 180
DNA polyplexes, caged   *I* 156

DNA-protein interactions   *II* 138–142
Domino dendrimers   *I* 59, 71
Domino dendrons, multi-triggered   *I* 88
Doxorubicin   *I* 4, 79
Drug delivery systems   *II* 1–65, 71–72
– polyion complex micelles   *II* 88–95
– polymer vesicles   *II* 95–96
– polymeric micelles *see* Polymeric micelles
Drug resistance   *II* 173–198
Drug targeting   *II* 1–65, 67–101
– cancer chemotherapy   *II* 103–121
– parenteral   *II* 11
– passive versus active   *II* 7–9
– tumour cells versus tumour vasculature   *II* 9–11
– tumour vasculature   *II* 42–50

Electrolyte homeostasis   *I* 14
Endotoxins   *I* 32
Enhanced permeability and retention effect   *II* 7, 8, 23, 72, 103–121
– modulation of   *II* 108–110
– theory and principles   *II* 106–108
EPR effect   *I* 4
Erythropoietin, PEGylation   *I* 121
Exotoxins   *I* 32

Fat binder, polymeric   *I* 49
Fenton reaction   *I* 21

Gene delivery   *I* 135, *II* 90–95
– device   *I* 62
Gene expression profiles   *II* 183–185
– transcriptional activation by synthetic polymers   *II* 185–189
Gene therapy   *I* 5
Genomic profiles   *II* 180–183
Glucosamine dendrimer   *I* 66
Glycocholate   26
Granulocyte colony stimulating factor   *I* 112
Growth hormone   *I* 115

Hemagglutination   *I* 37
Hemochromatosis   *I* 22
Hemoglobin   *I* 52
Heparanproteoglycans   *I* 141
HIV virus   *I* 40
HK polymer   *I* 139

Subject Index

HMG-CoA reductase inhibitors *I* 25
HPMA *I* 4
HPMA copolymer *II* 1–65
HPMA copolymer-1,5-diazaanthraquinone *II* 42
HPMA copolymer-antibody-doxorubicin conjugates *II* 26–27
HPMA copolymer-camptothecin *II* 28–31
HPMA copolymer-DACH platinate *II* 32–33
HPMA copolymer-doxorubicin-galactosamine *II* 22
HPMA copolymer-Gly-Phe-Leu-Gly-doxorubicin *II* 23–25
HPMA copolymer-Gly-Phe-Leu-Gly-doxorubicin-galactosamine *II* 25–26
HPMA copolymer-paclitaxel *II* 27–28
HPMA copolymer-platinate *II* 31–32
HPMA copolymer-TNP-470 (caplostatin) *II* 46–48
Human groth hormone *I* 115
Hydrazones *I* 154
Hydrophobic drugs, encapsulation of *II* 80–83
Hydroxamates *I* 23
Hydroxypropyl methacrylate *I* 159
Hyperkalemia *I* 14
Hyperphosphatemia *I* 15

IgG *I* 119
Immune response *II* 189–191
Immunocamouflage *I* 129
Inflammatory mediators *II* 110
Influenza virus inhibitors, polyvalent *I* 38
Inorganic ions *I* 14
Insulin, PEGylation *I* 120
Intercalators *II* 143–144
– direct imaging *II* 145–146
Interferons *I* 105, 109
Iron, sequestration *I* 21
Iron overload disorder *I* 21

LDLc *I* 25
Lectin *I* 67
Leptin (OB proteins), PEGylation *I* 121
Leukemia *I* 84
Ligand-receptor interactions *II* 164–166

Ligands *I* 31
Lipase inhibition *I* 48
Lipoprotein cholesterol *I* 25
Liposomal-PEG-Ala-Pro-Arg-Pro-Gly *II* 49–50
Lysosomotropic drug delivery *II* 19

Macromolecular drugs *II* 103–121
– intracellular uptake *II* 111
– quality of life *II* 114–115
– SMANCS *II* 111–113
McGrath adaptor unit *I* 70
Megakaryocyte growth and development factor *I* 114
Membrane proteins *II* 154–155
Mescaline-*N*-vinylpyrolidone *I* 3
Metronidazole *I* 32
Micelle *I* 2
Minor groove binders *II* 144
MOLT-3 leukemia *I* 84
Monoclonal antibodies *II* 18
Monovalent ions *II* 136–137
MRSA *I* 41
Multidrug resistance *I* 40
Multidrug resistance-associated proteins *II* 175
Multiple sclerosis *I* 44
Multi-prodrug *I* 59
Myelin based protein *I* 45

Naphthalenesulfonate-formaldehyde, HIV *I* 40
N-(2-hydroxypropyl)methacrylamide *see* HPMA
New chemical entities *II* 6
4-Nitroaniline *I* 72
Nucleic acids, delivery *I* 135

Obesity *I* 47
Oligolysines, disulfide cross-linking *I* 156
Orlistat *I* 47, 48
Oxidation therapy *II* 117

PA63 *I* 36
Paclitaxel conjugates *II* 33–35, 115
PAMAM *I* 60
Parenteral drug targeting *II* 11
Passive targeting *II* 7–9
PEGylated proteins *I* 4, 95
Phosphate binder therapy *I* 15

Phosphate ions, sequestration *I* 15
PEI-cholesterol *I* 148
Penicillin-G-amidase *I* 88
PEO-PPO-PEO *I* 145
Peptides, PEGylated *I* 95
P-glycoprotein *II* 175
PEG *II* 1–65
PEG-adenosine deaminase *II* 13–14
PEG-camptothecin *II* 36
PEG-DAC *II* 55–56
PEG-DAO *II* 54–55, 118
PEG-granulocyte-colony stimulating factor *II* 15–16
PEG-interferon-α *II* 16–17
PEG-*L*-asparaginase *II* 7, 14–15
PEG-paclitaxel *II* 37
PEG-XO *II* 54, 118
PEG-ZnPP *II* 55–56, 118
PEGylated-liposome-Raf mutant *II* 49
PEGylated-liposomes *II* 49–50
– liposomal-PEG-Ala-Pro-Arg-Pro-Gly *II* 49–50
– PEGylated-liposome-Raf mutant *II* 49
Peptide motifs *II* 45–46
Phenotype *II* 175–179
Phenotypic correction of immune response *II* 189–191
Plasma half-life *II* 103–121
Pluronic F68 *I* 52
Pluronic block copolymers *II* 176–177
Polyacetal-diethylstilboestrol *II* 40–42
Poly(amidoamine), dendrimers *I* 60, 142, 149, 151
Polyamines *I* 149
Poly(β-amino ester) *I* 152
Polycations, polyplexes *I* 140
Polydispersity *I* 12
Polyethyleneimines (PEI) *I* 62, 141, 147
– cyclodextrin-grafted *I* 148
– dextran-grafted *I* 147
– PEG-grafted *I* 147
Poly(ethyleneglycol) *I* 4, 62, 98, 145
Poly(ethylene oxide) *II* 71
Polyion complex micelles *II* 88–95
– gene delivery *II* 90–95
– properties of *II* 88–89
Poly(glutamic acid) *I* 4
Polyglycerol dendrimers *I* 63
Poly(lactide) *II* 73
Polylactide-hydroxyproline *I* 146

Poly-L-glutamic acid conjugates *II* 115–119
– camptothecin *II* 35–36, 116
– paclitaxel *II* 33–35, 115
Poly(lys-(AEDTP)) *I* 157
Polylysines *I* 65, 140, 146, 149
Polymer directed enzyme prodrug therapy *II* 50–52
Polymer-DNA complex *I* 2
Polymer-drug conjugate *I* 2
Polymer genomics *II* 192–194
Polymeric micelle *I* 2
Polymer-protein conjugate *I* 2
Polymer therapeutics *II* 1–65
– combinations *II* 50–56
– *see also individual compounds*
Polymer vesicles *II* 95–96
Polymer-coated surfaces *II* 191–192
Polymer-drug conjugates *II* 18–39
– alteration of signal transduction *II* 179–180
– angiogenesis inhibitors *II* 22
– anticancer agents *II* 20–21
– brain tumour implants *II* 38–39
– dextran-doxorubicin *II* 37
– and gene expression profiles *II* 183–185
– and genomic profiles *II* 180–183
– HPMA copolymer-antibody-doxorubicin conjugates *II* 26–27
– HPMA copolymer-camptothecin *II* 28–31
– HPMA copolymer-DACH platinate *II* 32–33
– HPMA copolymer-Gly-Phe-Leu-Gly-doxorubicin *II* 23–25
– HPMA copolymer-Gly-Phe-Leu-Gly-doxorubicin-galactosamine *II* 25–26
– HPMA copolymer-paclitaxel *II* 27–28
– HPMA copolymer-platinate *II* 31–32
– PEG-camptothecin *II* 36, 116
– PEG-paclitaxel *II* 37, 115
– phenotypic selectivity *II* 175–179
– poly-L-glutamic(PG)-camptothecin *II* 35–36, 116
– poly-L-glutamic(PG)-paclitaxel *II* 33–35, 116
– polymeric micelles *II* 37–38

Subject Index

Polymer-enzyme liposome therapy
  II 52–54
Polymer-protein conjugates *II* 11–18
– PEG-adenosine deaminase *II* 13–14
– PEG-granulocyte-colony stimulating
  factor *II* 15–16
– PEG-interferon-α *II* 16–17
– PEG-*L*-asparaginase *II* 14–15
– preclinical *II* 18
– styrene-co-maleic
  anhydride-neocarzinostatin *II* 17–18
Polymeric micelles *II* 37–38, 69–70
– blood circulation and tissue distribution
  II 78–80
– CDDP-incorporated micelles *II* 83–84
– drug delivery *II* 80–88
– encapsulation of hydrophobic drugs
  II 80–83
– intracellular location *II* 87–88
– properties of *II* 76–77
– stimuli-triggered drug release *II* 84–85
– surface-functionalized *II* 86–87
Polymethacrylates *I* 143
Polyorthoesters *I* 154
Polyphosphoesters *I* 154
Polyphosphoramidates *I* 154
Polyplexes *I* 2, 135, 140
– bioresponsive *I* 164
Polypropylene oxide, SCA *I* 52
Polypropyleneimines (PPI), dendritic
  I 150
Polysaccharide-enzyme interactions
  II 153
Polysaccharides *II* 150–153
– cellulose *II* 150
– dextran *II* 152–153
– starch grains *II* 150–152
Poly(styrene-4-sulfonate), HIV *I* 40
Poly(styrene sulfonic acid) *I* 34
Polyvalency *I* 12
Polyvalent interactions *I* 30
Potassium, sequestration *I* 14
Prodrug, dendritic *I* 78
Propranolol *I* 61
Prostacyclin antagonists *II* 110
Protein-DNA interactions *II* 138–142
Proteins
– DNA-protein interactions *II* 138–142
– mechanical properties *II* 162–164
– membrane *II* 154–155

– polymer-protein conjugates *II* 11–18
– single-force molecular interactions
  II 162
– on substrates *II* 157–160
– viral *II* 156–157
Proton sponge *I* 141, 146
Pyran copolymer *I* 3

Quality of life *II* 114–115
Quinone methide *I* 70

RBCs, PEGylation *I* 129
Reactive oxygen species *II* 54, 103–121
Receptors *I* 31
Renal failure *I* 15
Rheumatoid arthritis *I* 44
RNA *II* 146–150
– crystallization *II* 149
– force investigations *II* 149–150
– in situ synthesis *II* 147
– tectonics *II* 147–149
Rotavirus *I* 37

Sequestrant *I* 2, 13
Serum phosphate *I* 15
Shabat adaptor unit *I* 68
Sialic acids, influenza virus *I* 39
Sialyltransferase *I* 127
Sickle cell anemia *I* 22
– non-ionic surfactant *I* 52
Signal transduction *II* 179–180
– polymer-coated surfaces affecting
  II 191–192
Silanes
– aminopropyltriethoxysilane *II* 135
– DNA immobilisation *II* 135
Single-force molecular interactions
  II 162
– ligand-receptor interactions
  II 164–166
– mechanical properties of proteins
  II 162–164
SMANCS *II* 111–112
– clinical status *II* 112–113
– quality of life *II* 114–115
Solid tumours, characteristics *II* 106
Starburst dendrimers *I* 142
Starch grains *II* 150–152
Statins *I* 25
Stimuli-triggered drug release *II* 84–85

Stroke *I* 25
Styrene maleic anhydride *II* 7
Styrene-co-maleic
   anhydride-neocarzinostatin *II* 17–18
Super-stealth property *II* 96
Superoxide dismutase *II* 116
Surface-functionalized polymeric micelles
   II 86–87
Synthetic polyelectrolytes *II* 189–191

Taxol *I* 72
Thalassemia *I* 22
Thiol groups, PEGylation *I* 122
Thrombopoiesis *I* 114
Toxins, polymeric sequestrants *I* 32
Transferrin *I* 145
Transglutaminase, PEGylation *I* 125
Triacyl glycerides *I* 48
Tumour cell targeting *II* 9–11

Tumour vasculature, drug targeting
   II 9–11, 42–50
– delivery schedules and vehicles
   II 48–49
– HPMA copolymer-TNP-470 (caplostatin)
   II 46–48
– markers of angiogenesis *II* 43–45
– PEGylated-liposomes *II* 49–50
– peptide motifs *II* 45–46

Vancomycin *I* 32
– resistance *I* 41
Vascular endothelial growth factor *II* 72
Vascular pemeability factor *II* 72
Vaso-occlusion *I* 52
Viral attachment proteins *I* 37
Viral infections *I* 37
Viral proteins *II* 156–157
Viruses, artificial *I* 164